EENVIRO 2017 Workshop

Advances in Heat and Transfer in Built Environment

E3S Web of Conferences Volume 32 (2018)

Bucharest, Romania
24 November 2017

Editors:

C. Balan
M.C. Balan

ISBN: 978-1-5108-5878-7

Printed from e-media with permission by:

Curran Associates, Inc.
57 Morehouse Lane
Red Hook, NY 12571

Some format issues inherent in the e-media version may also appear in this print version.

Printed by Curran Associates, Inc. (2018)

For additional information, please contact EDP Sciences – Web of Conferences at the address below.

EDP Sciences – Web of Conferences
17, Avenue du Hoggar
Parc d'Activité de Courtabœuf
BP 112
F-91944 Les Ulis Cedex A
France

Phone: +33 (0) 1 69 18 75 75
Fax: +33 (0) 1 69 28 84 91

contact-edps@webofconferences.org

Additional copies of this publication are available from:

Curran Associates, Inc.
57 Morehouse Lane
Red Hook, NY 12571 USA
Phone: 845-758-0400
Fax: 845-758-2633
Email: curran@proceedings.com
Web: www.proceedings.com

TABLE OF CONTENTS

edp sciences

E3S Web of Conferences

Statement of Peer review

In submitting conference proceedings to E3S Web of Conferences, I certify to the Publisher that I adhere to the **Policy on Publishing Integrity** of the journal in order to safeguard good scientific practice in publishing.

1. All articles have been subjected to peer review administered by the proceedings editors.
2. Reviews have been conducted by expert referees, who have been requested to provide unbiased and constructive comments aimed, whenever possible, at improving the work.
3. Proceedings editors have taken all reasonable steps to ensure the quality of the materials they publish and their decision to accept or reject a paper for publication has been based only on the merits of the work and the relevance to the journal.

Title, date and place of the conference

"EENVIRO 2017 Workshop – Advances in Heat and Transfer in Built Environment"
24. Nov. 2017, Bucharest, Romania

Proceedings editor(s):

Prof. CORNELIU BALAN

Date and editor's signature 24. Nov. 2017

edp sciences

E3S Web of Conferences

Statement of Peer review

In submitting conference proceedings to E3S Web of Conferences, I certify to the Publisher that I adhere to the **Policy on Publishing Integrity** of the journal in order to safeguard good scientific practice in publishing.

1. All articles have been subjected to peer review administered by the proceedings editors.
2. Reviews have been conducted by expert referees, who have been requested to provide unbiased and constructive comments aimed, whenever possible, at improving the work.
3. Proceedings editors have taken all reasonable steps to ensure the quality of the materials they publish and their decision to accept or reject a paper for publication has been based only on the merits of the work and the relevance to the journal.

Title, date and place of the conference

EENVIRO 2017 Workshop - Advances in Heat and Mass Transfer in Built Environment

Bucharest, Romania

Proceedings editor(s): Mugur C. Balan

Date and editor's signature 24.11.2017

Experimental evaluation of passive cooling using phase change materials (PCM) for reducing overheating in public building

Abdullahi Ahmed[1, *], *Monica Mateo-Garcia*[1], *Danny McGough*[1] *Kassim Caratella*[2] *and Zafer Ure*[3]

[1]Centre for Built and Natural Environment, Faculty of Engineering, Environment and Computing, Coventry University, Sir John Laing Building, Much Park Street, Coventry, CV15FB, UK.
[2]Exergy, Coventry University Technology Park, Puma Way, Coventry, CV1 2TT, UK.
[3] Phase Change Material Products Limited, Unit 32, Mere View Industrial Estate, Yaxley, Cambridgeshire, PE7 3HS, U.K

Abstract. Indoor Environmental Quality (IEQ) is essential for the health and productivity of building users. The risk of overheating in buildings is increasing due to increased density of occupancy of people and heat emitting equipment, increase in ambient temperature due to manifestation of climate change or changes in urban micro-climate. One of the solutions to building overheating is to inject some exposed thermal mass into the interior of the building. There are many different types of thermal storage materials which typically includes sensible heat storage materials such as concrete, bricks, rocks etc. It is very difficult to increase the thermal mass of existing buildings using these sensible heat storage materials. Alternative to these, there are latent heat storage materials called Phase Change Materials (PCM), which have high thermal storage capacity per unit volume of materials making them easy to implement within retrofit project. The use of Passive Cooling Thermal Energy Storage (TES) systems in the form of PCM PlusICE Solutions has been investigated in occupied spaces to improve indoor environmental quality. The work has been carried out using experimental set-up in existing spaces and monitored through the summer the months. The rooms have been monitored using wireless temperature and humidity sensors. There appears to be significant improvement in indoor temperature of up to 5°K in the room with the PCM compared to the monitored control spaces. The success of PCM for passive cooling is strongly dependent on the ventilation strategy employed in the spaces. The use of night time cooling to purge the stored thermal energy is essential for improved efficacy of the systems to reduce overheating in the spaces. The investigation is carried within the EU funded RESEEPEE project.

1 Introduction

There is global concern relating to energy consumption in the built environment as one of the major contributors to greenhouse gas emissions blamed for causing climate change and its associated impacts. Majority of energy consumption in buildings is used for achieving and maintaining building indoor environmental quality. Significant proportion of the energy consumption in buildings is used for cooling and heating application [1]. Heating, ventilating and air conditioning (HVAC) systems have also been found to account for about 60% of the total energy consumed in buildings [2]. It is therefore essential that passive and low energy strategies are used to reduce building energy consumption for cooling and heating and for maintaining comfortable internal environment of buildings. One strategy that has been used for centuries to maintain comfortable internal conditions in building is the use of thermal mass strategy [3], historically sensible heat storage materials such as brick, concrete and earth have been used as part of the building fabric to provide the required building thermal mass. This is more difficult to achieve when retrofitting of existing buildings due to both the architectural and structural limitations in the case study building under review. This paper investigates the use PCM TUBE modules to increase thermal mass of an existing non-domestic building. The strategy is designed to improve the indoor comfort condition by reducing the peak daytime indoor temperatures. The benefit of the PCM TUBE modules is considered within existing building retrofit project. The benefit of building retrofit to increase the life of existing buildings is well documented, to avoid the unintended consequences, such as overheating, it is therefore essential that measures to improve IEQ are considered in the process low energy retrofit of buildings. This research will help in closing knowledge in this area.

2 Literature Review

* Corresponding author: abdullahi.ahmed@coventry.ac.uk

2.1. Risk of overheating in buildings

The risk of overheating in buildings is increasing and projected to increase under different future climatic conditions. Therefore, it is very important to adapt existing buildings to mitigate this risk and ensure sustainability and resilience of the buildings stock. The European heat wave in 2003 is a typical example of the potential devastating impact of persistent summer overheating caused by sustained elevated summer temperature. IPCC [4] shows that changes in urban temperature under observed and projected future climate change will lead to increase temperatures of between 2.4°C to 6.4°C at the end of 21st century.

Liu and Coley [5] concluded that under a changing climate, most of the locations exceeded the static overheating risk criterion recommended by CIBSE Guide A and suggested that to achieve comfortable indoor environment under changing climate, appropriate and sustainable adaptation is necessary for UK dwellings. Hamdy et al [6] evaluated the overheating risks in Dutch building stock built between 1964 and 2013 for four climate change scenarios, concluding that there is overheating risk in some poorly ventilated buildings and therefore most at risk under climate change, and suggested government actions through policy initiatives on adaptation measures to protect both existing and new buildings from the risk of overheating.

The term overheating is used to describe the excessive indoor temperature within buildings which can lead to discomfort and potential heat stress for occupants. Overheating often happens during warm weather in the summer, but it can happen in winter months too. The sudden elevation in indoor temperatures and prolonged periods of excess heat is a major challenge for building users and managers especially the vulnerable population such as unhealthy, elderly and young people [7]. The European heat wave in 2003 has caused increases in deaths in the UK and Europe and caused disruption to travel infrastructure. The METoffice [8] argued that climate change had at least doubled the risk of heat wave exceeding that of 2003 happening again.

Eicker [9] suggests that cooling of buildings can be achieved with different energy consumption from purely passive systems to high electrical energy consumption with systems such as chillers. To reduce the carbon emission associated with the built environment we should aim to mitigate the risk of overheating in existing buildings under current and potential future climatic conditions using passive and low-energy strategies that could improve comfort conditions in buildings with responsible use of energy. It is therefore important to identify and evaluate these potential passive solutions that can be used both in building adaptation and the construction of new buildings.

Building indoor temperature is primarily influenced by external climatic conditions, the internal gains from human occupancy, heat emitting equipment and building fabric efficiency and mass. To protect the building extreme diurnal temperature swings during hot periods, thermal mass strategy has historically been applied in hot climatic conditions to moderate absorb both solar and internal gains during the day and release it during night times when external temperatures significantly drop [10].

In building retrofit it is difficult to increase the thermal mass of the building with conventional sensible heat storage materials such as brick and concrete due to both structural and spacial limitations. Therefore, building retrofit materials that have high thermal capacity per unit volume are the preferred materials which can be installed safely without the need for too much structural upgrade to the building fabric and will allow for maximum space utilisation. Phase Change Materials (PCM) have high thermal capacity per unit volume compared to sensible heat storage materials such as brick and concrete, that have traditionally been applied in buildings, and work by raising the temperature of the materials. PCM on the other hand have both sensible heat storage capacity and significant amount of energy is exchanged during change of state of the materials [11].

2.2 Phase Change Material (PCM) Thermal Storage System

Phase Change Materials (PCM) work on the principle of latent heat storage and release during the melting & solidification process. When the material solidifies, large amount of latent heat of solidification is released to the ambient environment. On the other hand, when the material melts large amount of latent heat is absorbed from the ambient environment to aid the melting process.

PCM thermal storage systems have been used in building applications to improve both thermal comfort and energy efficiency. Auzebyet et al [12] used thermal modelling to evaluate the factors influencing the performance of PCM for reducing overheating in UK housing and concluded that there is significant potential for reducing overheating using PCM. Nghana and Tariku [13] investigated the impact of PCM on energy consumption and comfort in buildings under mild climatic conditions and found a reduction in peak temperature of about 0.6°K and increased the trough by 0.8°K.

PCM application has also been investigate under hot climatic conditions. Derraji et al [14] evaluated the application of PCM wall system for improving thermal comfort conditions in buildings under the hot climate of Algeria using Transient System Simulation Software (TRNSYS) and concluded that there is significant improvement in indoor condition in summer and winter. Solgi et al [15] also investigated the potential of using PCM and nigh ventilation for reducing cooling load in office buildings in hot Arid climate of Yazd, Iran. The study found a potential cooling load reduction of up to 7% which can significantly reduce cooling energy use.

Álvarez et al [16] investigated the application of PCM in conjunction with night time ventilation and investigated the influence of many factors affecting the effectiveness for PCM in night cooling applications and concluded that low utilisation of thermal energy stored, low convective heat transfer coefficient and limited contact area between PCM and room ambient air are some of the factors that need to be improved to ensure effective utilisation.

Souayfane et al [17] reviewed various applications of PCM for cooling application and concluded that there are some benefits of using PCM in buildings such as 'PCM can decrease energy consumption, shift the peak loads of cooling energy demand, decrease temperature fluctuations providing a thermally comfortable environment, and reduce the electricity consumption. Free cooling applications are effective when the diurnal temperature variations are large (up to 15'°C). Liu and Li [18] experimental investigates the impact of PCM on the performance of solar chimneys and shows that the PCM can 'reduce the air flow rate and air outlet temperature during charging period but increase them during discharging period compared with the solar chimney without PCM'. This shows that the ventilation rate will increase during the period when air movement is much desirable. Iten et al [19] in the review of air-PCM-TES application for free cooling and heating in the buildings found that there is potential for such systems to provide free-cooling and heating of buildings using auxiliary means rather than purely passive application.

In this research a purely passive approach is implemented to experimentally investigate the impact of PCM Thermal Energy Storage (TES) technologies and their effectiveness for moderating indoor thermal conditions. Because of the high diurnal temperature range of about 15°K in temperate climate conditions of the UK, the cool night-time ambient can be used to charge the latent heat capacity of various PCM solutions between +4°C and +117°C without the use of any refrigeration machinery by simply allowing the surrounding air to freeze these solutions. It is then possible later for the stored energy to be released back to the occupied space during daytime to handle the heat gains.

This technique is generally called Passive Cooling and it may enable the charging process to take place by means of free cooling, i.e. without running the chillers and as a result becomes a very economical and environmentally friendly system. Furthermore, PCM-TES opens new opportunities to explore heat balance for the existing and new systems, which could offer significant overall system efficiency improvements.

3 Methodology

The research was carried out using mixed methodology which includes experimental and modelling and quantitative user evaluation methods. The experiment was set-up in an educational space and three offices within a University Higher Education Building. The offices where selected on the South-west side of the building which received the highest solar radiation. Both the rooms with PCM and the control room were monitored. The experiment rooms and the control room-have been monitored using air temperature and humidity sensors. The monitored facilities are instrumented to monitor the indoor and outdoor climatic conditions. Data from the experimental period is analysed to understand thermal behaviour under different seasons based on the outdoor temperature and solar trend.

There are many building, and system modelling software tools used within industry and academia for the predictive analysis of building systems and their impact on energy and environmental performance. Each program has unique features in terms of modelling resolution, solution algorithms, intended target audience, modelling options, ease of use vs. flexibility [20]. IES Virtual Environment (IESVE) is an integrated suite of applications linked by a Common User Interface (CUI) and a single Integrated Data Model (IDM). This means that all the applications have a consistent "look and feel" and that data input for one application can be used by the others. The modular structure of the software allows for integrated building performance analysis in multiple domains (i.e. thermal, airflow and daylight) [21]. Crawley et al [22] critically evaluated twenty major software's such as EnergyPlus, ESP-r, ICE, and TRNSY. IESVE was categorized as one of the software's that has undergone the most rigorous validation studies with the most powerful modelling capabilities. IESVE software has therefore selected for estimating the cooling and heating demand of the case study spaces.

Apart from the experimental and modelling elements of the research, quantitative research methods have been used to collect user behaviour and satisfaction data that guides the selection of spaces for the intervention and for evaluating satisfaction with the quality of internal environment pre-and post PCM installation.

Description Case Study Building and Rooms

Figure 1 shows a model of the case study building which was used for the implementation of PCM. This case study building has a brick façade and 30% glazed areas. Window frame in this building is metal frame with 6mm single glazed. The structure is a concrete frame system.

Figure 1: Case Study Building Model

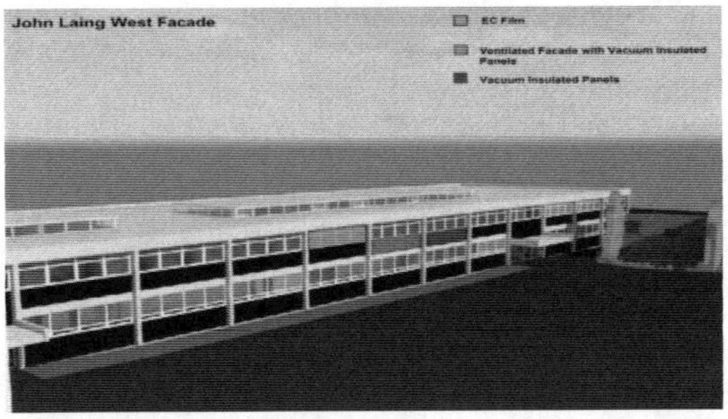

Figure 2: PCM Tube Retrofit

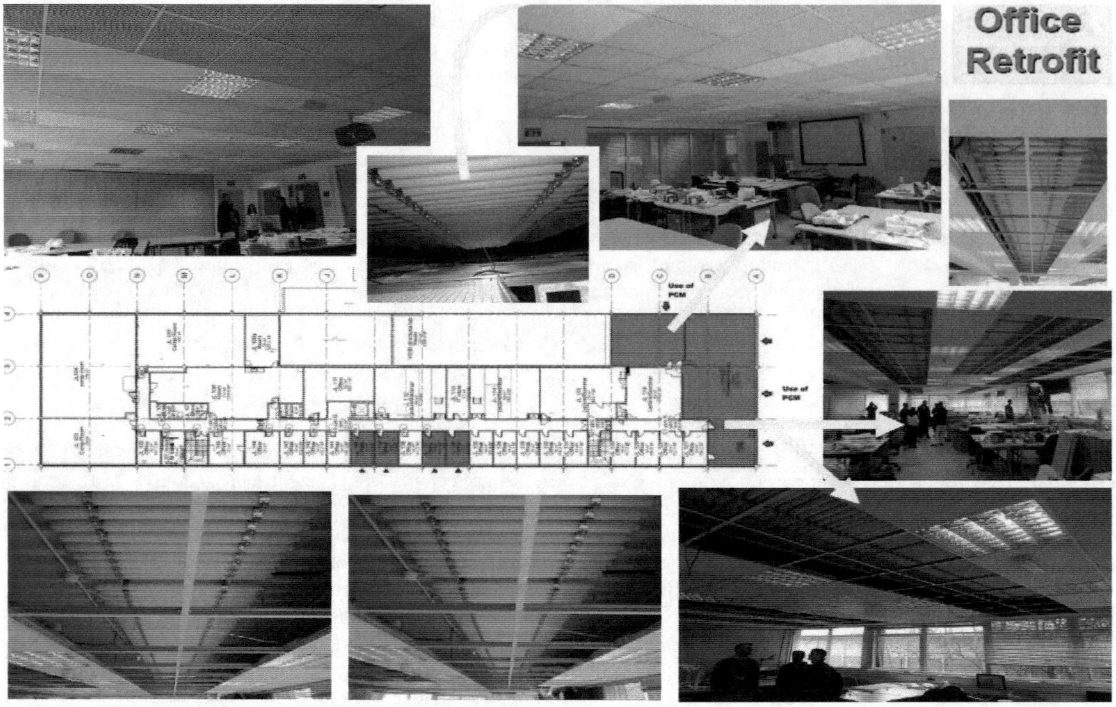

The IES model was used to evaluate the cooling and heating demand of various spaces within the building which was ultimately used in sizing the total number of PCM tubes required to meet the thermal demand of the rooms.

Figure 2 shows the offices and teaching spaces where the PCM was installed, and the installation process. The spaces have been selected based on an initial user satisfaction survey which suggests that occupants are not happy with the space due to complaints of overheating in summer. Both the offices and the Architecture studio are located on the south and south-westerly orientation which leads to significant solar gain that is one of the reasons for the elevated indoor temperature in summer. The Architecture studio is also used by students late in the evening. The students complained of significant drop in indoor temperature at night which suggests that there is significant diurnal temperature range because of both poor insulation and lack of building thermal mass to moderate the indoor temperature.

The PCM Tube passive cooling modules were installed behind ceiling tiles using standard 2" pipe brackets which was installed as secondary structure to support the weight of the PCM modules. The performance of the PCM tubes is dependent on the level of thermal interactions between the PCM tube and the room ambient air. This is essential for the convective heat transfer between the PCM tubes and the room ambient air which enables heat transfer from the room to the PCM and vice versa. The other important heat transfer mechanism that influences the performance of the PCM is the direct view between the PCM and the occupied space, which is essential for radiative heat transfer to take place [23]. As a result, the solid ceiling material has been replaced with perforated ceiling grills to enable both air circulation and the direct view between room and PCM, while maintaining the aesthetically pleasing view which is essential when considering internal interventions.

Table 1 and Table 2 show the estimate of number of PCM tubes based on the cooling load of the spaces for Architecture Studio and offices respectively. For the architecture studio, the cooling load estimated using IES virtual environment is used to calculate the cooling demand based on 12 hours occupancy, while for the offices 9 hours occupancy is used based on the average hours declared by occupants. 1m long PCM Tube will provide 0.145 kWh of thermal storage capacity [24], therefore, the total number of PCM tubes selected are 2211 and 143 for Architecture studio and the offices respective.

Table 1 Number of PCM Tubes for Architecture studio

Numbers of PCM Tubes for Architecture Studio	
Total cooling load (W)	25309.922
Total Cooling Load (kW)	25.310
Total Cooling demand (kWh) 12 hours occupancy	303.719
Tube/kWh	0.145
Number of Tubes	2211.075

https://doi.org/10.1051/e3sconf/20183201001

Table 2 Number of Tubes per office

Number of Tubes for Offices	
Total cooling load (W)	2189.84955
Total Cooling Load (kW)	**2.18984955**
Total Cooling demand (kWh) 12 hours occupancy	19.70864595
Tube/kWh	0.145
Number of Tubes	143.4789425

Results and Discussion

The indoor temperature of the spaces has been monitored from May to November 2016. Three offices were monitored: two with PCM Tubes and a control room. All the offices are free running rooms with no mechanical ventilation or cooling. Ventilation is by natural ventilation using openable windows. Figure 3 shows the indoor temperature for the three offices for the period May to October 2016. The blue and red lines show the rooms with PCM installation, while the red line is the temperature in the control room. During the study period there is consistent evidence that the rooms with PCM have more stable indoor internal conditions. The peak summer indoor temperature is consistently lower in the rooms with PCM compared to the control room.

Figure 3 Summer indoor temperatures for offices

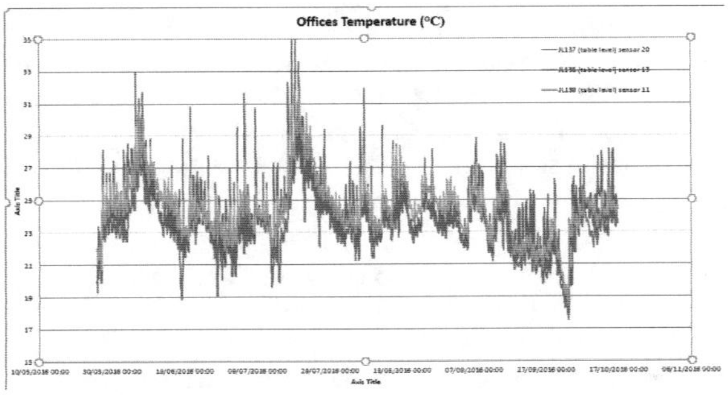

Figure 4 and Figure 5 also show the indoor temperature over a shorter cycle between 4th to 9th of June 2016 and 17th to 22nd of July 2016. The results again reveal significant benefit of PCM tube in supressing the peak indoor temperature during these periods. During the worst period of sustained high temperature, it shows that the rooms with PCM have high minimum temperatures of 22°C and 19.9°C compared to the control room temperature of 19.6°C. This suggests that the indoor diurnal temperature swing in rooms with PCM is much smaller than the control room, as suggested by the difference between maximum and minimum temperature of 6.8°C and 7.20 °C in rooms with PCM compared to 10.9°C in the control room. With air temperature being the main factor affecting thermal comfort, this reduction will have significant benefit for thermal comfort. Because of the increase in thermal mass, it is expected that there will be reduction in mean radiant temperature in the summer months will also be lower in rooms with PCM further, which will lower the operative temperature in the rooms with PCM.

Figure 4: Indoor temperature 05.06.2016 - 08.06.2016

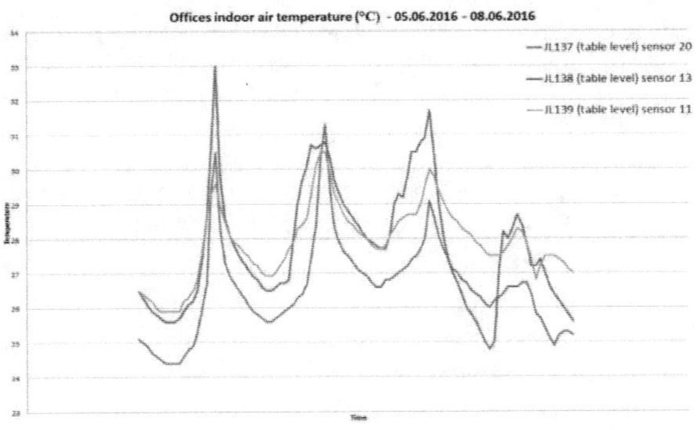

Figure 5 Indoor temperature: 17.07.2016 - 21.07.2017

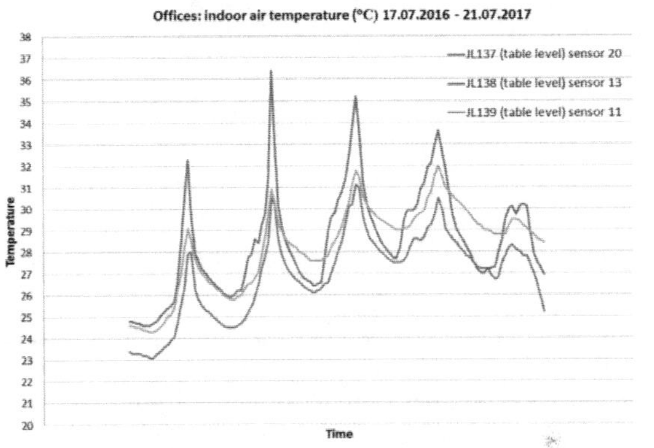

The absolute maximum temperature also shows significant benefit in the rooms with Phase Change Materials which have maximum value over the summer period of 32°C and 31°C compared to 36.4°C in the control room showing the temperature difference of up to 5K.

Even though the PCM tube shows significant potential for reducing indoor temperature during the summer, the current installation works on a 100% occupant control for opening windows during the day. For security reasons all the windows are closed overnight which means that not all the heat absorbed by the PCM during the day is released out of the room overnight. It is therefore expected that if the rooms were ventilated during the night as part of night ventilation strategy, the performance of the PCM will be much better than the current performance. The ventilation strategy can be made either in passive means or to install small fans that will be used to ventilate the building during cold periods of the night to remove the heat absorbed by the PCM during the day time. It is anticipated that this will further significantly improve the benefit of the PCM and further supress the maximum indoor temperature in room with PCM Tubes. The night ventilation strategy will be much better if the security situation can be overcome using fixed high-level windows.

User satisfaction questionnaires were deployed pre-and post-retrofitting of the PCM tubes. The results shown in Figure 6 shows that there is significant improvement in the user satisfaction after the PCM installation. There appear to be significant improvement in user perception with more occupants rating their indoor environment 4-7 in the 7-point scale compared to pre-retrofit period. It is therefore very conclusive evidence of PCM performance when looking at both objective and subjective measures of thermal comfort.

Figure 6: User Satisfaction

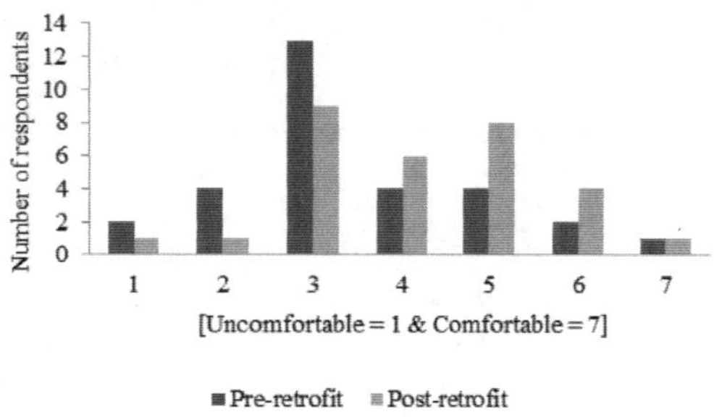

Conclusion

The paper investigates the benefit of PCM Tubes systems for reducing overheating in non-domestic buildings in the UK. The application of the PCM tube was based on a passive approach and relying on maximum occupant control of ventilation strategy such as opening and closing of windows. The study finds that it is possible to increase internal thermal mass in existing buildings through retrofitting of PCM tubes due to their thermal capacity per unit volume. The study also finds that the PCM tube can reduce peak indoor temperature by up to 5°K.

Despite the positive performance of the PCM tubes in this investigation, there is still potential to improve the performance of the system by implementing effective night ventilation. The main limitation of this study is the lack of effective night ventilation strategy. Further experimental study is in progress to evaluate the impact of the PCM when effective night ventilation strategy is implemented. There is also evidence through the user evaluation that the PCM has made a difference in terms of the perception of users for improving comfort conditions. The objective and subjective assessment of thermal comfort all suggests significant benefit to the application of PCM in the occupied spaces.

References

1. European Union, Directive 2012/27/EU of the European parliament and of the council efficiency Official Journal of the European Union, 25-Oct-2012, http://eurlex.europa.eu/LexUriServ/LexUriServ.do?uri=OJ:L:2012:315:0001:0056:en:PDF.
2. HughesBR, CalautitJK, GhaniSA. The development of commercial wind towers for natural ventilation: a review. Applied Energy 2012; 92: 606–627. http://dx.doi.org/10.1016/j.apenergy.2011.11.066.
3. Balaras, C. Heat Attenuation: in Passive Cooling of Buildings Ed. Santamouris, M and Asimakopoulous, D. (1996) Science Publishers, London.
4. IPCC Climate Change 2014: Impacts, Adaptation, and Vulnerability – Chapter 8 Urban Areas. (2014) Online: http://www.ipcc.ch/pdf/assessment-report/ar5/wg2/WGIIAR5-Chap8_FINAL.pdf. Accessed 10/10/2017.
5. Liu, C. and Coley, D. Overheating risk of UK dwellings under a changing climate. 6th International Building Physics Conference, IBPC 2015. Energy Procedia 78 (2015) 2796 – 2801 available online: http://www.sciencedirect.com/science/article/pii/S1876610215023607 accessed 10-10-2017
6. Hamdy. M., Carlucci, S., Hoes, P., and Hensen, J The impact of climate change on the overheating risk in dwellings – A Dutch case study. Building and Environment 122. Pages 307-323. (2017)
7. Zero Carbon Hub ASSESSING OVERHEATING RISK LEAFLET. Zero Carbon Hub. (2016) Online:http://www.zerocarbonhub.org/sites/default/files/resources/reports/ZCH-OverheatingLeaflet-3-Modelling-S.pdf accessed: 10/10/2017.
8. Metoffice. Too hot, too cold, too wet, too dry: Drivers and impacts of seasonal weather in the UK. (2014) Availableonline:https://www.metoffice.gov.uk/binaries/content/assets/mohippo/pdf/4/8/drivers_and_impacts_of_seasonal_weather_in_the_uk.pdf. Accessed 01/10/2017.
9. Eicker, U Passive Cooling Strategies: In: Low-Energy Cooling for Sustainable Buildings. Wiley, UK. (2009)
10. Balaras, C (1996) Heat Attenuation: in Passive Cooling of Buildings Ed. Santamouris, M and Asimakopoulous, D. Science Publishers, London.
11. Derradji, L., Errebaia F.B.,Amara, M., Effect of PCM in Improving the Thermal Comfort in Buildings. (2016) Energy Procedia 107 (2017) 157 – 161.
12. Auzeby, M., Wei,S. Underwood, C., Chen, C., Ling,H., Pan,S., Bobo N., Tindall, J., Buswell,R., Using phase change materials to reduce overheating issues in UK residential buildings. Energy Procedia 105 (2017) 4072 – 4077

13. Nghana, B., and Tariku, F.) Phase change material's (PCM) impacts on the energy performance and thermal comfort of buildings in a mild climate. Building and Environment: Volume 99, April 2016, Pages 221-238.
14. Derradji, L., Errebaia F.B.,Amara, M., Effect of PCM in Improving the Thermal Comfort in Buildings. Energy Procedia 107 (2017) 157 – 161.
15. Solgi, E., Fayaz, R., Kari, B.Z., Cooling load reduction in office buildings of hot-arid climate, combining phase change materials and night purge ventilation. Renewable Energy, Volume 85, January 2016, Pages 725-731
16. Álvarez, S., Cabeza, L.F., Ruiz-Pardo, A., Castell, A., Tenorio, J.S. Building integration of PCM for natural cooling of buildings. Applied Energy Volume 109, September 2013, Pages 514-522.
17. Souayfane,F. Fardoun, F. Biwole, P. Phase change materials (PCM) for cooling applications in buildings: A review. Energy and Buildings, Volume 129, 1 October 2016, Pages 396-431.
18. Liu, S. and, Li, Y. An experimental study on the thermal performance of a solar chimney without and with PCM. Renewable Energy, Volume 81, 2015, Pages 338–346.
19. Iten, M, Liu, S., and Shukla, A. A review on the air-PCM-TES application for free cooling and heating in the buildings. Renewable and Sustainable Energy Reviews. Volume 61, 2016, Pages 175-186.
20. Roel C.G.M. Loonen, Fabio Favoino, Jan L.M. Hensen & Mauro Overend. Review of current status, requirements and opportunities for building performance simulation of adaptive facades. Journal of Building Performance Simulation, (2016): DOI: 10.180/19401493.2016.1152303.
21. Roel C.G.M. Loonen, Fabio Favoino, Jan L.M. Hensen & Mauro Overend. Review of current status, requirements and opportunities for building performance simulation of adaptive facades. Journal of Building Performance Simulation, (2016): DOI: 10.1080/19401493.2016.1152303.
22. Crawley, D., J. Hand, M. Kummert, and B. Griffith. "Contrasting the Capabilities of Building Energy Performance Simulation Programs." 2008. Building and Environment 43 (4): 661–673. doi:10.1016/j.buildenv.2006.10.027.

[23] Moss.K.J. Heat and Mass Transfer in Buildings: Chaper 7 Heat Radiation. Second Editio. Taylor and Francis, Oxon. 2007.

[24] PCM Product Ltd. PlusICETM Phase Change Materials Passive Cooling. Available Online: http://www.pcmproducts.net/files/Passive%20Cooli ng%20Catalogue.pdf

E3S Web of Conferences **32**, 01002 (2018) https://doi.org/10.1051/e3sconf/20183201002
EENVIRO 2017

Determination of thermal and acoustic comfort inside a vehicle's cabin

Alexandra Ene (Angelescu)[1], *Tiberiu* Catalina[1*], and *Andreea* Vartires[1]

[1]Technical University of Civil Engineering, Faculty of Building Services, Romania

Abstract. Thermal and acoustic comfort, inside a vehicle's cabin, are highly interconnected and can greatly influence the health of the passengers. On one hand, the H.V.A.C. system brings the interior air parameters to a comfortable value while on the other hand, it is the main source of noise. It is an intriguing task to find a balance between the two. In this paper, several types of air diffusers were used in order to optimize the ratio between thermal and acoustic interior comfort. Using complex measurements of noise and thermal comfort parameters we have determined for each type of air diffuser the sound pressure level and its impact on air temperature and air velocity.

1 Introduction

Nowadays, many people spend a significant proportion of their time travelling and there is an increasing demand for comfort, in both private and public transport [1].
Noise and vibrations have an important influence on a customer's perception of vehicle quality and interior cabin noise levels are key criteria when choosing a vehicle [2]. Further on, the HVAC system represents one of the most important parameters that influences the interior comfort of a vehicle's cabin. On one hand it represents necessary equipment in order to obtain thermal comfort while on the other hand, it represents the most important parameter that influences negatively the acoustic comfort, due to air flows from the multitude of the HVAC settings. Different supply air velocity and temperature on the car chamber has a great influence on air distribution characteristics [3].
Commonly, vehicle occupants choose to operate the vehicle ventilation system in the "recycle" mode to prevent outdoor-polluted air entering the vehicle's cabin and by that raising the concentration of CO_2 [4].
The thermal comfort sensation is assured by the factors that depend on the heat exchange between the human body and the ambient environment [5].
The ventilation and air conditioning system consists of a fan, a mixing unit (formed from various equipment) and air ducts that lead to the places where the air enters the cabin, including the front grids, the grids from the lower/feet area, for both the front and back passengers, demisting and defrosting actions, and other combinations depending on the type and construction of the chosen car [6].

2 Experimental campaign

2.1 Acoustic measurements

The determinations were performed on a 2012 Dacia Duster vehicle, in an enclosed room in order to maintain external background noise as low as possible. The tested vehicle presents different positions through which air is introduced into the cabin, as shown in Fig. 1. The grids numbered with "1" have a demisting role. The four ones, numbered with "2", introduce the air towards the face of the occupants and the ones numbered with "3" can introduce air at the bottom part of the cabin where the passenger's feet should be.

Fig. 1. Dacia Duster vehicle's interior

* Corresponding author: tiberiu.catalina@gmail.com

https://doi.org/10.1051/e3sconf/20183201002

The fan presents four different gears through which the airflow introduced into the cabin can be changed, as shown in Fig. 2. The car also presents five types of combinations which can be chosen for introducing the air into the cabin, formed from a combination of the three positions shown in Fig. 1.

- Position (a): the air is introduced for demisting (fig.1, grids "1");
- Position (b): the air is introduced for demisting and the lower/feet area (fig. 1, grids "1" and "3");
- Position (c): the air is introduced for the lower/feet area (fig.1, grids "3");
- Position (d): the air is introduced towards the face and for the lower/feet area (fig.1, grids "1" and "3");
- Position (e): full towards the face (fig.1, grids "2").

Fig. 2. Methods for adjusting the H.V.A.C. system

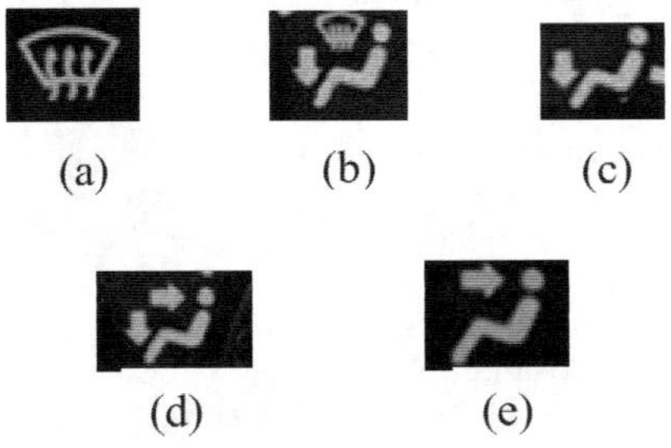

Fig. 3. Typical vehicle HVAC system operations

2.1.1. Equipment used in the study

The sound pressure levels were determined by means of a sound level meter *Bruel&Kjaer 3560-B-020,* precision class 1. The sonometer presents five receivers giving the ability to measure noise level from five different points, simultaneously.

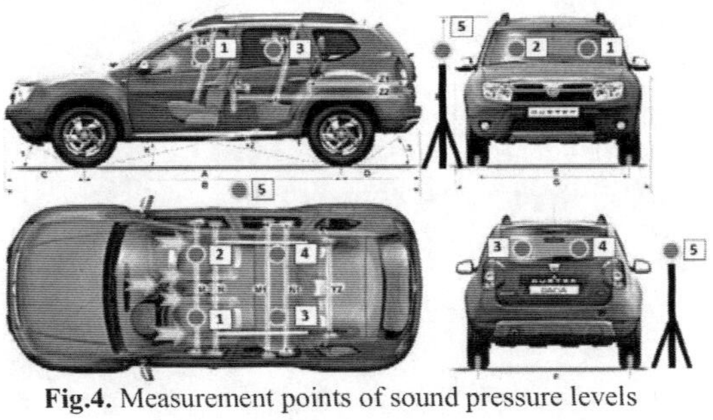

Fig.4. Measurement points of sound pressure levels

The receivers were placed on each of the points occupied by a passenger, as shown in Fig. 4. Point "1" represents the driver's position, point "2" the front passenger, near to the driver's seat, point "3" the passenger from the back of the chauffeur and the "4" point corresponds to the back passenger who seats diagonally from the driver's chair. The exterior sound pressure level was constantly measured by receiver "5" placed outside the car at a distance of about 0,5 m.

The experimental campaign was conducted for five different air vents. The first one used was the initial air diffuser of the car, the second one – air diffuser spherical type one (prototype 1), third one – air diffuser spherical type two (prototype 2), fourth - air diffuser propeller type (prototype 3), fifth - classic air diffuser type (prototype 4).

First of all, the sound pressure level with the ventilation system off was measured, then the measurements were conducted for every fan gear, for every position of air inlets and also replacing the originals air grids with the four new prototypes. The sound pressure level was recorded for 20 seconds in each mentioned case.

The sound pressure level resulted exclusively from the ventilation system as a difference between the sound pressure level measured with the ventilation system on (for the mentioned cases) and the sound pressure level measured with the ventilation system off.

The results were downloaded on the computer by means of the software "Labshop" and the data was imported into an Excel document.

The air flows for every fan gear were determined by placing the Testo 480's anemometer at the air inlet. The air flows resulted for every fan gear are: Gear 1=169.5 m^3/h, Gear 2=251.0 m^3/h, Gear 3=343.2 m^3/h, Gear 4=451.3 m^3/h.

Fig.5. Photographs taken while measurement campaign was going

2.1.2. Results of the acoustic experimental campaign

The results display the fact that by using the demisting air vents, the interior sound pressure level is highest, up to 5 dBA higher than the others, representing a value which can be detected by the human ear. The air vents present a higher contribution on the sound pressure level because of the air hitting solid surfaces. Furthermore, air entering the cabin in full face mode represents the second highest sound pressure level. The sound pressure level differences between air vents is overall small, the highest one being of 2 dBA.

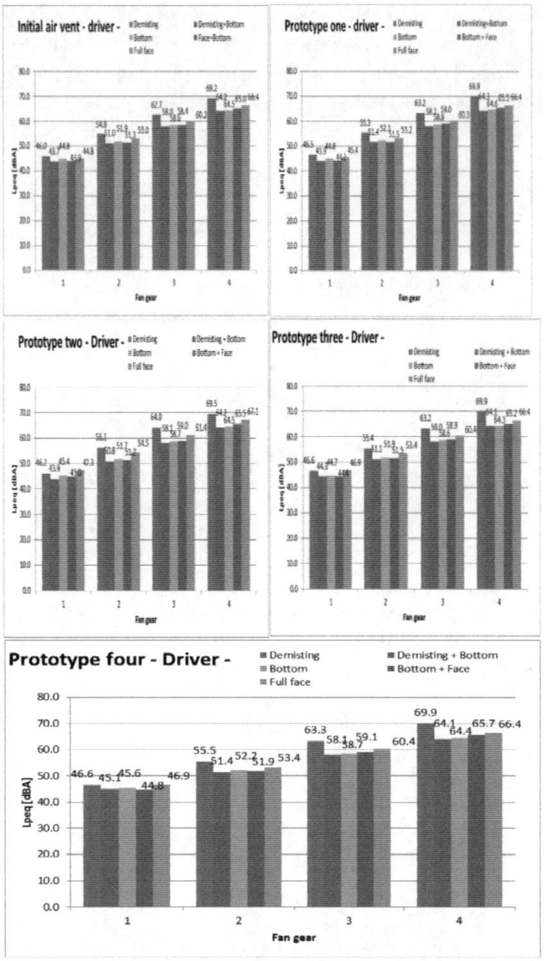

Fig. 6. Sound pressure level for driver position with different air vents

The results showed also a rise of 48.2% of the sound pressure level for gear 4 of the fan, comparing with a rise of the air flow of 166% from gear 1 of the fan. Further on, the graphics show a difference up to 5 dBA sound pressure level between the front and the back passengers' positions for every type of air vent analyzed.

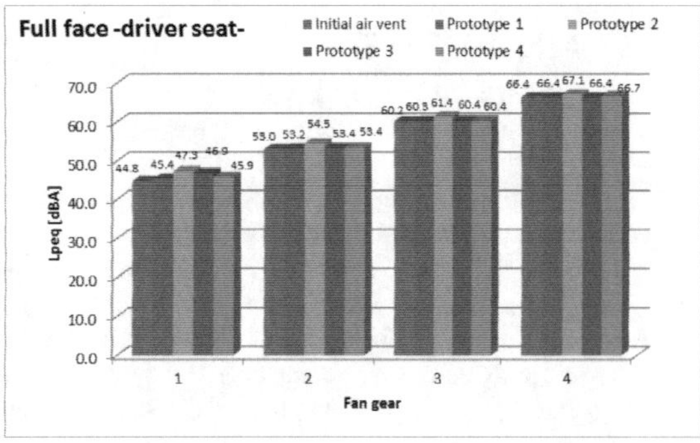

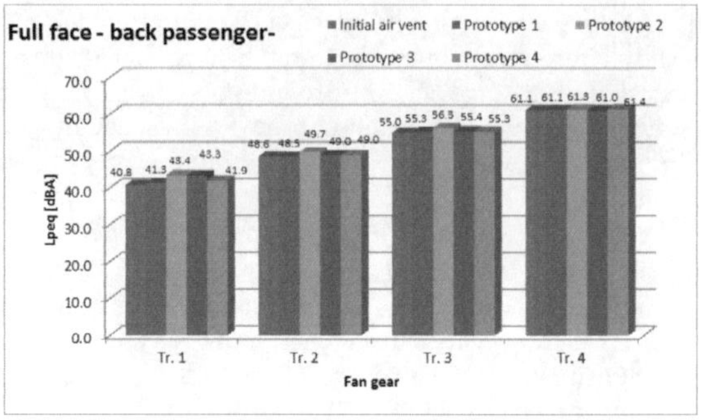

Fig. 7. Sound pressure level for driver and back passenger's position with different air vents

2.2. Thermal comfort measurements

The ventilation and air conditioning system presents the role of maintaining the interior air parameters in comfort conditions for passengers. The main factors which affect the human thermal comfort depend on four physical aspects which vary with the environment:

- Air temperature;
- Relative humidity;
- Mean radiant temperature;
- Air velocity.

The interior environment of a vehicle is affected by a large number of parameters in which can be included the indoor and outdoor temperature, solar radiation, geometrical characteristics and materials, reflection coefficient and also thermal mass of the interior materials. All these factors slow the calculation method and the predictability of the thermal comfort. For this reason in the present paper, the analyzed situation is the one in which a part of these factors were eliminated with the purpose of a more accurate determination of the interior comfort, without the external intervention of individuals presenting factors which are difficult to predict.

This measurement campaign was conducted in an indoor room with the purpose of eliminating the direct solar radiation. Also, the same model of vehicle, presented before, was used.

2.2.1. Equipment used in the study

For this experimental campaign, two identical equipment Testo 480 were used, which contain multiple devices for measuring the IAQ, for calculation and analyzing PMV/PPD, air temperature, air velocity etc.

https://doi.org/10.1051/e3sconf/20183201002

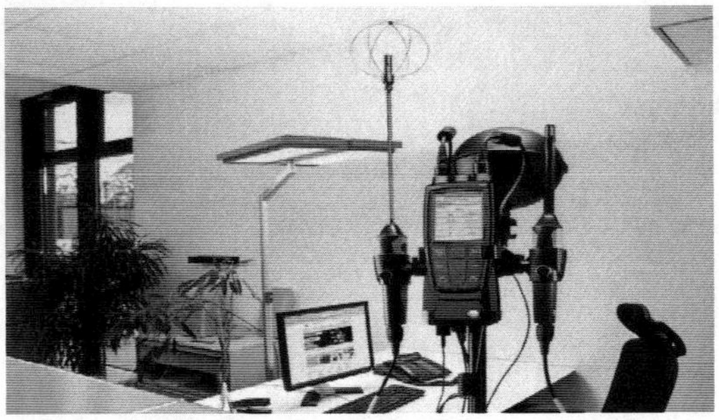

Fig. 8. Equipment used for the comfort determinations Testo 480

The determinations were conducted in an enclosed space in order to eliminate various factors like direct solar radiation. The two Testo 480 equipment were placed in the car (one on the driver's seat, and the second one on the back passenger's seat) as shown in fig. 9.

The HVAC system was set in order to introduce fresh air from the outside, with the recycling mode off, fan gear two and the air conditioning on. Every determination lasted ten minutes and consisted in 600 measurements. For every new phase, it was needed to get inside the vehicle and save the data. Also in order to determine the PMV and PPD indices, cloth isolation was set at 0.6 clo and the metabolism 1 met. During the experimental campaign the air temperature raised from 30°C to 34°C.

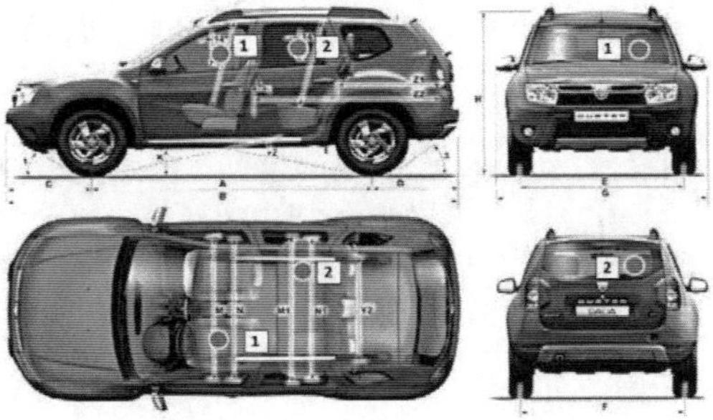

Fig. 9. Measurement points for the comfort analysis

2.2.2. Results of the thermal comfort experimental campaign

The results showed a significant air temperature difference between the front and the back passengers' position. This difference can lead up to 2°C when using the full face inlets and the combination of face and lower air grids. Furthermore, the indoor air temperature raises from about 23 °C (for the air diffusers propeller type) to 30 °C at the end of the measurement campaign (for the car's initial air diffusers).

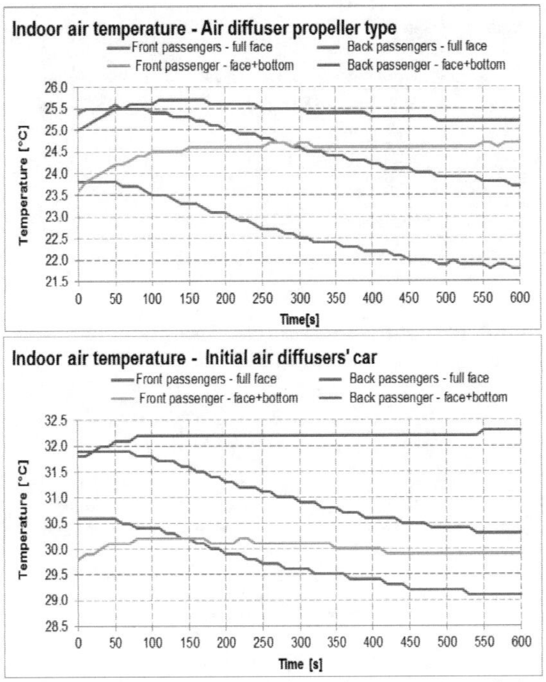

Fig. 10. Variation of the indoor air temperature during measurement campaign

2.2.3. Results of the second thermal comfort experimental campaign

Due to the fact that indoor air temperature rose drastically through the experimental campaign, it was decided to retake the measurements. Position and time of the measurement remained the same as in the first campaign, the only difference being that the car was placed outdoor, screened away from the direct solar radiation. However, the outdoor air temperature raised 2 °C while the measurement campaign took place.

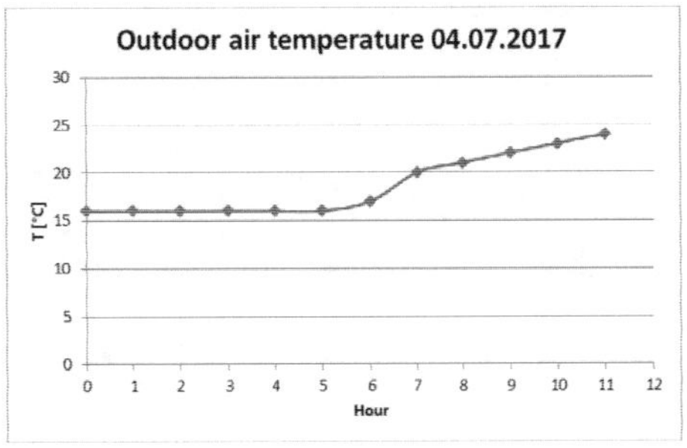

Fig. 11. The evolution of the outdoor air temperature in the day of the measurement campaign

Further on, the air conditioning system was turned off due to cold temperatures outside.

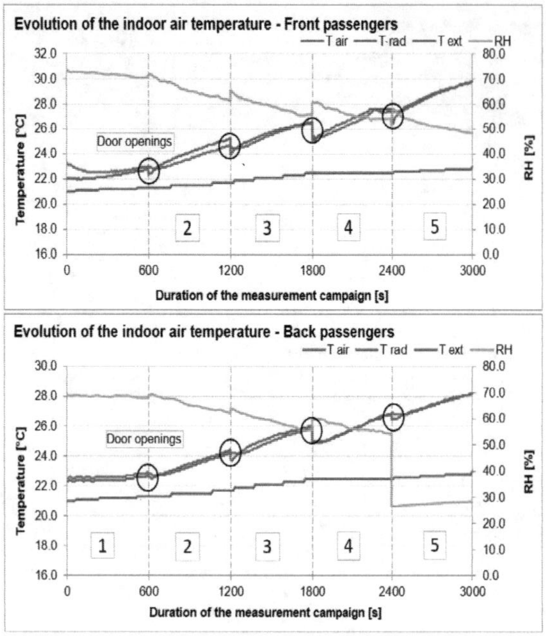

Fig. 12. The evolution of the indoor air temperature

As Fig. 12 shows, the indoor air temperature raises drastically during the measurement campaign, while the outdoor temperature remains constant. The vertical lines mark the measurement sets. If at the beginning of the measurement, indoor air temperature was about 22.5°C and the outdoor air temperature 21°C, at the end of the campaign the interior air temperature in the cabin's vehicle reaches 29.9°C even though the outside air temperature doesn't exceed 23°C.

Likewise, the door openings were highlighted with a circle. Regarding the back side of the car, the air temperature differences can still be identified.

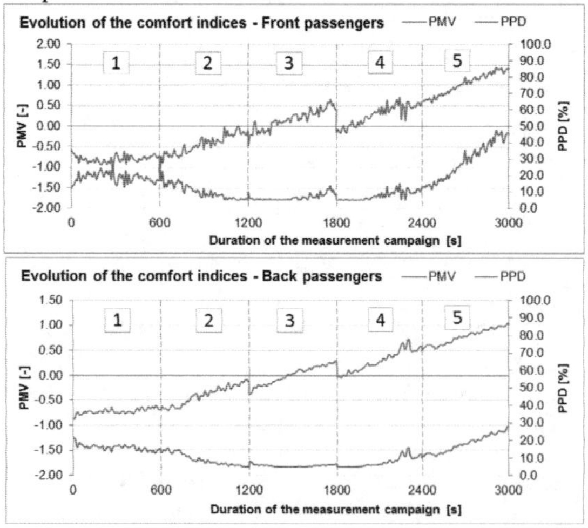

Fig. 13. The evolution of the PMV and PPD indices

Simultaneously with the raise of the air temperature the PMV and PPD indices are progressing, thus, if at the beginning of the measurement campaign PMV has the value of -0.5, at the end of the measurements the value does not exceed 1.

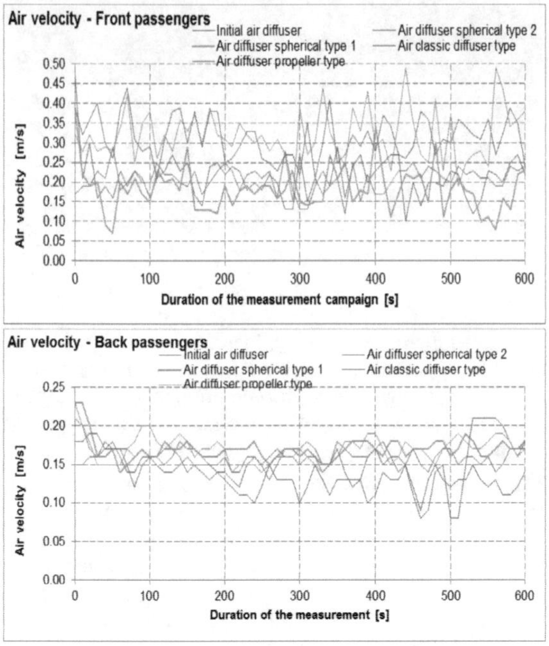

Fig. 14. Air velocity introduced by different air vents

Due to the fact that the temperature inside the vehicle varies excessively, the only way to compare the air vents is through air velocity. As it can be seen in Fig. 14, the initial air vents of the car can introduce an air velocity up to 0.45 m/s, representing the highest values. Analyzing the average values of the air velocity, it can be concluded that air diffusers spherical type 1 determine the lowest value at about 0.183 m/s. Regarding the back passengers, they would have preferred the use of the air diffuser spherical type 2 which can introduce a decrease down to 17%, unlike the initial air diffuser of the car.

3 Conclusion

For the acoustic comfort it can be concluded that:

- The sound pressure levels coming from the HVAC system are higher than the ones coming from the engine;
- The acoustic differences between air vents are very low;
- In order to reduce the noise level, a damper is needed.

As for the thermal comfort inside the vehicle's cabin, it can be concluded that:

- The air temperature inside the cabin is higher than the outside air temperature;
- The air diffusers spherical type introduce the lowest values for air velocity;
- The air temperature inside the cabin increases simultaneously with the raise of the operating time of the vehicle, giving the need of using the air conditioning even though the outside temperature is at a comfortable value;

– The vehicle presents a difference of 2 °C between front and back passengers.

Acknowledgements

This work was supported by a grant of the Romanian National Authority for Scientific Research, UEFISCDI, project number PN-II- PT- PCCA-2013- 4-0569.

References

1. Gameiro da Silva M. C., *Measurements of comfort in vehicles*, Measurement Science and Technology, (2002)
2. Watkins S., Wang X., Mavuri S. P., *An Investigation and Bench Marking of Vehicle HVAC Cabin Noise*, SAE TECHNICAL PAPER SERIES (2008)
3. Yang H. s.a., *The Analysis on the Effect of Passenger car Air Conditioning and Distribution with Different Inlet Parameters*, International Conference on Information Sciences, Machinery, Materials and Energy, (2015)
4. Wells B., McDonald S.,*Vehicle Cabin Air Quality Monitor for Fatigue and Suicide Prevention*, SAE TECHNICAL PAPER SERIES, (2000)
5. Simion M. s.a., *Factors which influence the thermal comfort inside of vehicles*, Sustainable Solutions for Energy and Environment, EENVIRO - YRC 2015, (2015)
6. Angelescu A. s.a., *Acoustic measurements inside a vehicle with different air prototype diffusers,* Romanian Journal of Acoustics and Vibration, Volume XIV (2017)

Air solar collectors in building use - A review

Andrei-Stelian Bejan[1], *Abdelouhab* Labihi[2], *Cristiana* Croitoru[1,*] and *Tiberiu* Catalina[1]

[1]Technical University of Civil Engineering Bucharest, Faculty of Building Services, 021414 Pache Protopopescu Blvd., Bucharest, Romania
[2]Faculty of Sciences and Techniques, Cadi Ayyad University, LP2M2E Laboratory, Marrakech, Morocco

Abstract. In the current energy and environmental context it is imperative to implement systems based on renewable energy sources in order to reduce energy consumptions worldwide. Solar collectors are studied by many years and many researchers are focusing their attention in order to increase their efficiency and cost-effectiveness. Water solar collectors are often implemented for domestic hot water, heating or industrial processes and already have a place on the market. A promising system which is not yet widely known is represented by air solar collectors that could represent an efficient way to use the solar energy with a lower investment cost, a system that can be used in order to preheat the fresh air required for heating, drying, or to maintain a minimum temperature during winter. This paper presents a comprehensive literature review on air solar collectors used mainly in buildings, acting as a solar wall. Air solar collectors are roughly classified into two types: glazed and opaque. The present study comprises the solar collector classification, applications and their main parameters with a special focus on opaque solar collectors.

1 Introduction

The renewable sources can provide low-cost energy consumption solutions when using passive systems. Among these renewable sources, the use of systems which are using solar energy is largely acknowledged, being easy to implement and efficient from the accessibility point of view in the zones with solar potential. Such systems can respond to the European Directives concerning energy performance of buildings [1] which impose significant reduction of the energy consumption. Generally, building sector consumes 35.3% from the total energy demand [2, 3]. In this context the use of renewable energies is an attractive solution for fulfilling the two requirements: indoor quality and energy efficiency. Solar systems integrated in the building can use inertial elements where the energy is stored, or the solar energy is used to heat the fluid: air or water. Thermal systems can be classified into four important categories: systems that use a massive wall to store heat (known as *solar walls or Trombe walls* depending on some characteristics), systems that use water to store the heat, systems that use eutectic materials (or phase changing materials) and *solar collectors*.

Thermal solar collectors are a cost-effective method of using the solar energy as a renewable source of energy. These solutions have been used and studied since earliest times and can be classified into solar collectors using water and solar collectors using air. Solar water collectors are a more often an option for domestic hot water preparation or for preheating the thermal agent required for heating. Solar thermal air collectors are used to preheat the fresh air required for heating, drying, or to maintain a minimum temperature during winter. Unlike water collectors, the air collectors have the advantage that there is no risk of freezing and also, these systems have a lower environmental impact [4]. According to Goyal et al. [5], solar thermal air collectors can provide a temperature of up to 65 °C, indicating a real interest in building use. The survey of literature allows us to be aware of the huge possibilities of such devices in energy recovery from the Sun.

A Transpired Solar Collector (TSC) or air solar collector is made of metal cladding with perforations, installed at a certain distance from a building wall, thus creating a cavity through which the air is circulating. The metal cladding is heated by the solar radiation from the sun and ventilation fans create negative pressure in the cavity, extracting the solar heated air through the perforated panel. The air is generally taken off at the top of the wall (due to air temperature gradients in the cavity) ensuring that all of the produced solar heat is collected and then distributed in the building via the ventilation system. The heat transfer between the fluid and the metal is enhanced depending on the **flow's characteristics and other external parameters**. Studies such as the ones of Van Decker et al. [6], Gunnewieck et al. [7, 8] are very interesting, given the information regarding the direct possibilities of improvement of these perforations and since the **form of the orifices can be an important factor**. The literature indicates the interest in using such technologies and **the implementation of a special geometry** for the perforations would conduct to a heat

* Corresponding author: cristiana.croitoru@utcb.ro

transfer increase, but no significant studies have been published [9].

2 Classification

Depending on the construction characteristics, solar collectors can be classified in several ways. According to Lai et al. [10] solar air collectors can be grouped into:
• Opaque solar collectors
• Transparent or glazed solar collectors.

Opaque solar collectors may be of several types: solar collectors in which the absorber is plane, solar collectors in which the absorber is corrugated, solar collectors where the absorber is perforated, or solar collectors in which the opaque absorber is protected by a layer of glass. Glazed solar collectors can also be divided into several categories [11], respectively the Trombe wall (absorbing element having the role of thermal mass), solar collectors with a plane absorber. Both types of collectors can contain several coils to increase heat transfer and both types of collectors can be with or without heat storage media (inertial materials).
In figure 1, Nkwetta et al. [12] propose a way of classifying solar collectors according to their heat storage capacity.

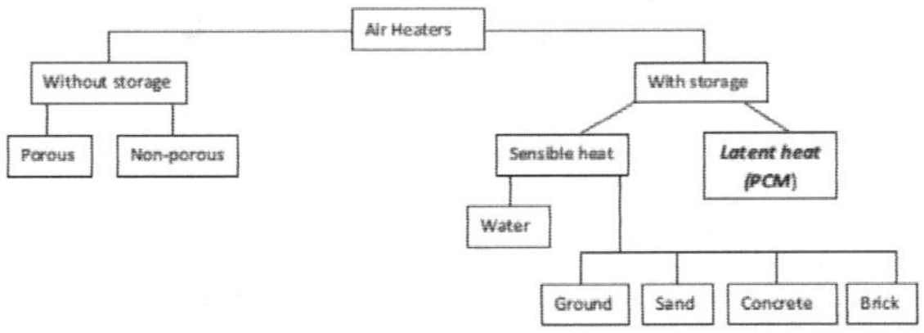

Fig. 1 Classification of solar thermal collectors according to the capacity to store the thermal energy according to [12]

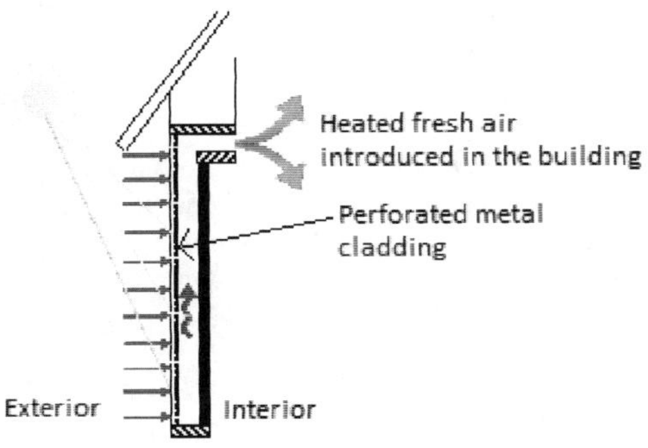

Fig. 2 Opaque solar thermal collector (perforated) model [13]

Regardless of their type, the functioning principle of the solar collectors is the same. Solar radiation is captured by an absorber that releases the heat to the circulating air through the formed cavity. In the case of an opaque solar collectors, the heat is usually captured by a dark metal panel or perforated panel (Figure 2), then it is ceded to the airflow which is then collected at the top of the collector and introduced into the room by a fan (in winter or during transition). If there are inertial elements integrated, the heat accumulated during the daytime can be later released at night. During daytime, in the summer season, the solar collector can function as a buffer space that reduces heat input through the building envelope or as a ventilated facade with the same role.

Regarding glazed solar collectors, the phenomenon is similar. Figure 3 represents a classic Trombe wall. The system is made up of a coating material (usually glass) that causes the greenhouse effect and allows the sun's radiation to penetrate (having the advantage of better lighting). Incidental solar radiation is captured by an absorbent layer with or without inertial elements, and then is released to the convective airflow. During the transition period and winter, the hot air can be used to heat the spaces or provide the fresh air needed. Accordingly to Hami et al. [11], Trombe walls and glazed solar collectors are the most used types of collectors.

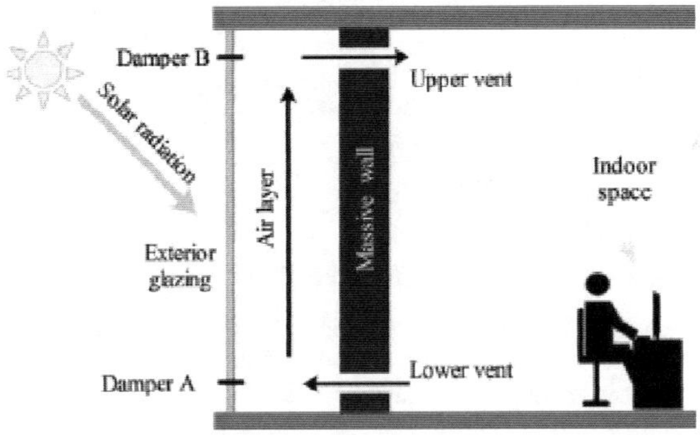

Fig. 3 Solar glazed solar collector model (Trombe wall) [14]

In figure 4, Shukla et al. [9] highlights the types of air solar collectors that have been studied so far (Figure 4). Figure 4a shows an opaque solar collector consisting of an absorbent element, the insulating wall and the cavity through which the air circulates. Figure 4b shows a glazed solar collector consisting of a glass layer, the cavity through which the air circulates and an insulating wall on which the absorbent part is mounted. Figure 4c shows a collector similar to the previous, but with two glass sheets. Figure 4d shows an opaque solar collector in which the absorbent element is protected by a glass layer which releases the heat to the circulated air. In this case, the absorbent element can also be perforated (similar to Figure 4f). Figure 4e shows a multi-coil solar collector used to enhance thermal transfer, while Figure 4f shows an opaque solar collector in which the absorbent element is perforated and releases heat to the air penetrating into the cavity through the orifices. In each of the cases presented, the insulating wall may be replaced by a heat storage material (sensitive or latent).

Perforated opaque solar collectors typically have a lower implementation cost and higher efficiency when used in heating systems, fresh air preheating systems, or drying systems [15] and may also have lower lifetime costs [4]. Opaque solar collectors are typically used for applications where the required temperature is lower [16] and the side winds are not very strong [17].

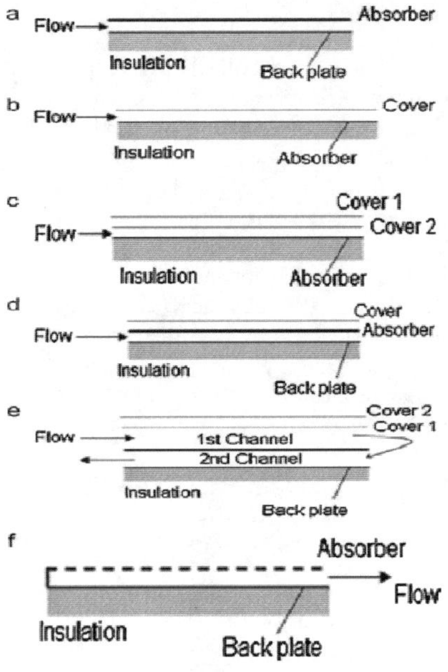

Fig. 4 Solar collector types [9]

According to different studies [12, 14, 15, 18-22], solar air collectors can be used for the following applications, with enormous potential to reduce energy consumption and operating costs [19]:

- buildings heating
- maintaining a low functioning temperature in industrial areas (for heating and drying of industrial halls)
- drying food
- preheating necessary fresh air
- heating and drying of greenhouses
- improving the efficiency of photovoltaic systems
- reducing energy consumption for cooling.

According to Leon and Kumar [19], solar air collectors have often been used for drying, using collectors with surfaces ranging from 90 to 1300 m^2 which have provided an air temperature between 27-70 ° C. In terms of building implementation, solar air collectors have been installed [9, 12, 14, 18]: in external walls (such as solar façades or ventilated facades), in the roof (as a ventilated roof) or as individual systems mounted on the roof, facade or exterior of the building. Another interesting solution involves the use of solar air for heating the air or other thermal agent for heat pumps [23, 24]. This way the performance coefficient of the air-to-water heat pumps can be improved. Accordingly to Ciriminna et al. [21] the main obstacle to the implementation of solar collectors with air is low market reputation and difficult access to this type of technology.

3 Applications

Perforated opaque solar collectors are often used in buildings due to their low cost of implementation and efficiency [25]. They are already implemented in large buildings in Canada, US or Europe and have the potential to replace glazed solar collectors [19], the implementation degree having an upward trend [23]. Accordingly to Gunnewiek et al. [8] they are especially recommended when direct outdoor air heating is desired and can achieve efficiency up to 72%.

Paya-Marin [26] has carried out an economic study on the implementation of opaque solar collectors. The author compared two types of opaque solar collectors, taking into account the installation and operating costs: collector without perforations (the air being oriented towards the back of the absorbent plate) and with a perforated collector. The payback period of the perforated collector is approximately 4 years, while in the case of the collector without perforations, the payback period is approximately 3 years (with an investment cost of 41% lower). However, the solar collector with perforations has a higher efficiency with 28%, this aspect being underlined by Belusko et al. [27]. The presence of perforations within the absorber plate allows recovery of lost energy by convection and radiation at the surface of the plate [28, 29]. Accordingly to Chan et al. [29] the efficiency of the perforated opaque collectors is even 50% higher than the flat ones without perforations.

4 Functioning parameters and energy efficiency of opaque solar collectors

According to several studies [14, 25], the important parameters defining the operation of perforated opaque solar collectors are:

- the pitch between the perforations (and the porosity of the plate)
- the diameter of the perforations and the shape of the perforations
- plenum size (airflow cavity)
- absorption coefficient of the absorbent element
- ambient temperature
- air density
- kinematic viscosity
- thermal conductivity of the air
- solar radiation incident on the solar collector
- air flow and fan speed
- wind speed (lateral or front)
- the temperature of the absorbent plate
- the temperature of the exhaust air.

All these parameters are included in the calculation of the parameters that define the thermal performance of the collector, namely: the efficiency of the heat exchange (εHX), efficiency of the solar collector (η) and temperature difference between the introduced air and the exhaust air (ΔT).

Accordingly to Cirimina et al. [21] the most important parameter of an opaque solar collector with perforations is the heat exchange efficiency and its value depends on: the diameter and the pitch of the perforations, the thickness of the absorbent plate and the material from which it is made. The thermal efficiency of air solar collectors could be improved by using a series of baffles, obstacles or fins that increase the heat exchange surface and increase the uneven flow of the flow [27]. Goyal et al. [5] states that such a solution has a low investment cost and can increase collector efficiency by 10-15%. A comprehensive bibliographic study on the types of absorbent elements that can be used to improve thermal transfer was performed by Razak et al. [30]. In general, the materials used to make the absorbing element are steel, copper, aluminium and plastic and its performance is influenced by porosity, geometry, thickness, heat exchange surface and material-specific properties.

The equations describing the energy balance and the characteristics of the perforated solar collectors are described by Leon and Kumar [19]. They also carried out a parametric study in which they assessed the influence of certain parameters on a perforated solar collector with the following characteristics: flow velocity 0.02-0.03 m/s, incident solar radiation 400-900 W/m^2, ambient temperature 30 °C, wind velocity 1.2 m/s, pressure loss through the absorbent plate 25-80 Pa, 120 mm cavity, 12-24 mm perforation pitch, 0.8-1.55 mm perforation diameter and a steel absorbent plate with an absorption coefficient of 0.95 and emissivity coefficient of 0.85. According to the hypotheses taken into account, the absorbent plate and the opposite wall plate have the uniform (isothermal) temperature on the surface. The results found are interesting and can determine the choice of the type of solar collector:

- the temperature difference between the inlet and exhaust air decreases as the air flow increases, the pitch between the holes increases and the diameter of the holes increases
- the smaller the perforation diameter and the pitch between, the higher the heat exchange efficiency (but the pressure losses increase);
- the higher the solar radiation and the lower the air flow, the higher the air temperature
- the higher the absorption coefficient of the plate, the higher the resulting temperature (the emissivity coefficient has a lower impact).

Perforated opaque solar collectors have been used since the '90s to heat the air [14] and can lead to good performances [9]. An experimental study of Wang et al. [25] highlights the performance of an opaque solar collector with circular perforations. The collector walls are thermally insulated and there are no inertial elements embedded. The collector was positioned outdoor and subjected to external temperatures between 11-21 °C and solar radiation values between 400-1000 W/m^2. Several plates with circular perforations with a diameter between 1 mm and 2 mm and a pitch of 20 mm to 30 mm were tested in the study. The airflow rates used were between 80 m^3/h and 200 m^3/h, and the resulting velocities were between 0.011 m/s and 0.028 m/s. According to the study, the solar radiation should be higher than 400 W/m^2 to reduce the radiant heat losses of the absorbing element to the exterior. The obtained results show a difference in temperature between the inlet and outlet air which varies between 10.55 and 31.16 ° C (the higher the air velocity, the lower the temperature achieved), a collector efficiency of 46.83-74.93% (efficiency increases with increasing velocity), respectively a heat transfer efficiency between 61.99-82.32% (which decreases with increasing velocity).

The ratio between the perforated collector height and its thickness should be less than 1/50[8]. According to [14, 19, 21, 31] the cavity through which the air circulates in the case of perforated collectors has a thickness of 5-30cm, the perforation surface varies between 0.5-2% of the total surface of the absorbent plate and it must not necessarily be of metal but may also be of plastics with low thermal conductivity [32]. However, the metal plates are isothermal in contrast to the rest of the materials, but this does not influence too much the performance of the collectors [19]. Vaziri et al. [33] have made a perforated collector in which they replaced the metal plate with plexiglass of 3 mm thick and have achieved superior efficiency when using plastic.

Plate emissivity is an important factor in influencing the efficiency of solar collectors. An extensive bibliographic study [34] on types of absorbent plates demonstrates that low emissivity can increase annual solar thermal output by up to 40%. The same aspect is highlighted by Li et al. [35], the thermal transfer being more efficient when using an absorbent plate with a high absorption coefficient and low emissivity (the latter having a less significant impact).

A numerical study of Van Decker et al. [6] for a perforated opaque collector shows that about 62% of the thermal energy is released to the air in front of the absorbent plate, 28% in the orifice and about 10% behind the plate, and the thermal transfer efficiency decreases as the suction velocity, the pitch between the holes and the diameter of the holes increase and increases with increasing wind velocity and plate thickness.

Belusko et al. [27] has been studying the implementation of an opaque solar collector with a curved flat absorbent plate in the roof of a building. The air is aspirated at the bottom, recovers the heat from the corrugated board, and then it is introduced into the building. To improve heat transfer, the author proposes to make perforations in this plate, the air being entrained through the orifices even by the air driven behind the plate (creating a series of small jets). This process (also called "jet impingement") has improved the thermal efficiency of the solar collector by 21%. If the diameter of the holes increases, the impact of this process decreases, and the thermal efficiency of the collector decreases from 69.3% to 66.4%. Li et al. [36] proposes the use of flat opaque solar collectors, demonstrating that their efficiency is about 15% higher than that of corrugated opaque solar collectors.

Another study performed by Cordeau and Barrington [28] is showing the implementation of a perforated solar collector on a barn (see the principle in figure 5).

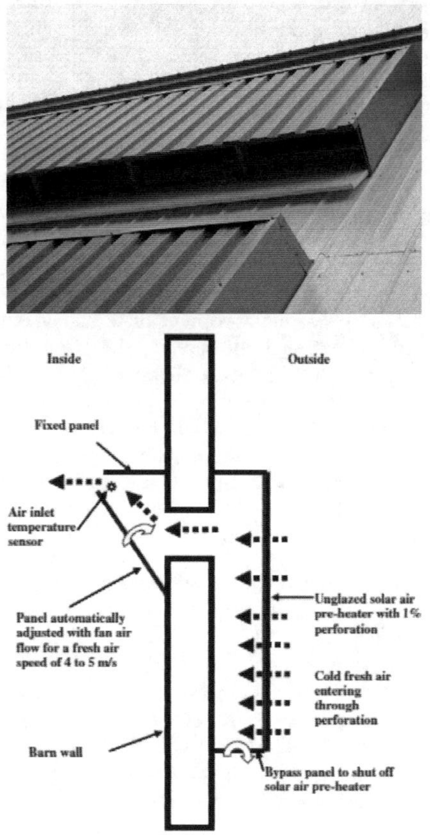

Fig. 5 Perforated solar collector in real conditions: photo and principle diagram [28]

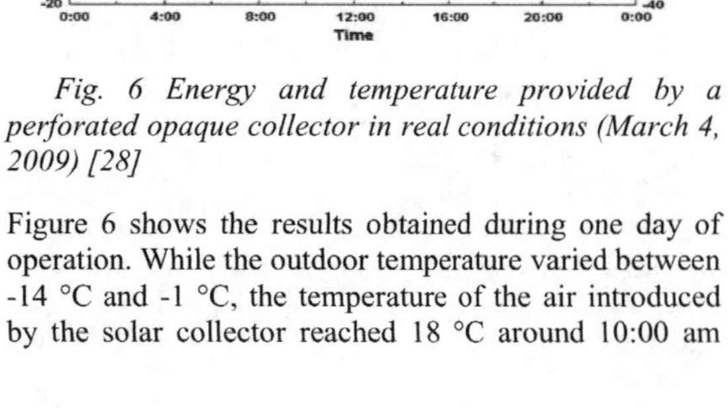

Fig. 6 Energy and temperature provided by a perforated opaque collector in real conditions (March 4, 2009) [28]

Figure 6 shows the results obtained during one day of operation. While the outdoor temperature varied between -14 °C and -1 °C, the temperature of the air introduced by the solar collector reached 18 °C around 10:00 am and during the day, about 532MJ were saved for heating of the building (147 kWh).

For less than 2 m/s wind velocity, average efficiency was 65%, and at high velocities (over 7 m/s), efficiency dropped below 25%. Even under these conditions, the perforated solar collectors used in Cordeau and Barrington study have reduced the building's operating costs in the cold season by \$ 14.8 / m^2.

The front or side wind significantly influence the efficiency of the solar air collectors accordingly to Gunnewiek et al. [8] and could reverse the flow direction (reverse flow effect), causing large pressure losses. According to the authors, the minimum suction velocity of the collector (airflow through the perforated plate relative to the total perforation surface), if there is no wind, it should be at least 0.0125 m/s. If the side wind has a velocity of 5 m/s, the minimum suction velocity should be greater than 0.017 m/s; if the wind blows on the front, the minimum suction velocity should be up to 0.039 m/s if the angle of incidence is 45 °. Collins and Abulkhair [37] have carried out experimental studies on a solar collector with perforations for wind velocities between 0.5-2 m/s and perforation velocity between 0.01-0.04 m/s and concluded that the efficiency of the heat exchange of the absorbent plate decreases with increasing suction velocity. To be able to carry out the numerical and experimental studies on solar collectors, Chan et al. [29] have made a number of assumptions: the radiant heat transfer to the surface of the plate is constant, convection losses to the environment are negligible, there is no inverse flow effect, and air properties at plenum flow are the same. Perforated solar collectors have better efficiency for uneven flows at low speeds [7].

Accordingly to Zhang et al. [14, 20], in the absence of wind, the perforated solar collector is more efficient with 13% resulting an air temperature increase of 3.4 °C in comparison with a perforated collector protected with a glass sheet (this reduces the effect of solar radiation) but at wind velocities greater than 3 m/s the efficiency of the latter becomes superior. However, adding a glass sheet brings, besides the advantages mentioned, the benefit of better control (closure can be done through a flap) and a higher thermal insulation (thermal buffer).

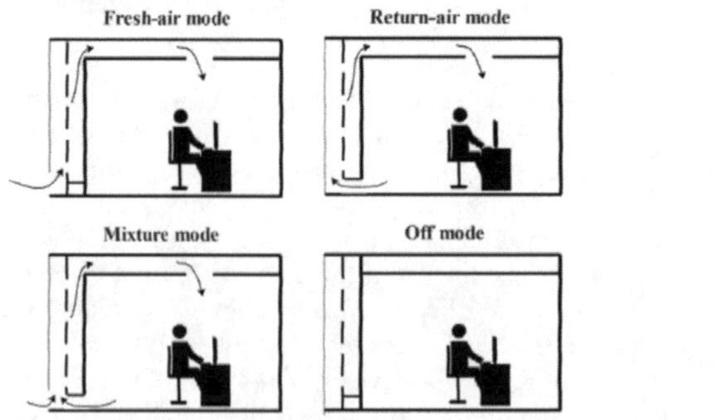

Fig. 7 Operating modes of a opaque solar collector with perforations protected with a glass sheet [20]

The system is shown in figure 7, and the experimental results on a house in China show that indoor average temperature can be raised from 12.12 °C to 16.17-18.19 °C. An alternative to this solution that can be used for space heating is proposed by Zheng et al. [38] in a complex parametric study on a perforated solar collector protected with a glass layer closed at the bottom (the ventilated air is aspirated from inside, figure 8), and the temperature difference between the inlet air and the exhaust from the collector was up to 23 ° C in case of solar radiation of 1200 W/m^2.

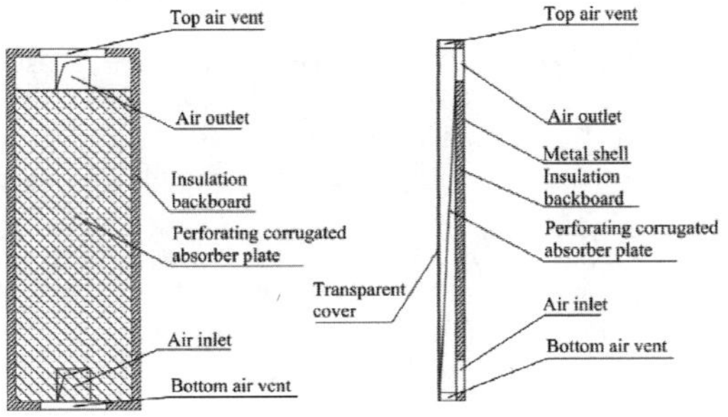

Fig. 8 Solar Collector perforated protected by a sheet of glass with no communication to exterior [38]

An opaque solar collector with high efficiency perforations was proposed by Croitoru et al. [13, 39]. The study proposes the use of lobed geometries instead of circular ones as perforations in classical solar collectors. Figure 9 shows the types of orifices studied. The lobed geometries used are 15% more efficient for flows between 80-220 m^3/h /m^2 and can lead to an exhaust air temperature up to 2 °C higher. Unconventional lobed shape of the orifices causes a more complex flow which improves the heat transfer by 40% [13].

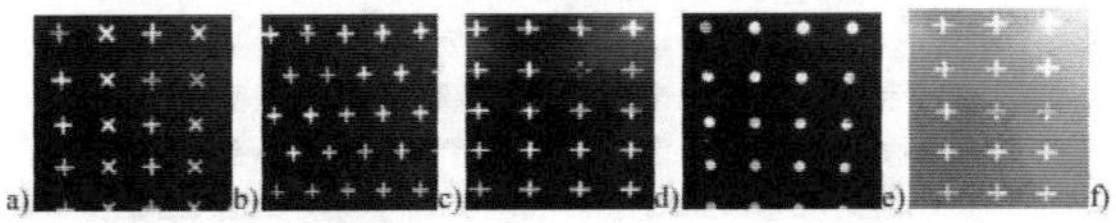

Fig. 9 Types of geometries for the orifices on the absorber [39]

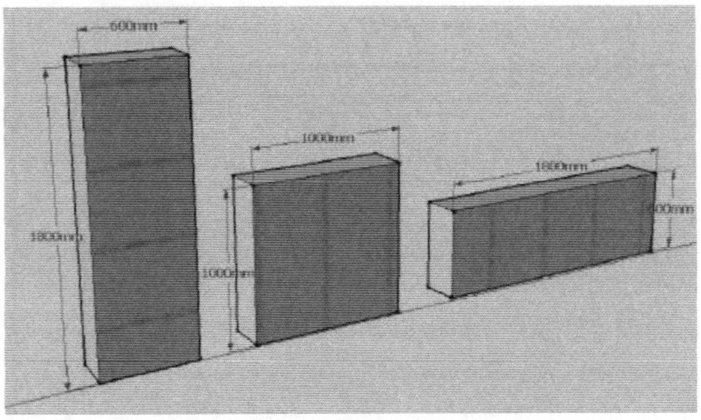

Fig. 10 Solar collectors with same area of 1m^2 [40]

An interesting study was conducted by Perisoglou and Dixon [40] which compared experimentally three solar collectors with a surface of 1m^2, but with different geometries (Figure 10). The study was conducted for an air duct velocity ranging from 1.5-5.5 m/s and the results show efficiencies ranging between 19-71%, respectively, temperature differences between inlet and outlet ranging from 5-22 °C, while the collector placed horizontally had the highest efficiency. The authors also studied the position of the outlet in the collector and found that a 100 mm positioning from the top is better than a positioning in the middle of the collector (the air can pass through the absorbent plate).

Accordingly to Hollick [41] opaque perforated collectors combined with solar cells can be made. A classic collector can reduce annual energy consumption for heating with 500-1000 kWh/m^2 of the collector, and combined with photovoltaic cells, it can produce around 50-100 kWh/m^2 of electricity per year. In addition, the efficiency of the photovoltaic system may increase due to cell ventilation, and the efficiency of the collector may increase due to the heat from the photovoltaic cells.

Regarding the financial aspects, same author, Hollick, in another study [42] presents a commercially perforated solar collector which can determine a temperature difference between the inlet air and the outlet air of about 30-35% (the system is located at 25cm from the building wall). Such a collector was implemented in a building covering a 335 m^2 wall, covering 28% of the ventilation air demand (195700 kWh), which translates into annual savings of $ 4860 (for Canada's climate) and payback time of about 6 years. [43]. An interesting numerical and experimental study was conducted by Eryener and Akhan [44] on a 720 m^2 air collector mounted in Turkey. The solar collector used supplies an air treatment system and has very good efficiency on sunny winter days, bringing down monthly costs by up to 62%. Compared to classical heating solutions, perforated solar collectors have lower operating and running costs over their lifetime and the potential for reducing energy consumption for space heating in cold climates is very high [38].

5 Conclusions

Air solar collectors are widely studied and in the last years these technologies are gaining more and more attention from the researchers. This paper presents a literature review on air solar collectors acting as a solar wall with focus on transpired solar collectors. The existing applications are summarized and the performances and benefits are discussed. The study presents all the types of air solar collectors used and makes a classification based on the several applications found. Moreover, the main parameters of opaque solar collectors are presented and some particular cases are emphasized. From the literature survey we can observe that transpired solar collectors are cost-effective systems that could be used in order to gather energy from the Sun. Researchers are focusing their attention especially on the materials used, flows, collector geometry, dimensions and pitch of the orifices, but there are only a few studies regarding the impact of the orifices geometry. Also, we can observe a lack of studies about coupling the air solar collectors with thermal energy storage materials, which are used mainly in Trombe walls, but no study was found regarding the implementation of thermal energy storage (sensible or latent heat storage/phase changing materials) in transpired solar collectors. This strategy could enhance the overall efficiency of the collectors by lagging the thermal energy stored during the periods when solar radiation is available.

This work was supported by a grant of the Romanian National Authority for Scientific Research, CNCS – UEFISCDI, project number PN-III-P2-2.1-PED-2016-1154.

References

1. Energy, E.P.a.C.o., *The Directive on the energy performance of buildings 2002/91/EC.*
2. IEA, *Renewables for heating and cooling: untapped potential. France: OECD/IEA;*. 2007.
3. Chan, H.-Y., S.B. Riffat, and J. Zhu, *Review of passive solar heating and cooling technologies.* Renewable and Sustainable Energy Reviews, 2010. **14**(2): p. 781-789.
4. Reichl, C., et al., *Comparison of modelled heat transfer and fluid dynamics of a flat plate solar air heating collector towards experimental data.* Solar Energy, 2015. **120**: p. 450-463.
5. Goyal, R.K., G.N. Tiwari, and H.P. Garg, *Effect of thermal storage on the performance of an air collector: A periodic analysis.* Energy Conversion and Management, 1998. **39**(3): p. 193-202.
6. Van Decker, G.W.E., K.G.T. Hollands, and A.P. Brunger, *Heat-exchange relations for unglazed transpired solar collectors with circular holes on a square or triangular pitch.* Solar Energy, 2001. **71**(1): p. 33-45.
7. Gunnewiek, L.H., E. Brundrett, and K.G.T. Hollands, *Flow distribution in unglazed transpired plate solar air heaters of large area.* Solar Energy, 1996. **58**(4–6): p. 227-237.
8. Gunnewiek, L.H., K.G.T. Hollands, and E. Brundrett, *Effect of wind on flow distribution in unglazed transpired-plate collectors.* Solar Energy, 2002. **72**(4): p. 317-325.
9. Shukla, A., et al., *A state of art review on the performance of transpired solar collector.* Renewable and Sustainable Energy Reviews, 2012. **16**(6): p. 3975-3985.
10. Lai, C.-M. and S. Hokoi, *Solar façades: A review.* Building and Environment, 2015. **91**: p. 152-165.
11. Hami, K., B. Draoui, and O. Hami, *The thermal performances of a solar wall.* Energy, 2012. **39**(1): p. 11-16.
12. Nkwetta, D.N. and F. Haghighat, *Thermal energy storage with phase change material—A state-of-the art review.* Sustainable Cities and Society, 2014. **10**: p. 87-100.
13. Croitoru, C.V., et al., *Thermodynamic investigation on an innovative unglazed transpired solar collector.* Solar Energy, 2016. **131**: p. 21-29.
14. Zhang, T., et al., *The application of air layers in building envelopes: A review.* Applied Energy, 2016. **165**: p. 707-734.
15. Dymond, C. and C. Kutscher, *Development of a flow distribution and design model for transpired solar collectors.* Solar Energy, 1997. **60**(5): p. 291-300.
16. Molineaux, B., B. Lachal, and O. Guisan, *Thermal analysis of five unglazed solar collector systems for the heating of outdoor swimming pools.* Solar Energy, 1994. **53**(1): p. 27-32.
17. Gao, L., H. Bai, and S. Mao, *Potential application of glazed transpired collectors to space heating in cold climates.* Energy Conversion and Management, 2014. **77**: p. 690-699.
18. Alkilani, M.M., et al., *Review of solar air collectors with thermal storage units.* Renewable and Sustainable Energy Reviews, 2011. **15**(3): p. 1476-1490.
19. Leon, M.A. and S. Kumar, *Mathematical modeling and thermal performance analysis of unglazed transpired solar collectors.* Solar Energy, 2007. **81**(1): p. 62-75.
20. Zhang, T., et al., *A glazed transpired solar wall system for improving indoor environment of rural buildings in northeast China.* Building and Environment, 2016. **98**: p. 158-179.
21. Rosaria Ciriminna, F.M., Mario Pecoraino, Mario Pagliaro, *Solar Air Heating and Ventilation in Buildings: A Key Component in the Forthcoming Renewable Energy Mix.* Energy Technology, 2017. **5**: p. 1-9.
22. Paya-Marin, M.A., et al., *Large scale test of a novel back-pass non-perforated unglazed solar air collector.* Renewable Energy, 2015. **83**: p. 871-880.
23. Brown, C., et al., *Transpired Solar Collector Installations in Wales and England.* Energy Procedia, 2014. **48**: p. 18-27.
24. Januševičius, K., et al., *Validation of Unglazed Transpired Solar Collector Assisted Air Source Heat Pump Simulation Model.* Energy Procedia, 2016. **95**: p. 167-174.
25. Wang, X., et al., *A simplified method for evaluating thermal performance of unglazed transpired solar collectors under steady state.* Applied Thermal Engineering, 2017. **117**: p. 185-192.
26. Paya-Marin, M.A., *Chapter 5 - Solar Air Collectors for Cost-Effective Energy-Efficient Retrofitting*, in *Cost-Effective Energy Efficient Building Retrofitting.* 2017, Woodhead Publishing. p. 141-168.

27. Belusko, M., W. Saman, and F. Bruno, *Performance of jet impingement in unglazed air collectors.* Solar Energy, 2008. **82**(5): p. 389-398.
28. Cordeau, S. and S. Barrington, *Performance of unglazed solar ventilation air pre-heaters for broiler barns.* Solar Energy, 2011. **85**(7): p. 1418-1429.
29. Chan, H.-Y., et al., *Thermal Analysis of Flat and Transpired Solar Facades.* Energy Procedia, 2014. **48**: p. 1345-1354.
30. Razak, A.A., et al., *Review on matrix thermal absorber designs for solar air collector.* Renewable and Sustainable Energy Reviews, 2016. **64**: p. 682-693.
31. Chan, H.Y., J. Zhu, and S. Riffat, *Heat Transfer Analysis of the Transpired Solar Facade.* Energy Procedia, 2013. **42**: p. 123-132.
32. Arulanandam, S.J., K.G.T. Hollands, and E. Brundrett, *A CFD heat transfer analysis of the transpired solar collector under no-wind conditions.* Solar Energy, 1999. **67**(1–3): p. 93-100.
33. Vaziri, R., M. Ilkan, and F. Egelioglu, *Experimental performance of perforated glazed solar air heaters and unglazed transpired solar air heater.* Solar Energy, 2015. **119**: p. 251-260.
34. Hall, R. and J. Blower, *Low-emissivity Transpired Solar Collectors.* Energy Procedia, 2016. **91**: p. 56-63.
35. Li, X., C. Li, and B. Li, *Net heat gain assessment on a glazed transpired solar air collector with slit-like perforations.* Applied Thermal Engineering, 2016. **99**: p. 1-10.
36. Li, S., et al., *Airflow and thermal analysis of flat and corrugated unglazed transpired solar collectors.* Solar Energy, 2013. **91**: p. 297-315.
37. Collins, M.R. and H. Abulkhair, *An evaluation of heat transfer and effectiveness for unglazed transpired solar air heaters.* Solar Energy, 2014. **99**: p. 231-245.
38. Zheng, W., et al., *Thermal characteristics of a glazed transpired solar collector with perforating corrugated plate in cold regions.* Energy, 2016. **109**: p. 781-790.
39. Croitoru, C., et al., *Thermal Evaluation of an Innovative Type of Unglazed Solar Collector for Air Preheating.* Energy Procedia, 2016. **85**: p. 149-155.
40. Perisoglou, E. and D. Dixon, *Experimental Monitoring of Different Dimensions of Transpired Solar Collectors.* Energy Procedia, 2015. **70**: p. 111-120.
41. Hollick, J.C., *Solar cogeneration panels.* Renewable Energy, 1998. **15**(1–4): p. 195-200.
42. Hollick, J.C., *Unglazed solar wall air heaters.* Renewable Energy, 1994. **5**(1–4): p. 415-421.
43. Hollick, J.C., *World's largest and tallest solar recladding.* Renewable Energy, 1996. **9**(1): p. 703-707.
44. Eryener, D. and H. Akhan, *The Performance of First Transpired Solar Collector Installation in Turkey.* Energy Procedia, 2016. **91**: p. 442-449.

Experimental investigation of the charge/discharge process for an organic PCM macroencapsulated in an aluminium rectangular cavity

Andrei – Stelian Bejan[1,*], *Abdelouhab* Labihi[1,3], *Cristiana Verona* Croitoru[1,2], *Tiberiu* Catalina[1,2], *Hassan* Chehouani[3] and *Brahim* Benhamou[4,5]

[1]Technical University of Civil Engineering Bucharest, Faculty of Engineering for Building Services, 021407 Bucharest, Romania
[2]CAMBI Research Center, 021407 Bucharest, Romania
[3]Laboratory of Processes, Metrology, Materials for Energy and Environment (LP2M2E), Faculty of Sciences and Techniques, Cadi Ayyad University, Marrakech, Morocco
[4]Laboratoiry of Renewable Energies and Energy Efficiency (EnR2E), National Center of Studies and Research on Water & Energy (CNEREE), Cadi Ayyad University, Marrakech, Morocco
[5]Energy Process Research Group at LMFE, Faculty of Sciences Semlalia, Cadi Ayyad University, Marrakech, Morocco

Abstract. Buildings sector has one of the highest potential regarding the reduction of greenhouse gases emissions, as being responsible for more than 40% of energy consumption worldwide. This is why, in order to achieve indoor thermal comfort, it is mandatory to use energy-efficient systems. Materials acting as thermal energy storage (TES) represents one of the most effective strategy that can be implemented and nowadays, many studies are focusing their attention on latent heat storage, respectively on phase changing materials (PCM) which can embed a large embed a high quantity of energy, unlike classic materials acting as thermal mass. This purpose of this paper is to experimentally investigate the charge and discharge processes for an organic PCM (RT35 paraffin) macroencapsulated in an aluminium rectangular cavity which was placed first in a horizontal position and after in a vertical position. After several experimental campaigns conducted we determined that the vertical position enhance the heat transfer because of the natural convection which occurs inside the cavity. Therefore, the charging time is lower in case of the vertical cavity and the temperature measured inside and on the surface is higher.

1 Introduction

1.1 Context and motivation

According to the United Nations Environment Programme [1] building sector has the highest potential regarding the reduction of greenhouse gases emissions around the world, with relative low investment costs. Moreover, the building sector is responsible for more than 40% of energy consumption and 33% of greenhouse gases emissions worldwide.

Furthermore, all the members of the United Nations (196 states) have, after the COP21 conference (which took place in 2015 in Paris, France), a common goal: to limit the global warming at 2°C above preindustrial era [2] which imposes even more stringent requirements and regulations regarding energy consumption in buildings, buildings envelope and buildings systems.

In Europe, more than 50% of the energy consumption in buildings is due to the HVAC systems that ensure the indoor thermal comfort and indoor air quality [3]. Moreover, an important percentage of the total amount of energy consumption is due to the fresh air needed by the buildings occupants and approximately 20-50% of the energy consumption for heating and cooling is due to the building envelope [4]. Also, all these values percentages have an ascending trend in all the developed countries [5].

In order to reduce energy consumptions and CO_2 emissions determined by the buildings sector it is very important to implement innovative materials and energy efficient systems using renewable energy sources.

1.2 Latent heat thermal energy storage (LHTES)

Thermal energy storage (TES) materials represent one of the most promising methods of reducing energy consumptions in buildings by incorporating them into passive systems and active systems [6, 7].

TES materials can be classified by taking into account the storage method as: sensible heat storage, latent heat storage and chemical heat storage [8]. One of the most promising materials used in order to store latent heat are the phase changing materials (PCMs) because they can store 5-14 times more energy than classical materials [9, 10]. According to Soares et al. [9], PCMs integrated in building envelope or building systems could determine the following benefits:

* Corresponding author: andrei.s.bejan@gmail.com

• Lower energy consumption for heating and cooling
• Lower operating costs and CO_2 emissions
• Lower heating and cooling loads (undersized systems)
• Better performance of building envelope
• Better thermal comfort (fewer overheating hours, lower temperature variations).

Phase changing materials is a TES medium that draws the attention of many researchers over the last decade [11, 12] and according to Navarro et al. [6] these materials can improve the overall efficiency of systems using renewable energy sources by storing the surplus energy and using it during the periods when the source is not available (e.g. during night-time in case of solar radiation).

There are several types of PCMs studied in literature [6, 7, 13, 14]: organic PCMs, inorganic PCMs (salt hydrates) and eutectics. The organic PCMs (usually paraffin) are one of the most used in the building sector because their characteristics: large storage capacity, phase change temperature within the comfort limits, market availability, low-cost and because they are non-toxic. According to many researchers, the most important properties that must be taken into account when choosing the type of PCM are [15-17]: phase change temperature (near to the operating value), high thermal conductivity, large storage capacity, low volume variations, good thermal stability in time, high specific heat etc..

PCMs can be microencapsulated an macroencapsulated [18]. In case of microencapsulation the bulk material is embedded in capsules with a very small diameter (less than 1mm), while in case of macroencapsulation the bulk materials is embedded in different types of containers (e.g. metal panels, spheres, cylinders, tubes etc.) [19].

1.3 Charge/discharge problem

According to the literature the main problem in case of organic PCM is that the thermal conductivity has low values (approximately 0.2 W/mK for paraffin) [17]. There are many strategies that can be used in order to improve the heat transfer [13, 15, 17]: the use of materials with high thermal conductivity (graphite, aluminium or copper matrixes, metal particles, carbon fibres), the use of metal structures for macroencapsulation (metal spheres, metal rings, metal fins inside or outside the cavity, rectangular or cylindrical tubes) and the use of metal containers for microencapsulation (small iron spheres, metal cylinders).

According to Iten and Liu [20], in order to choose the proper container it is important to know: details about the application and available space, boundary conditions and the contact material. Same authors emphasise that usually the metal containers are used in order to improve the heat transfer (especially aluminium, steel and copper) and moreover, the rectangular containers are being used often because of easy manufacturing and high storage density of PCM (up to 90% of the total volume). Not only the geometry and material is important when choosing the PCM and related container, but also the position relative to the vertical direction [21] which has a determinant role regarding the thermal transfer processes and can improve the natural convection issues inside the cavity [22].

This paper aims to study in an experimental manner the charge and discharge process for an organic PCM macroencapsulated in an aluminium rectangular cavity which can be used in different building applications and especially in solar air collectors, building walls, or building systems.

2 Experimental setup

2.1. Materials and methods

Figure 1 shows the aluminium rectangular tube which was used as a container for the PCM. The tube has the following dimensions: 800x60x20mm (2mm thickness) and it was filled with approximately 0.9l of paraffin (macroencapsulation). After this stage it was sealed with special glue, painted in black and covered with 2.5mm of insulation on all sides (except the one exposed to the heat source). In order to cope with the volume changes during the phase change, one very small hole was made in one corner of the tube.

Fig. 1. Aluminium rectangular tube: a-empty, b-filled with PCM and prepared for experimental study

Fig. 2. Experimental setup: 1-vertical tube, 2-lamps for vertical tube, 3-horizontal tube and 4-lamps for horizontal tube

The experimental setup considers two rectangular tubes: one mounted vertical and one mounted horizontal. Every bar is heated by two halogen lamps (500W each) placed on a stand at 500mm distance, which ensures a uniform heat distribution on each tube.

The organic PCM used for the experimental study was bulk paraffin RT35 with the phase changing temperature around 35°C. All thermal properties of the RT35 are presented in table 1.

Table 1. PCM RT35 properties

Data	Value	Unit
Melting area	29-36	°C
Congealing area	36-31	°C
Heat storage capacity	160	kJ/kg
Specific heat capacity	2	kJ/kgK
Density solid (at 15°C)	0.86	kg/l
Density liquid (at 45°C)	0.77	kg/l
Heat conductivity	0.2	W/mK

2.2 Measurement procedure

In order to study the charge/discharge process in case of RT35 PCM macroencapsulated in the rectangular cavity, two aluminium bars were used. Inside each tube, three K-type thermocouples with accuracy of 0.2°C were mounted at 200mm distance, in order to measure the temperature in the cavity in different points x20, x40 and x60 for the horizontal bar and y20, y40 and y60 for the vertical bar). Another two thermocouples were placed on the surface of each aluminium bar in order to measure the surface temperature in the middle of the container. One more thermocouple was used to measure the ambient temperature.

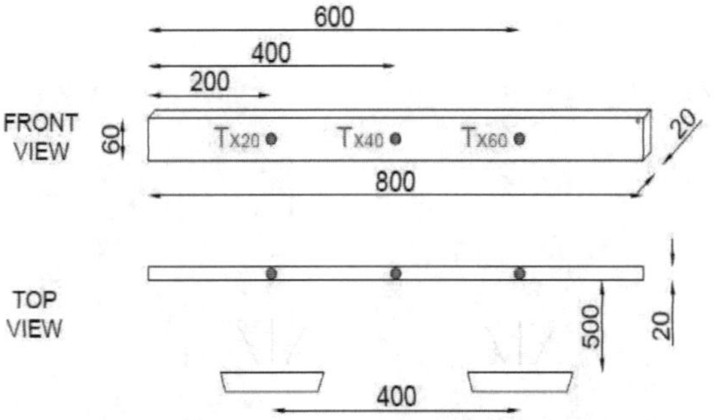

Fig. 3. Dimensions and placement of the sensors – horizontal tube

The measurements were conducted for 21 hours, temperature values were recorded every 5 minutes and all the data was collected using a very precise data logger. Moreover, we used the FLIR-E40 thermal imaging camera in order to measure the surface temperature after two hours of charge and two hours of discharge (correlation with thermocouple and the temperatures inside the cavity). Both devices are presented in figure 5.

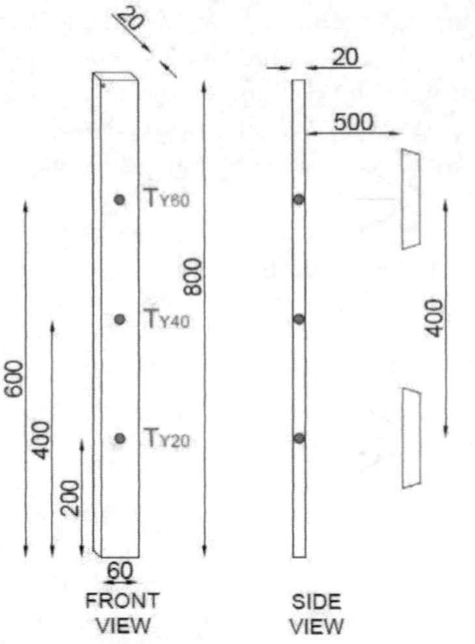

Fig. 4. Dimensions and placement of the sensors – vertical tube

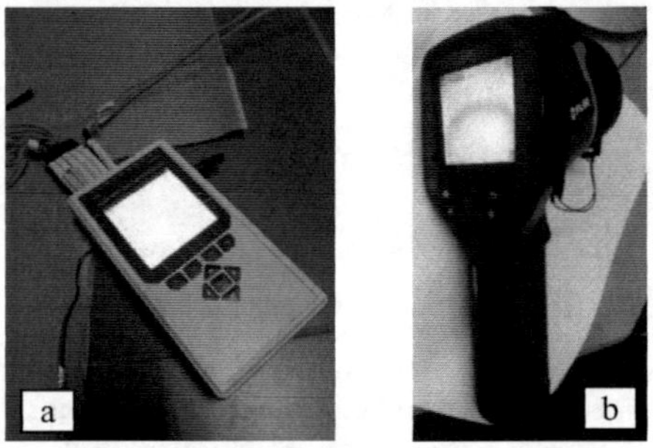

Fig. 5. Data logger (a) and thermal imaging camera (b)

3 Results

3.1 Analysis of charge/discharge

Following the experimental studies conducted we obtained very interesting results. The measurements were made in similar conditions for both vertical and horizontal rectangular aluminium containers with PCM.

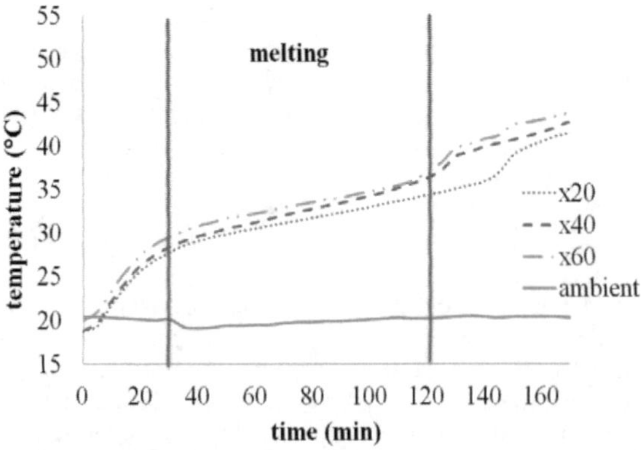

Fig. 6. Charging – horizontal rectangular tube

Figure 6 shows the charging process for the horizontal rectangular tube filled with organic PCM. The

https://doi.org/10.1051/e3sconf/20183201004

process has three stages: solid state (storing sensible heat), mixture (storing latent and sensible heat while melting) and liquid state (storing sensible heat). The melting period starts after 30 minutes and it takes approximately 90 minutes to change phase from solid to liquid mainly because of thermal conduction. During the charging period, the PCM transforms to liquid state after 120 minutes and reaches the maximum temperature of 43.6°C after 170 minutes. Because of the enhanced thermal conductivity determined by the aluminium material, the heat is evenly distributed to PCM, with minimal heat losses. This phenomenon determines only small differences between the values measured by the thermocouples in different points in the rectangular cavity (up to 1.8°C).

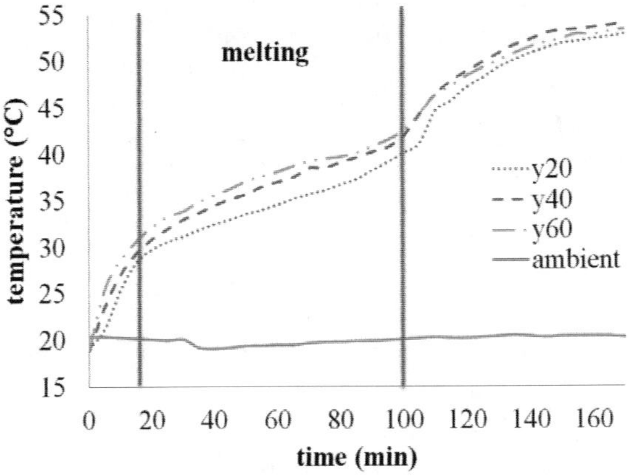

Fig. 7. Charging – vertical rectangular tube

Figure 7 shows the charging process in case of the vertical rectangular tube filled with organic PCM, experimentally studied in the same conditions of ambient temperature and radiation (from the heat source) as the horizontal one. However, the results emphasise important differences. The melting stage starts after 18 minutes, much faster than in the first case and it takes approximately 80 minutes to change phase from solid to liquid. During the charging period the PCM transforms to liquid state after 98 minutes, reaches 43.6°C after 110 minutes (60 minutes faster) and after 170 minutes achieves the maximum temperature of 54.6°C (11°C more). In case of the vertical cavity, because of the high thermal conductivity of aluminium and the evenly distributed heat he PCM begins to melt near the cavity walls and all the liquid paraffin goes in the upper part of the rectangular cavity. This is why we can observe even 3.5°C difference between T_{Y20} and T_{Y60}. This stratification inside the cavity is due to the fact that, besides the conduction phenomena, an important role is fulfilled by the natural convection inside the cavity, as mentioned before in different studies [20, 21, 23].

Figure 8 shows the discharging process for the horizontal rectangular tube. The process has three stages: liquid state (releasing sensible heat), mixture (releasing sensible and latent heat while solidifying) and solid state (releasing sensible heat, stabilizing to ambient temperature). The sensible heat during the liquid state is discharged quickly and after only 15 minutes the PCM begins to solidify. The solidification lasts for about 165 minutes, although the phenomenon is not so obvious (the evolution of the discharging curve changes around the 350th minute of the experimental campaign).

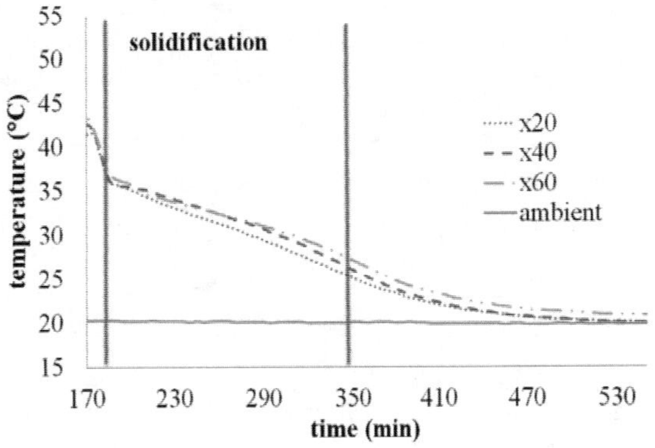

Fig. 8. Discharging – horizontal rectangular tube

Figure 9 shows the discharging process in case of the vertical rectangular tube, also in the same conditions of ambient temperature as in the previous case. The sensible heat during the liquid state is release slower because the temperature of PCM is higher, but after 30 minutes the PCM begins to solidify. The stratification in case of discharging is no longer clear as in the case of charging process because the PCM begins to solidify near to the aluminium wall. Because of the good conductivity of aluminium the solidification inside the cavity is uniform until the moment when temperature measured by thermocouple T_{Y20} drops suddenly (275th minute) which means that an important amount of solid PCM gathered in the lower part of the vertical cavity and this phenomena enhance the convective transfer within the cavity. The solidification process lasts for about 135 minutes, although the phenomenon is not so obvious, similar with the previous case.

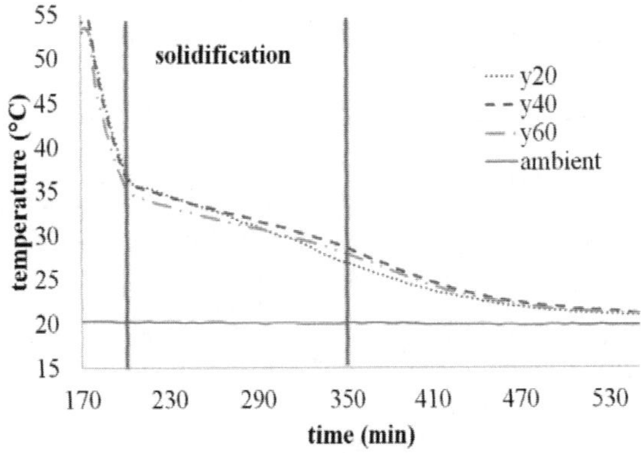

Fig. 9. Discharging – vertical rectangular tube

3.2 Analysis of surface temperature

Because of a lower charging period in case of the vertical cavity, the temperature on the surface of aluminium rectangular tube is also higher. This determines also an enhanced convective transfer between the vertical PCM filled bar and ambient air that can explain the fast heat release in the first stage of the discharging process.

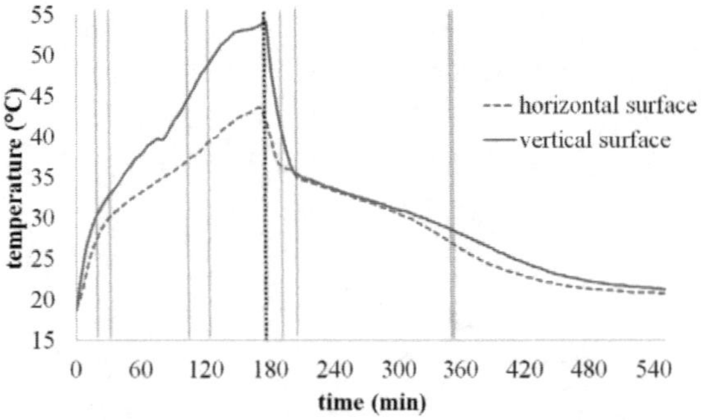

Fig. 10. Variation of surface temperature during charging and discharging

This phenomena is emphasised by figure 10 and also by thermal images presented in figure 11 and figure 12.

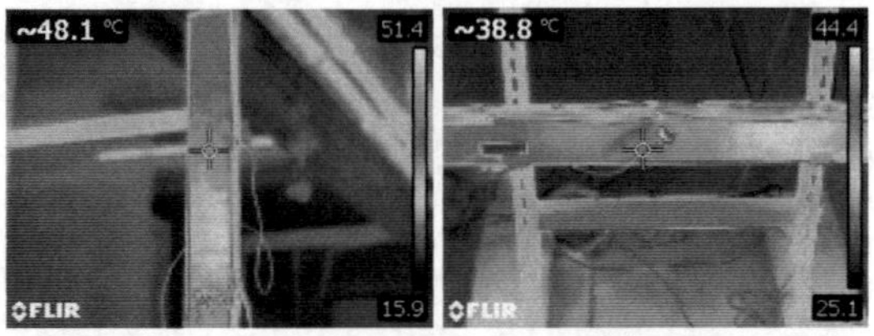

Fig. 11. Thermal images after two hours of charging

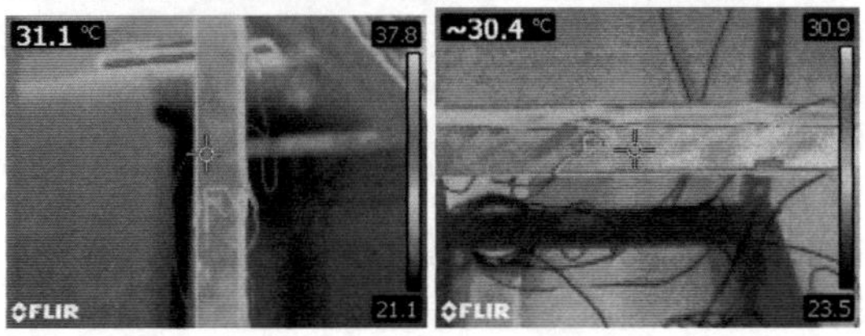

Fig. 12. Thermal images after two hours of discharging

4 Conclusions

In the current context (high energy consumptions and CO_2 emissions determined by the building sector) it is essential to implement highly efficient systems using renewable energy sources and innovative materials. Nowadays, many researches focus their attention on the latent heat thermal energy storage, such as phase changing materials which can determine important benefits regarding achieving indoor thermal comfort with minimum energy consumption.

This paper aims to evaluate in an experimental manner the charging and discharging phenomena for an organic PCM (RT 35 paraffin) macroencapsulated in an aluminium rectangular cavity. After several experimental studies conducted we can conclude that:

• The charging time in case of a vertical aluminium tube filled with PCM is lower, unlike the horizontal position

• The stratification inside the vertical cavity is easy to observe during the charging period, but not so obvious during the discharge process

• During charging process the thermal transfer in enhanced in case of the vertical tube due to the convective transfer (in addition to conductive transfer which is predominant in the case of the horizontal tube)

• The aluminium distributes uniform the heat towards the PCM because of high conductivity

• After 170 minutes of charging the maximum temperature inside the vertical cavity reaches 54.6°C, unlike 43.6°C in case of the horizontal cavity

• The 43.6°C temperature is achieved after only 110 minutes in case of vertical cavity (60 minutes faster)

• The discharging period is longer than the charging period, similar with other studies from the literature

• The discharging period is similar in both cases and the convective transfer is not so clear during this stage

• The surface temperature is higher in case of vertical cavity which enhance the convective transfer between the aluminium bar and ambient air.

Further studies will be conducted in order to better understand the complex phenomena which occurs inside the aluminium rectangular cavity filled with organic PCM. The charge and discharge will be also studied in transparent cavities with different thicknesses in order to assess the impact of geometry. Other types of phase changing materials with different phase changing temperatures will be investigated. Moreover, the studied bars will be used in several building elements and systems in order to assess the impact of PCM implementation in terms of energy savings and improvement of indoor thermal comfort. One the main studies which will be conducted by authors of the present paper will investigate the implementation of thermal energy storage in a transpired solar collector (with air), which acts as a solar wall. In this case, the implementation of latent heat storage can improve the overall efficiency of the system by lagging the solar energy stored during the periods when the solar radiation is available (e.g. during day-time) and using it during periods when solar radiation is not available (e.g. during night-time, when the heat will be slowly released to the air).

This work was supported by a grant of the Romanian National Authority for Scientific Research, CNCS – UEFISCDI, project number PN-III-P2-2.1-PED-2016-1154.

References

1. UNEP, *Buildings and Climate Change Summary for Decision-Makers*. (United Nations Environment Programme - Sustainable Buildings & Climate Initiative, 2009)
2. G. Rajgor, Renewable Energy Focus **17**, 175-177 (2016)
3. M. Saffari, A. de Gracia, S. Ushak, and L.F. Cabeza, Energy and Buildings **112**, 159-172 (2016)
4. T. Zhang, Y. Tan, H. Yang, and X. Zhang, Applied Energy **165**, 707-734 (2016)

5. P. Kumar and L. Morawska, Environmental Science & Technology **47**, 7591−7592 (2013)
6. L. Navarro, A. de Gracia, S. Colclough, M. Browne, S.J. McCormack, P. Griffiths, and L.F. Cabeza, Renewable Energy **88**, 526-547 (2016)
7. L. Navarro, A. de Gracia, D. Niall, A. Castell, M. Browne, S.J. McCormack, P. Griffiths, and L.F. Cabeza, Renewable Energy **85**, 1334-1356 (2016)
8. T. Khadiran, M.Z. Hussein, Z. Zainal, and R. Rusli, Renewable and Sustainable Energy Reviews **57**, 916-928 (2016)
9. N. Soares, J.J. Costa, A.R. Gaspar, and P. Santos, Energy and Buildings **59**, 82-103 (2013)
10. A. Kylili and P.A. Fokaides, Journal of Building Engineering **6**, 133-143 (2016)
11. C. Liu, C. Wang, Y. Li, and Z. Rao, RSC Advances **7**, 7238-7249 (2017)
12. S.N. Al-Saadi and Z. Zhai, Renewable and Sustainable Energy Reviews **21**, 659-673 (2013)
13. S.A. Memon, Renewable and Sustainable Energy Reviews **31**, 870-906 (2014)
14. S.E. Kalnæs and B.P. Jelle, Energy and Buildings **94**, 150-176 (2015)
15. A. Waqas and Z. Ud Din, Renewable and Sustainable Energy Reviews **18**, 607-625 (2013)
16. V.V. Tyagi and D. Buddhi, Renewable and Sustainable Energy Reviews **11**, 1146-1166 (2007)
17. D. Zhou, C.Y. Zhao, and Y. Tian, Applied Energy **92**, 593-605 (2012)
18. Y. Konuklu, M. Ostry, H.O. Paksoy, and P. Charvat, Energy and Buildings **106**, 134-155 (2015)
19. L.F. Cabeza, A. Castell, C. Barreneche, A. de Gracia, and A.I. Fernández, Renewable and Sustainable Energy Reviews **15**, 1675-1695 (2011)
20. M. Iten and S. Liu, Energy Conversion and Management **77**, 608-627 (2014)
21. A. Baïri, E. Zarco-Pernia, and J.M. García de María, Applied Thermal Engineering **63**, 304-322 (2014)
22. A.I.N. Korti and F.Z. Tlemsani, Journal of Energy Storage **5**, 177-186 (2016)
23. A. Labihi, F. Aitlahbib, H. Chehouani, B. Benhamou, M. Ouikhalfan, C. Croitoru, and I. Nastase, Applied Thermal Engineering **126**, 305-314 (2017)

Impact of the formaldehyde concentration in the air on the sink effect of a coating material

Anne-Lise Tiffonnet[1,*], *Céline* Tourreilles[2], and *Thierry* Duforestel[2]

[1] LUSAC, Université de Caen Normandie, Saint-Lô, France
[2] EDF Département EnerBAT, EDF Lab Les Renardières, France

Abstract. This study aims to characterize, from a numerical modelling, the sorption behaviour of a material (a plasticised flooring material) when it is exposed to a pollutant commonly encountered in indoor environments (formaldehyde). It deals with the influence of the pollutant concentration in the room air on the sink effect of the material. The numerical simulations are based on a macroscopic modelling using experimental test results obtained elsewhere. The consequences on the room inertia are also discussed, and analogies between mass transfer and heat transfer are highlighted.

1 Introduction

In the fields of energy efficiency and indoor air quality of buildings, the characterisation of the thermal and solutal behaviour of the used construction and coating materials is essential. The prediction of the temperatures of the air and the walls of a room allows to evaluate the thermal comfort. At the same time, the prediction of the pollutants concentrations allows to appreciate the risks for the occupants' health.

When dealing with pollutant transfers between air phase and walls, many studies represent these transfers like global processes and walls are often considered as reversible sinks. Two coefficients called "adsorption rate constant" and "desorption rate constant" are used to describe respectively the pollutant transfer from de room air to the wall and from the wall to the room air.

This study deals with the influence of the concentration level of an indoor air pollutant (formaldehyde) on the sorption capacity of a material used as typical flooring material (plasticised flooring material).

2 Description of the macroscopic model

The most widely used model for predicting either the VOC emission from materials or the adsorption and desorption cycles of airborne contaminants is the one-sink model initially suggested by Tichenor and co-workers [1]. It considers that at any time, the mass flux on the material surface is the difference between the adsorption and the desorption rates of the material. According to Langmuir's theory, the former is assumed to be proportional to the contaminant concentration in the bulk air of the room, while the latter is assumed to depend upon the amount of contaminant adsorbed on the material surface.

Therefore, the governing equations for the indoor concentration C (kg.m^{-3}) and the surface mass M (kg.m^{-2}) of a compound, respectively, are (fig. *1*):

$$V\frac{dC}{dt} = Q(C_{ext} - C) - Sk_aC + Sk_dM \quad (1)$$

$$\frac{dM}{dt} = k_aC - k_dM \quad (2)$$

where C_{ext} (kg.m^{-3}) is the pollutant concentration in the inlet air, V (m^3) the room volume, Q (m^3.s^{-1}) the air exchange flow through the room, k_a (m.s^{-1}) the adsorption rate coefficient, k_d (s^{-1}) the desorption rate constant, and S (m²) the material surface area exposed to the bulk air. It means that increased surface areas due to surface roughness and internal pores of the material, as well as the effects of bulk mass transport and diffusive transport processes, are implicitly lumped into the adsorption and desorption rate coefficients.

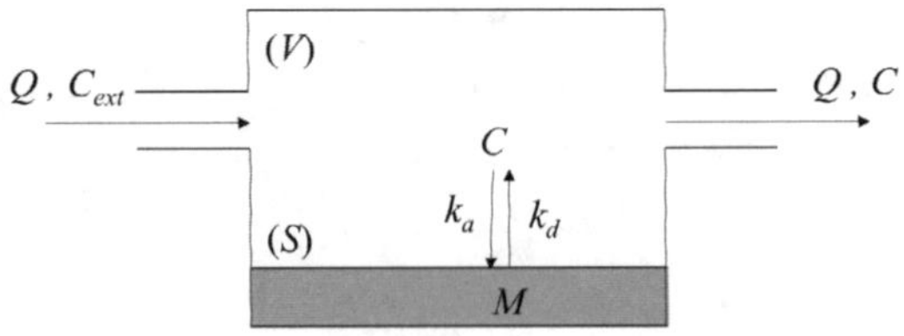

Fig. 1. Schematic representation of pollutant transfers.

3 Sorption capacity of a material

3.1 Experimental characterisation

In order to characterize the sorption capacity of the studied material, tests in experimental chamber were carried out.

* Corresponding author: anne-lise.tiffonnet@unicaen.fr

https://doi.org/10.1051/e3sconf/20183201005

The experimental device is based on an existing equipment previously used to characterize sorption phenomena by static experiments [2]. This equipment was then adapted to characterize the same phenomena by dynamic experiments. This experimental device is described in details elsewhere [3].

Once the material is placed in the chamber and the climate parameters (temperature and humidity) are stabilised, a known amount of pollutant is injected in the supply air of the chamber. The polluted air then flows continuously during the adsorption test period. Then, the comparison between the pollutant concentration in the supply air of the chamber and the exhaust one leads to determine the pollutant mass which is adsorbed by the material.

Once the steady state is reached (i.e. when we cannot observe any variation of the concentration in the air of the chamber), the pollutant injection is stopped. The supply air of the chamber is now a clean air. This is the desorption phase. Comparing the profiles of the obtained concentrations with and without material allows to determine the pollutant mass which is desorbed by the material.

Experiments were carried out with a plasticised flooring material exposed to formaldehyde under dry air (measured relative humidity is 8%) at a constant temperature equals to 19°C. The pollutant injection was imposed during 3 hours (adsorption phase). Measurements were realised continuously during 6 hours (adsorption + desorption periods) with sampling points every 3 minutes. The loading factor of the material (i.e. the ratio between the material area exposed to the room air and the chamber volume) was 0.628 m^{-1}. Figure *2* shows the profile of the resulting concentration in the chamber air. For this test, the injection corresponded to an equilibrium concentration of 174 $\mu g.m^{-3}$. Without material, concentration would be comparable to the injection scheme one.

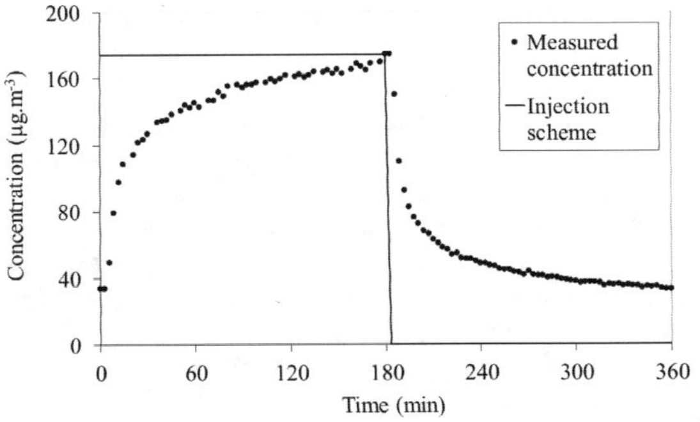

Fig. 2. Injection scheme and resulting concentration profile in the chamber air.

3.2 Identification of the model coefficients

This part presents the method used for the k_a and k_d coefficients identification. The principle is explained from the results of the experiment described above.

The insertion of equation (2) in equation (1) leads to, with $C_{ext} = C_\infty$:

$$\frac{dC}{dt} = \frac{Q}{V}C_\infty - \frac{Q}{V}C - \frac{S}{V}\frac{dM}{dt} \tag{3}$$

The gaseous phase concentration variation is calculated by using the following finite differences scheme:

$$\frac{dC}{dt} = \frac{C^{t+\Delta t} - C^t}{\Delta t} \tag{4}$$

Where C^t is the concentration at the time t, $C^{t+\Delta t}$ is the concentration at the next time and Δt is the time lapse.

The mass variation within the material is, from equation (3):

$$\frac{dM}{dt} = -\frac{V}{S}\left(\frac{dC}{dt} + \frac{Q}{V}\left(C - C_\infty\right)\right) \tag{5}$$

Here it is important to note that the concentration C_∞ is the steady state concentration (i.e. the concentration which is reached at the end of the adsorption phase) in these equations. The pollutant mass inside the material is calculated any time by the equation (6) with, as the initial condition, a null initial mass in the sink (material): $M_0 = 0$.

$$M^{t+\Delta t} = \left(\frac{dM}{dt}\right)\Delta t + M^t \tag{6}$$

where M^t is the mass in the sink at time t, and $M^{t+\Delta t}$ is the mass at the next stage.

Equation (2) can be rewritten as such:

$$\left(\frac{dM}{dt}\right)/M = k_a C/M - k_d \tag{7}$$

It is then possible to graphically determine the k_a and k_d coefficients from the draw of the following function:

$$\left(\frac{dM}{dt}\right)/M = f(C/M) \tag{8}$$

The coefficients of the straight line in the transient state give directly the k_a and k_d coefficients, as shown in figure *3*. Thus we obtained for the adsorption phase: $k_a = 8.65\ 10^{-2}\ m.s^{-1}$ et $k_d = 9.00\ 10^{-4}\ s^{-1}$. These values allow to compute the concentration and to compare it to the measured one (fig. *4*).

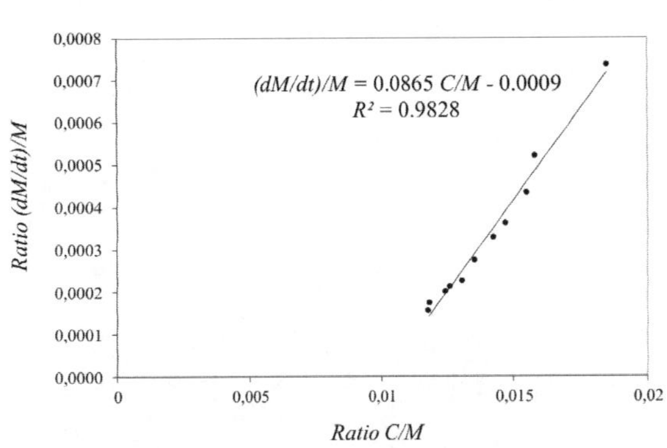

Fig. 3. Identification of k_a and k_d coefficients.

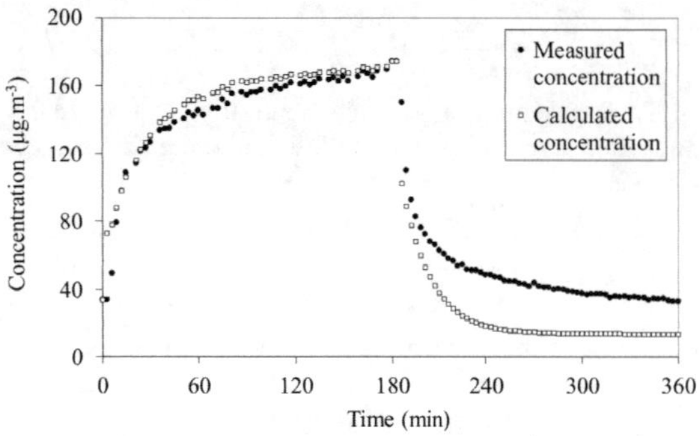

Fig. 4. Profiles of the measured and calculated chamber air concentrations.

The k_a and k_d values relevance is checked recovering the following relation in steady state:

$$k_a C_\infty = k_d M_\infty \quad \text{or} \quad \frac{k_a}{k_d} = \frac{M_\infty}{C_\infty} \qquad (9)$$

It is checked here because the same value is found for the two ratios (96,1 m).

Theoretically, the « k_a - k_d » model being based on total reversibility of the phenomena, the coefficients must be the same for the two phases (adsorption and desorption). The drawn in fig. *4* leads to the same conclusions as Singer and co-workers [4]: there is a larger difference between the measured and calculated concentration values for the desorption period than the adsorption period, coefficients being determined from the adsorption period.

To characterize the desorption period with other values of k_a and k_d coefficients, these values must check the equality (9). The values are obtained by iterative calculations. For the studied case, the coefficients are: $k_a = 4.88\ 10^{-3}$ m.s^{-1} and $k_d = 3.30\ 10^{-4}$ s^{-1}. The new drawn of the evolution of the computed concentration is given figure *5*. Once again, a difference between measured values and computed values is observed for desorption period, but it is lower than previously. These observations match those by Singer and co-workers [4].

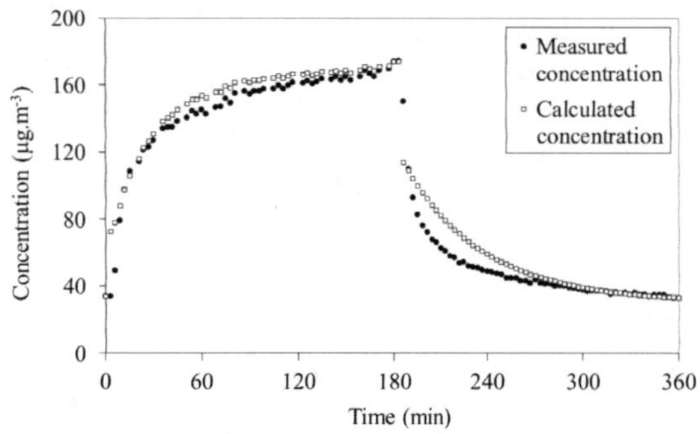

Fig. 5. Profiles of the measured and calculated chamber air concentrations with two different « k_a - k_d » values.

Other tests were carried out in the experimental chamber with the same environmental conditions (same temperature, same relative humidity) and the same material loading factor but for two other equilibrium concentration levels (71 and 118 µg.m^{-3}). The durations of the injection and desorption periods were also the same as previously (3 hours for each one). The exploitation of these tests, as previously explained, leads to obtain other k_a and k_d values.

Table *1* gives all the obtained values, which match data from literature [1,4,5,6,7,8].

Table 1. Values of the model coefficients.

	Adsorption		Desorption	
C (µg.m^{-3})	k_a (m.s^{-1})	k_d (s^{-1})	k_a (m.s^{-1})	k_d (s^{-1})
71	6.84 10^{-2}	1.70 10^{-3}	1.22 10^{-2}	3.30 10^{-4}
118	9.58 10^{-2}	1.30 10^{-3}	1.69 10^{-2}	5.30 10^{-4}
174	8.65 10^{-2}	0.90 10^{-3}	4.88 10^{-3}	3.30 10^{-4}

No conclusion can be drawn from these results. To compare them, it will be necessary and even required to neglect: 1) the fact that initial concentrations are not null in adsorption phase, and 2) the fact that the steady-state concentrations are not null in desorption phase. Thus, a dimensionless concentration is defined and used thereafter.

4 Results exploitation

4.1 Reduced concentration

Like the reduced temperature occurs in heat transfer, the reduced concentration is defined by the following expression:

$$C^* = \frac{C(t) - C_\infty}{C_0 - C_\infty} \qquad (10)$$

where C_0 is the initial concentration and C_∞ is the steady state concentration (equilibrium concentration in the room air).

The value of this dimensionless concentration ranges from 0 to 1. The closer it is to 1, the closer the concentration is to the initial concentration. It means a total adsorption of the injected pollutant. Or rather, the closer the reduced concentration is to 0, the closer the concentration is to the injected one. In this case, it means that the adsorption is very low.

Figure *6* shows the reduced concentration evolutions for the three studied concentration levels (formaldehyde injection concentrations: 174, 118, et 71 µg.m^{-3}) during the adsorption phase. The fact of being able to distinguish the three curves indicates a dependence on the pollutant concentration of the sorption capacity of the material. One can see that the adsorption capacity seems to be growing up along with the concentration.

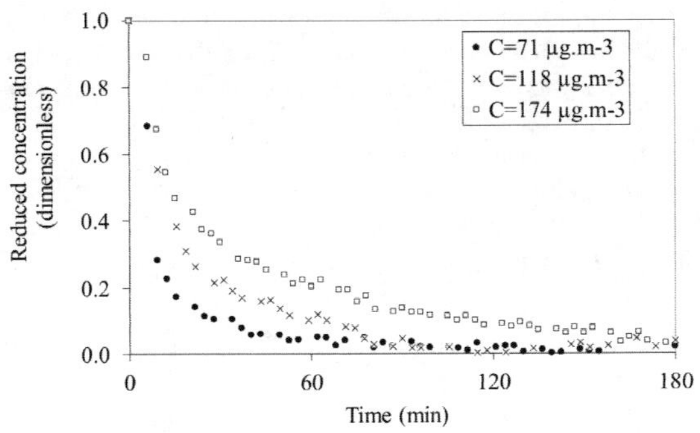

Fig. 6. Profile of the reduced concentration during the adsorption period, for the three studied concentration levels.

Equation (1) can be rewritten as:

$$\frac{dC}{dt}+\left(\frac{Q}{V}+k_a\frac{S}{V}\right)C=\frac{Q}{V}C_\infty+k_d\frac{S}{V}M \quad (11)$$

The solution to this equation without second term takes the following form:

$$C=A\exp\left(-\left(\frac{Q}{V}+k_a\frac{S}{V}\right)t\right) \quad (12)$$

A specific solution, corresponding to the steady state, is, with $dC/dt=0$:

$$C=\frac{\frac{Q}{V}C_\infty+k_d\frac{S}{V}M_\infty}{\left(\frac{Q}{V}+k_a\frac{S}{V}\right)}=C_\infty \quad (13)$$

Thus, the solution to equation (11) is the sum of the two previous results:

$$C=A\exp\left(-\left(\frac{Q}{V}+k_a\frac{S}{V}\right)t\right)+C_\infty \quad (14)$$

where A is determined from the initial condition of the adsorption phase: $A=C_0-C_\infty$.

Consequently, the reduced concentration profile can be expressed as:

$$C^*=\exp\left(-\left(\frac{Q}{V}+k_a\frac{S}{V}\right)t\right) \quad (15)$$

or:

$$C^*=\exp\left(-\frac{t}{\tau}\right) \quad (16)$$

where τ is the time constant (s).

Here, the analogy with heat transfer is highlighted. In the case of thermally thin media within which temperature is considered as uniform, the evolution of the reduced temperature is given exactly by the same expression.

When dealing with the desorption period, equation (1) can be rewritten as follows (with $C_{ext}=0$):

$$\frac{dC}{dt}+\left(\frac{Q}{V}+k_a\frac{S}{V}\right)C=k_d\frac{S}{V}M \quad (17)$$

As previously, the form of the solution of the equation without second term is given by expression (12). When the steady state is reached, the room air concentration would be equal to the background level (or initial concentration of the adsorption period). At the beginning of the desorption period the concentration is the concentration reached at the end of the adsorption period. The integration constant is obtained from this initial condition: $A=C_\infty-C_0$.

Thus, the reduced concentration profile takes the form of the following expression:

$$C^*=1-\exp\left(-\left(\frac{Q}{V}+k_a\frac{S}{V}\right)t\right) \quad (18)$$

4.2 Time constant and inertia

The time constant of the phenomenon, τ, is such as:

$$\frac{1}{\tau}=\frac{Q}{V}+k_a\frac{S}{V} \quad \text{or} \quad \tau=\frac{V}{Q+k_aS} \quad (19)$$

It is possible to evaluate this constant by drawing the function $Ln(C^*)=f(t)$.

The time constant values fo the three studied concentration levels and obtained from the adsorption period are indicated in table *2*. Time constant increases as concentration increases. For a given loading factor, the higher the concentration is, the longer the time to reach equilibrium is.

The time constant is characteristic of inertia. The higher the time constant is, the larger inertia is. A material is considered as thermally inert when the energy it can store during a time lapse is significant. Otherwise it is qualified as reactive.

In the same way, a material would be considered as a pollutant buffer if the mass it could adsorb is significant. The profile of the mass in the sink for the three studied concentration levels is presented in figure *7*. The resulting adsorbed masses during the adsorption periods were computed from the time integration of the solution of equation (2).

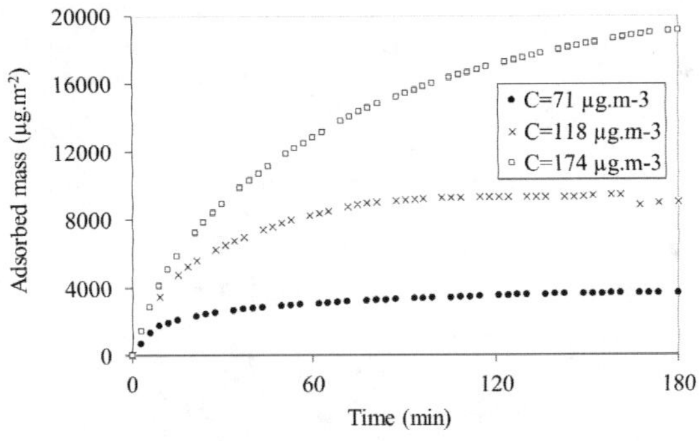

Fig. 7. Evolution of the adsorbed mass on the sink during the adsorption period, for the three concentration levels.

Equation (2) can be rewritten:

$$\frac{dM}{dt} + k_d M = k_a C \qquad (20)$$

The form of the solution of the same equation without second term is:

$$M = B\exp(-k_d t) \qquad (21)$$

and a specific solution corresponding to the equilibrium is: $M = M_\infty$.

Elsewhere, the constant is determined from the initial condition of the adsorption period: $B = -M_\infty$. Thus:

$$M = M_\infty(1 - \exp(-k_d t)) \qquad (22)$$

The time integration of this last equation leads to the total mass (m) of pollutant adsorbed during the time of the adsorption period. The values for the three studied levels of pollutant concentration are also given in table *2*. These results confirm the previous conclusions.

Table 2. Time constant and adsorbed mass.

C (μg.m^{-3})	τ (min)	m (μg)
71	12.05	78.5
118	18.87	234.3
174	22.22	434.3

5 Conclusion

Apart from the notion of reduced concentration – analogous to the reduced temperature one – there are many analogies between heat and mass transfers [9]. Notions like time constant and inertia are analogous: the more a material occurs as a wall component of the room the more it gives inertia to this room. The larger the material surface is exposed to the room air, the more important the sink effect due to the material is: the room air concentration is smaller during the adorption period and the pollutant reemission is slower during the desorption period.

Here we studied a synthetic non-porous material, and we used a « k_a - k_d » model which is based on the hypothesis that concentrations on the material surface and room air are in equilibrium. Transfers inside the materials were neglected. This modelling is analogous to the heat transfer one for thin material – which does not display any temperature gradient (temperature is uniform) – exposed to an external temperature variation.

This simplification leads to some discrepancies between the measured data (results from experimental tests) and the computed one (results from the model). The k_a and k_d coefficients would depend on the material and the air flow on the surface of this material.

It also depends on the pollutant transfers (mainly diffusion [10]) which take place within the material after the adsorption or before the desorption. Thus, the knowledge of the solubility constant for the studied pollutant/material couple and the diffusion coefficient of the pollutant within the material would be useful in order to evaluate the lack of precision of the simplified model which has been used here.

We wish to thank the laboratory LaSIE of the University of La Rochelle for the use and the exploitation of the experimental results.

References

1. B. Tichenor, Z. Guo, J. Dunn, L. Sparks, M. Mason. The Interaction of Vapor Phase Organic Compounds with Indoor Sinks. *Indoor Air*, **1**, pp. 23-35, (1991)
2. A.L. Tiffonnet. Contribution à l'analyse de la Qualité de l'Air Intérieur : Influence des transports de Composés Organiques Volatils (COV) entre les parois et l'ambiance. PhD thesis - Université de La Rochelle (2000)
3. C. Tourreilles. Qualification énergétique et sanitaire des systèmes d'épuration intégrés aux réseaux de ventilation. PhD thesis - Université de La Rochelle (2015)
4. B. Singer, A. Hodgson, K. Guevarra, E. Hawley, W. Nazaroff. Gas-phase organics in environmental tobacco smoke. Effects of smoking rate, ventilation, and furnishing level on emission factors. *Environmental Science & Technology*, **36**, pp. 846-853, (2002)
5. Y. An, J. Zhang, C Shaw. Measurements of VOC adsorption / desorption characteristics of typical interior building materials. *HVAC&R Research : International Journal of Heating, Ventilating, Air-Conditioning and Refrigerating Research*, **5**, issue 4, pp. 297-316, (1999)
6. W. Bouhamra, A. Elkilani. Investigation and modeling of surface sorption / desorption behavior of volatile organic compounds for indoor air quality analysis. *Environmental Technology*, **20**, issue 5, pp. 31-45, (1999)
7. R. Jorgensen, T. Dokka, O. Bjorseth. Introduction of a Sink-Diffusion Model to Describe the Interactions between Volatile Organic Compounds (VOCs) and Material Surfaces. *Indoor Air*; **10**, pp. 27-38, (2000)
8. D. Won, R. Corsi, M. Rynes. Sorptive Interactions between VOCs and Indoor Materials. *Indoor Air*, **11**, pp. 246-256, (2001)
9. A. Damian, A.L. Tiffonnet, P. Blondeau, F. Allard. Mise en évidence de l'effet des parois sur les concentrations intérieures en polluants – Analogie avec le transfert thermique, *Actes du Forum des Associations du Génie Civil et Urbain*, Lyon, pp. 125-132, (2000)
10. P. Blondeau, A.L. Tiffonnet; F. Allard, F. Haghighat. Physically-based modeling of the material and gaseous contaminant interactions in buildings: models, experimental data and future developments. *Advances in Building Energy Research*, **2**, pp. 57-94, (2008)

A Numerical Analysis of the Air Distribution System for the Ventilation of the Crew Quarters on board of the International Space Station

Florin Bode[1,2], Ilinca Nastase[2], Cristiana Verona Croitoru[2], Mihnea Sandu[2], Angel Dogeanu[2]

[1]CAMBI Research Center, Technical University of Civil Engineering Bucharest, 021414 Bucharest, Romania
[2]Technical University of Cluj Napoca, Department of Mechanical Engineering 400020 Cluj - Napoca, Romania

Abstract. Quality of life on the International Space Station (ISS) has become more and more important, since the time spent by astronauts outside the terrestrial atmosphere has increased in the last years. The actual concept for the Crew Quarters (CQ) have demonstrated the possibility of a personal space for sleep and free time activities in which the noise levels are lower, but not enough, compared to the noisy ISS isle way. However, there are several issues that needs to be improved to increase the performance of CQ. Our project QUEST is intended to propose a new concept of CQ in which we will correct these issues, like the noise levels will be lower, more space for astronaut, increased thermal comfort, reduce the CQ total weight, higher efficiency for the air distribution, personalized ventilation system in CQ for the crew members in order to remove CO_2 from the breathing zone. This paper presents a CFD study in which we are comparing the actual and a proposed ventilation solution for introducing the air in CQ. A preliminary numerical model of the present configuration of the air distribution system of the Crew Quarters on board of the ISS, shows the need for an improved air distribution inside these enclosures. Lower velocity values at the inlet diffuser, distributed over a larger surface, as well as diffusers with improved induction would appear to be a better choice. This was confirmed through the development of a new model including linear diffusers with a larger discharge surface. In this new configuration, the regions of possible draught are dramatically reduced. The overall distributions of the velocity magnitudes displaying more uniform, lower values, in the same time with more uniform temperatures. All these observations allow us to consider a better mixing of the air inside the enclosure.

1 Introduction

Space is one of the most extreme environments imaginable. Outside ISS, the wall temperature will vary from 121°C (Sun facing side) to -157°C (dark side)[1]. Due to the fact that the time spent by astronauts on ISS has increased over the years, quality of life has become a matter that needs to be improved. Furthermore, future space stations and long range transportation space ships will demand for more comfortable life conditions for both crew members and travellers. Over the last two-and-a-half decades, The International Space Station's Environmental Control and Life Support System (ECLSS) has developed in size, complexity, and capability continuously from the beginning [2]. ISS crew members are living and working in a high-risk environment, and their comfort plays a major role in ISS proper operation, because a rested crew member operating in a comfortable environment is less prone to human error.

Nowadays, on the ISS operates several personal Crew Quarters (CQ) for the crew members [3-5]. The most advanced form all CQ are four of them which are installed in the Harmony Node (also known as Node 2) [6]. Functionally, the role of a CQ is to provide a quiet private place in which crew personnel can sleep or relax in the free time (Figure 1 and Figure 2).

However, there are several issues that needs to be improved to increase the performance of CQ. NASA and ESA admitted that on ISS, higher noise levels, CO2 and dust accumulation are frequently occurring issues [7, 8]. From our observations we believe that the current CQ design can be improved in order to reduce the noise levels, increase the personal space for astronaut, increased thermal comfort, reduce the CQ total weight, higher efficiency for the air distribution, personalized ventilation system in CQ for the crew members in order to remove CO_2 from the breathing zone. These improvements are the objective of the QUEST research project developed by our research team.

One of the goals of our project is to understand the flows occurring in the actual design of air diffusion system of the existing CQ on the ISS. Our team has experience on studying the air flow through different types of diffusers. This paper presents a CFD study in which we are comparing the actual and a proposed ventilation solution for introducing the air in CQ [9-11].

In the last five decades, extensive investigations and experiments involving human subjects have resulted in methods for predicting the degree of thermal discomfort of people exposed to a certain environment. One of the most well-known and widely accepted method is Fanger's "Comfort Equation" and his practical concepts

* Corresponding author: ilinca.nastase@utcb.ro

 https://doi.org/10.1051/e3sconf/20183201006

of "Predicted Mean Vote" and "Predicted Percentage of Dissatisfied" [12].

Fanger's hypothesis relies on the thermal balance between the heat produced, consumed and transferred to the environment when the human body finds itself in a state of neutrality. The most important parameters that can influence the thermal comfort state were considered to be the metabolic rate, the clothing thermal resistance, the air temperature, velocity, humidity, pressure and turbulence intensity [12-15].

Figure 1 : Crew Quarters installed on ISS Harmony node [16]

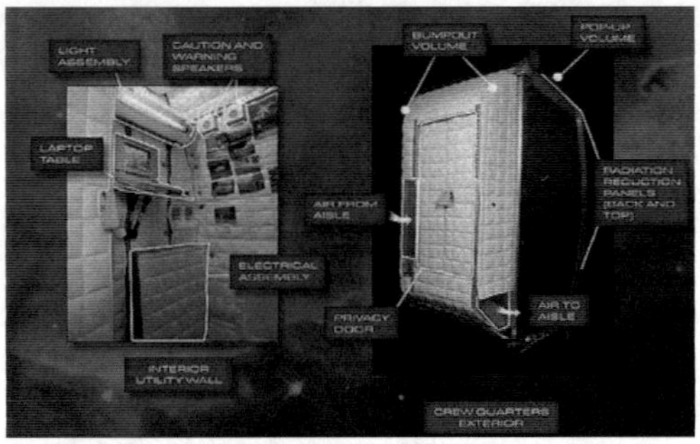

Figure 2: CQ in its actual structure [3]

When the heat balance equation proposed by Fanger is satisfied, the heat generated by the human body is dissipated without having an increase or a decrease in the human body [12]. PMV index values are between -3 and 3. These values quantify the average opinion of a group of subjects on the state of comfort. Associated with this parameter, is the index PPD, indicating the percentage of occupants in thermal discomfort.

Fanger and Christensen [17] proposed correlations for the local turbulence intensity of the indoor air flow with the thermal discomfort sensation in an index called "Draught Rate" (DR). The Draught Rate is an index that depends on the mean velocity and temperature but also on the air turbulence intensity value. DR values are divided in four regions of values corresponding to a classification of the indoor ambiance regarding its comfort level. This way, a values of $DR \leq 15\%$ are associated to the zones with a high standing quality of the ambiance, values of $15\% \leq DR \leq 20\%$ are corresponding to a good quality of the ambiance, values $20\% \leq DR \leq 25\%$ are corresponding to an acceptable quality of the ambiance and values of $DR > 25\%$ correspond to unacceptable conditions to regions where the sensation of draught might create a serious discomfort. More information on thermal comfort models for indoor spaces and vehicles can be found in [18].

The CQ total volume is around 2.1 m^3 and is limited to the size of a standard US rack volume [6, 19, 20]. The actual ventilation system provides adjustable airflow at three different flow rates. The CQ ventilation system's main purpose is to lower the carbon dioxide concentration, which in high levels represents an asphyxiation hazard. Another role of the CQ ventilation system is to remove the heat generated in the CQ which is composed from crew member's metabolic heat (100-132 W) and the electronics heat (around 153 W) [6, 19, 20]. The primary ventilation requirements for the CQ are: 0.42-5.1m^3/min of airflow and exhaust air velocity lower than 1.2 m/s.

2 MATERIAL AND METHOD

The CQ geometrical model along with the virtual manikin was built using the SolidWorks software in Figure 3. Unstructured tetrahedral meshes were generated for both studied cases. The actual ventilation solution study of the CQ (Figure 3a) was already presented in [21]. This configuration of the air diffuser is composed by four rectangular regions, each of them having independent vertical and horizontal guiding vanes [6, 19, 20]. The proposed ventilation solution is presented in Figure 3b. The main difference between the presented solutions in Figure 3 is the replacement of the axial fan solution with a tangential fan solution.

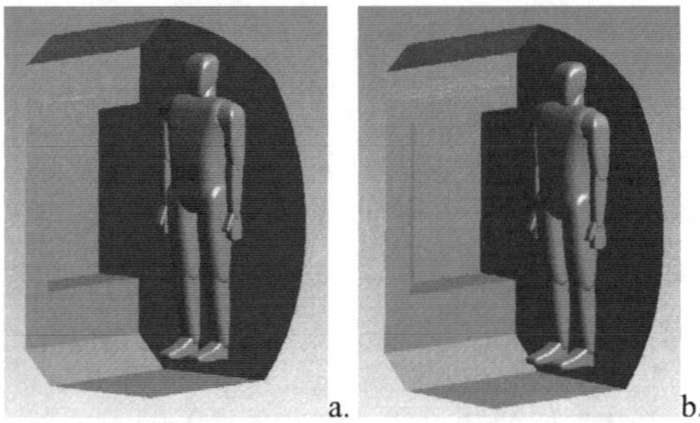

a. b.

Figure 3: Studied geometrical model a. Actual ventilation solution - Case 1 b. Proposed ventilation solution - Case 2

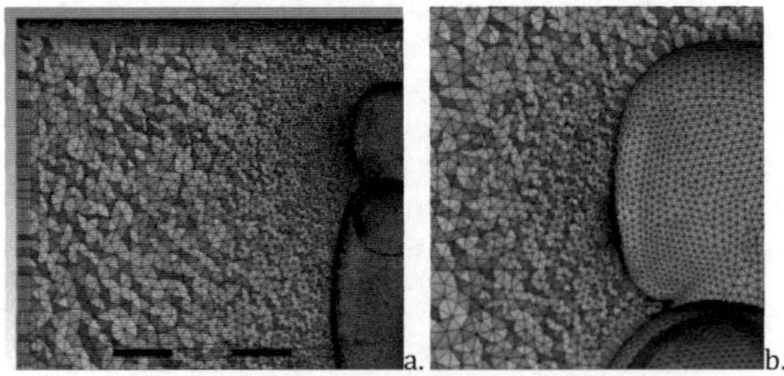

a. b.

Figure 4: Computational mesh for case 2
a. General view b. Close-up view

For the proposed ventilation solution (Case 2), a new mesh was generated consisting of 5.8 million elements. A boundary layer of 5 layers was generated (Figure 4 a and b).

Ansys Fluent was the software used for the numerical modelling. For the pressure-velocity coupling, Fluent provides four segregated types of algorithms (SIMPLE, SIMPLEC, PISO, Fractional Step) and a coupled one. For this study we used coupled algorithm which offers some advantages over the segregated approach because it contains a robust and efficient single phase implementation for steady-state flows, with superior performance compared to the segregated algorithms [22]. A second order upwind scheme was used to calculate the convective terms in the equations, integrated with the finite volume method. SST k-ω was the turbulence model used for the numerical simulation due to our experience in using this model for indoor air circulation [23]. The reason is that in SST k-ω turbulence model, for flows, both the boundary layer as well as the free stream are considered.

For the boundary conditions, we imposed three mass flow rates corresponding to the three positions of the fan controller: low, medium and high, respectively 108 m^3/h, 138 m^3/h and 196 m^3/h [20]. The Reynolds numbers at the exit of the air vents based on streamwise mean velocity and on the equivalent diameter (De_1= 1.144m) were Re_1= 4462, Re_2=5692, and Re_3=8098. The inlet turbulence intensity was imposed at: 5.6%, 5.43% and 5.19%, being calculated using the empirical relation proposed by Jaramillo [24]. The present configuration of the air diffuser is presented in [6, 19, 20] and is composed by four rectangular regions, each of them having independent vertical and horizontal guiding vanes. Inspired by [20] we considered that the air flow is deflected equally towards right and left in the horizontal plane with a 45° angle respectively, and in the same time the air flow is deflected equally towards the 'up' and 'down' in the vertical plane with a 45° angle as well.

For the cases 1, 2 and 3 we have considered that the air flow is deflected right and left on the horizontal and up and down on the vertical with a 45° angle [20]. Case 1, 2 and 3 are corresponding to the flow rates of 108 m^3/h, 138 m^3/h and 156 m^3/h. For the cases 4 and 5 the air was deflected 45° angle down, through an improved linear diffuser. This case is associated with an idea of a first level improvement of the air distribution system in the CQ. This way, we propose to keep the actual configuration of the inlet and the exhaust ducting parts but renounce to the ducting and abatements that are situated along the door of the CQ. This will be possible using transverse flow (cross-flow) fans. The inlet and the outlet grilles inside the CQ cabin will be kept at the same locations. The inlet air grille to the Node 2 cabin will be placed on the 'upper' chamfered part of the bump-out while the exhaust air grille will be kept at the same position on the lateral flat side of the bump-out.

In the case 4 we tested the medium flow rate of 138 m^3/h, and in the case 5 we tested the high flow rate of 156 m^3/h. We considered to present here only the two highest flow rates that were recorded to produce problems in terms of local air speed discomfort. The inlet air temperature was in all cases 18°C corresponding to the air taken from Node 2. The heat flux from the walls was equivalent to 150 W (imposed wall temperature was 22 °C). On the different zones of the virtual manikin inside the CQ we imposed the temperatures from Table 1.

Table 1. Temperatures imposed on the manikin surface

t	t_{head}	t_{neck}	t_{torso}	$t_{shoulders}$	t_{arms}	$t_{forearms}$	t_{hand}	t_{thigh}	t_{leg}	t_{feet}
°C	36	35	34	34	33	32	30	32	30	28

3. RESULTS AND DISCUSSION

Figures 5, 6 and 7 are presenting the distributions of the velocity magnitude, air temperature, of the in-plane vectors, and of the predicted percentage of dissatisfied regarding the draught sensation in the coronal plane of the virtual manikin in the first three cases (cases 1, 2 and 3). These three cases are corresponding to the current situation of the air distribution system in the CQ on board of the ISS.

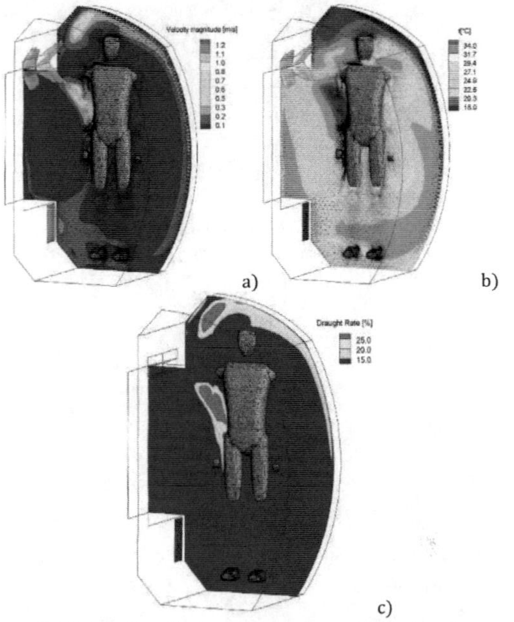

Figure 5: Distributions in the coronal plane of the virtual manikin – case 1 - 108m^3/h : a) velocity magnitude, b) air temperature, c) DR – predicted percentage of dissatisfied regarding the draught sensation

On the velocity magnitudes distributions in the coronal plane we superimposed iso-values of the velocity magnitude in the entire cabin corresponding to 0.6 m/s. On the temperature distributions in the coronal plane we superimposed iso-values of the temperature in the entire cabin corresponding to 18°C. Figures 8 and 9 are presenting the same distributions for the cases 4 and 5 corresponding to the medium flow rate and the proposed new air distribution strategy.

For the first three cases, the global pattern of the flow inside the CQ is changing dramatically with a relatively slight variation of the inlet velocity. In the region of the head and of the chest, fairly high values of the velocity magnitude from the point of view of limitations imposed by the thermal comfort standards, suggest that a sensation of thermal discomfort might occur. This is confirmed by the distributions of the draught rate.

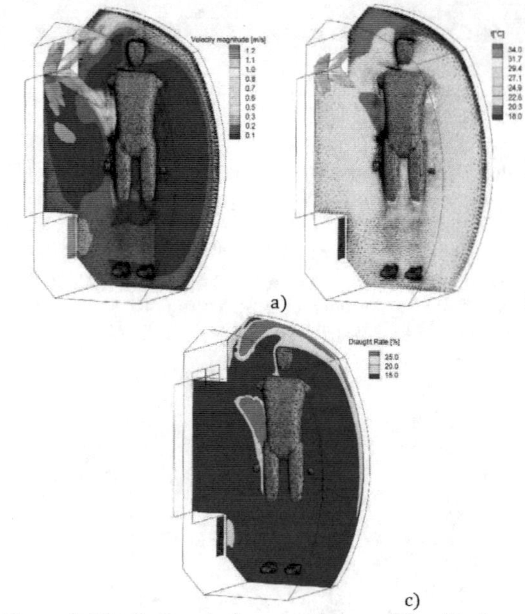

a) b) c)

Figure 6: Distributions in the coronal plane of the virtual manikin – case 2 - 138m^3/h : a) velocity magnitude, b) air temperature, c) DR – predicted percentage of dissatisfied regarding the draught sensation

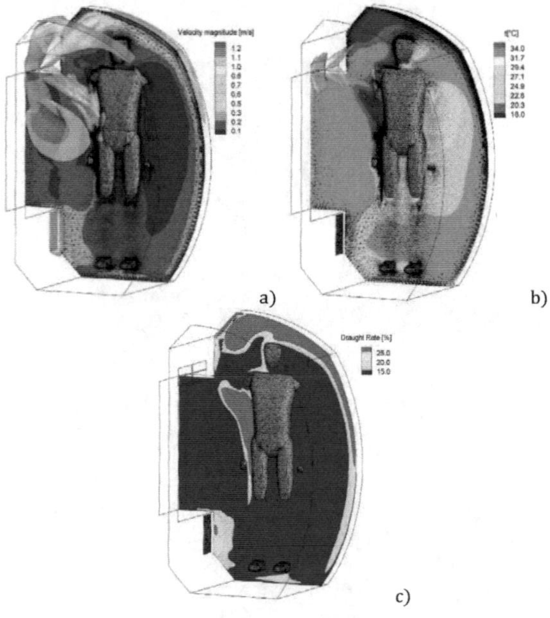

a) b) c)

Figure 7: Distributions in the coronal plane of the virtual manikin – case 3 - 196m3/h: a) velocity magnitude, b) air temperature, c) DR – predicted percentage of dissatisfied regarding the draught sensation

The regions of possible draught are dramatically reduced in the case of the new linear diffusers. The flow is introduced with lower mean velocities over a larger surface, and the flow is oriented away of the head of the occupant. The overall distributions of the velocity magnitudes in Figure 8 and 9 are displaying lower values, more uniform than in the previous cases. In the same time we could observe more uniform temperature distributions inside the CQ. All these observations allow us to consider a better mixing of the air inside the enclosure.

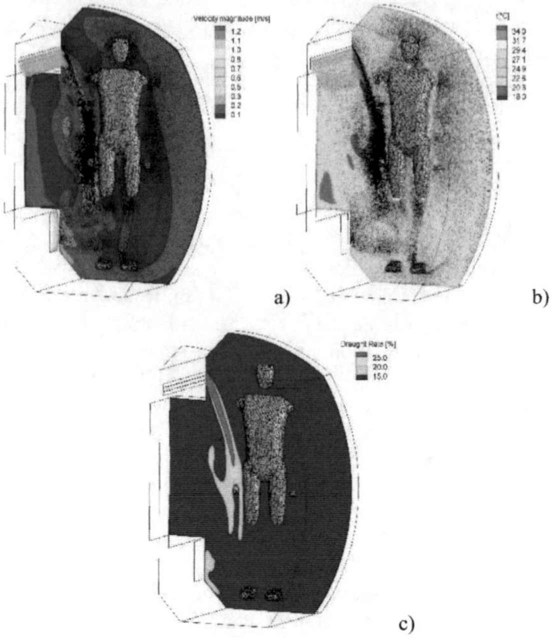

a) b) c)

Figure 8: Distributions in the coronal plane of the virtual manikin – case 4 - 138m^3/h: a) velocity magnitude, b) air temperature, c) DR – predicted percentage of dissatisfied regarding the draught sensation

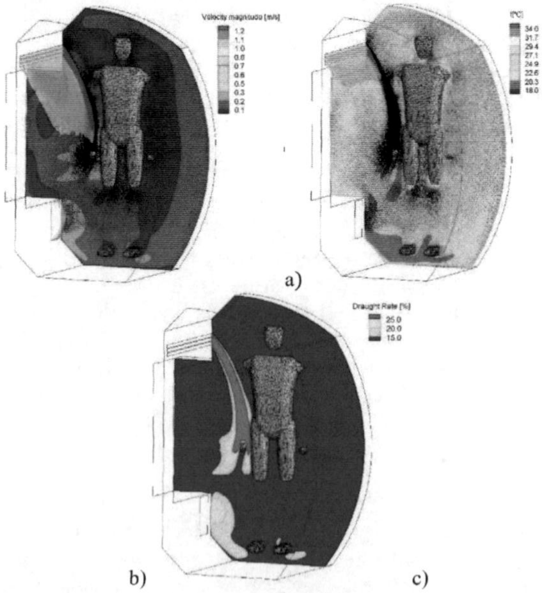

a) b) c)

Figure 9: Distributions in the coronal plane of the virtual manikin – case 4 - 138m^3/h: a) velocity magnitude, b) air temperature, c) DR – predicted percentage of dissatisfied regarding the draught sensation

CONCLUSIONS

From our preliminary numerical study representing the testing of a model of the present configuration of the air distribution system of the Crew Quarters on board of the ISS, it appears that the need for an improved air distribution inside these enclosures is obvious. This way, lower velocity values at the inlet diffuser, distributed over a larger surface, as well as diffusers with improved induction would appear to be a better choice. This was confirmed through the development of a new model including linear diffusers with a larger discharge surface.

In this new configuration, the regions of possible draught are dramatically reduced. The overall distributions of the velocity magnitudes displaying more uniform, lower values, in the same time with more uniform temperatures. All these observations allow us to consider a better mixing of the air inside the enclosure.

Other further improvements, like a supplementary PV ventilation system, can aid in diminishing the issues with CO2 accumulation during the sleeping hours. A combination between the general ventilation system in the CQ module and a personalized ventilation system with an adjustable air diffuser will allow us to obtain good air quality near the crew member in any conditions and further improve thermal comfort by covering the need for fresh air without having to lower the overall temperature inside the CQ module. This configuration has to be tested next.

This work was supported by a grant of the Romanian space agency ROSA, QUEST - Advanced air diffusion system of the crew quarters for the ISS and deep space habitation systems, STAR-CDI-C3-2016-577

References

1. NASA, https://science.nasa.gov/science-news/science-at-nasa/2001/ast21mar_1. 2001.
2. Robert M. Bagdigian, et al. *International Space Station Environmental Control and Life Support System Mass and Crewtime Utilization In Comparison to a Long Duration Human Space Exploration Mission. International Conference on Environmental Systems ICES*. 2015. Seattle,
3. Schlesinger, T.P., B.R. Rodriguez, and M.A. Borrego, *International Space Station Crew Quarters On-Orbit Performance and Sustaining Activities*. AIAA Journal, 2013.
4. Broyan, J.L., S.M. Cady, and D.A. Welsh, *ISS Crew Quarters Ventilation and Acoustic Design Implementation*. AIAA Journal, 2010.
5. *NASA Reference guide to the ISS* https://www.nasa.gov/pdf/508318main_ISS_ref_guide_nov2010.pdf. 2010.
6. Fairburn, S. and S. Walker, *'Sleeping With the Stars'–The Design of a Personal Crew Quarter for the International Space Station*. 2001, SAE Technical Paper.
7. Matty, C.M., *Overview of Carbon Dioxide Control Issues During International Space Station/Space Shuttle Joint Docked Operations*. American Institute of Aeronautics and Astronautics, 1976.
8. https://www.newscientist.com/article/dn9379-noisy-iss-may-have-damaged-astronauts-hearing/.
9. Amina Meslem, et al., *Optimization of a Lobed Perforated Panel Diffuser - A Numerical Study of Orifice Arrangement*. The International Journal of Ventilation, 2012. **11**(3): p. 16.
10. Bode, F., A. Meslem, and C. Croitoru, *Numerical simulation of a very low Reynolds cross-shaped jet*. Journal Mechanika 2013. **19**(5): p. 6.
11. Bode, F., et al., *The influence of the Inlet angle of vehicle air diffuser on the thermal comfort of passengers*. ENERGY and ENVIRONMENT (CIEM), 2017 p. 442-446.
12. Fanger, P.O., ed. *Thermal Comfort-Analysis and Applications in Environmental Engineering*. ed. C.D.T. Press. 1970.
13. *ISO 7730 - Ergonomics of the thermal environment*. 2005.
14. Djongyang, N., R. Tchinda, and D. Njomo, *Thermal comfort : A review paper*. Renewable and Sustainable Energy REviews, 2010. **14**: p. 2626-2640.
15. Nilsson, H.O., *Thermal comfort evaluation with virtual manikin methods*. Building and Environment, 2007. **42**(12): p. 4000-4005.
16. NASA, https://spaceflight.nasa.gov/gallery/images/station/crew-26/html/iss026e012169.html. 2010.
17. Fanger, P.O. and N.K. Christensen, *Perception of draught in ventilated spaces*. Ergonomics, 1986. **29**(2): p. 215 - 235.
18. Croitoru, C., et al., *Thermal comfort models for indoor spaces and vehicles—Current capabilities and future perspectives*. Renewable and Sustainable Energy Reviews, 2015. **44**(Supplement C): p. 304-318.
19. Adams, C. *Four Legs in the Morning: Issues in Crew-Quarter Design for Long-Duration Space Facilities*. in *28th International Conference on Environmental Systems, Danvers, MA*. 1998.
20. Broyan, J.L., M.A. Borrego, and J.F. Bahr. *International Space Station USOS Crew Quarters Development*. in *08ICES-0222*. 2008.
21. Florin Bode, et al., *Analysis for an improved concept of ventilation system for the Crew Quarters on board of the International Space Station*, in *6th CEAS Air & Space Conference*. 2017: Bucharest.
22. Ansys Inc, https://www.sharcnet.ca/Software/Ansys/17.0/en-us/help/flu_th/flu_th_sec_uns_solve_pv.html.
23. Danca, P., et al., *On the Possibility of CFD Modeling of the Indoor Environment in a Vehicle*. Energy Procedia 2017. **112**: p. 656-663.
24. Jaramillo, J.E., et al., *Numerical study of plane and round impinging jets using RANS models*. Numer. Heat Transfer Part B 2008. **54**: p. 213-237.

E3S Web of Conferences **32**, 01007 (2018) https://doi.org/10.1051/e3sconf/20183201007
EENVIRO 2017

Thermal expansion coefficient determination of polylactic acid using digital image correlation

Adrian - Ioan Botean[1,*]

[1]Mechanical Engineering Department, Mechanical Faculty, Technical University of Cluj – Napoca, Romania

Abstract. This paper aims determining the linear thermal expansion coefficient (CTE) of polylactic acid (PLA) using an optical method for measuring deformations called digital image correlation method (DIC). Because PLA is often used in making many pieces with 3D printing technology, it is opportune to know this coefficient to obtain a higher degree of precision in the construction of parts and to monitor deformations when these parts are subjected to a thermal gradient. Are used two PLA discs with 20 and 40% degree of filling. In parallel with this approach was determined the linear thermal expansion coefficient (CTE) for the copper cylinder on the surface of which are placed the two discs of PLA.

1 Introduction

In this study we aim to determine the linear thermal expansion coefficient (CTE) [1] of the polylactic acid (PLA) using the digital image correlation method.

Polylactic acid (PLA) is a thermoplastic material made from renewable resources such as corn starch or sugar cane. It is biodegradable and difficult to recycle. PLA does not present hazardous components and is considered compatible with the environment. PLA can come into contact with food and is a non-toxic thermoplastic material. [2], [3]

It is a strong and rigid material, making it more complicated to use for parts that require joining. PLA is used as a matrix for some fiber – reinforced composites that are 100% biodegradable. [4] Polylactic acid is currently used in the biomedical field, from which various elemens such as stents are made.

Depending on the molecular weight and degree of crystallinity, polylactic acid can be used in various applications. For example, if biaxially oriented the material can maintain its stability to 130-150 °C. It also has very good optical properties, with excellent machinability and excellent behaviour both at twisting and bending. [4]

The material can be configured to provide different mechanical properties depending on the chosen technological processing. This process can be by injection, molding, foil extrusion, blowing, or firing using the best known manufacturing methods and equipment. [4]

The advantage of polylactic acid is the low – temperature 3D printing capability.

The 3D printer is defined as a mechanism thet makes physical objects using input file information obtained through a three – dimensional graphics software, such as SolidWorks. [5] Through 3D printing, mechanism with a fully functional role can be created from just one technological process, without the need for other post – production assembly methods.

Fused Deposition Modeling (FDM) – extrusion medeling is the most widely used manufacturing technology because of its simplicity and affordability (Fig.1). [6]

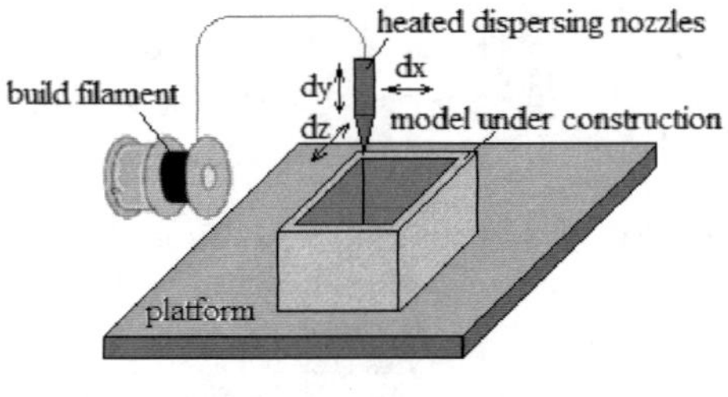

Fig. 1. Extrusion modeling.

Fig. 2. Filament made from PLA (polylactic acid).

This method is used in modeling, prototyping, but also in production applications. It works with a software application where the 3D model is sliced, in the first phase, in cross – sectioned sections called layers. The printing process is carried out by passing a PLA filament (Fig.2) through an extruder that heats it to the melting point, after which the layer is applied uniformly over the layer until the 3D model conforms to the CAD file. [6] The head or extruder is heated to melt the filament made

* Corresponding author: boteanadrianioan@yahoo.com

from PLA. It moves both horizontally and vertically under the action of a numerical control mechanism. It is controlled directly by the CAM printer application. In motion, the extruder dposits a thin plastic string that, by cooling, sticks to the previous layer so that the shape of the desired pattern is created. [7]

With the help of the digital image correlation method (DIC), it is possible to characterize, in real time, the coefficient of linear thermal expansion (CTE) of PLA. It can be a reliable tool for measuring thermal deformations, especially for plastics material with a high CTE. [7], [8]

2 Experimental part

Materials and methods. To determinate CTE for PLA two disks werw made using the 3D CREATR printer based on the Fused Deposition Modeling (FDM) method. The first disk has a filling degree of 20% (d=29.8 [mm], h=3.118 [mm]) and the second disk has a filling degree of 40% (d=29.2 [mm], h=2.952 [mm]), according to Fig.3.

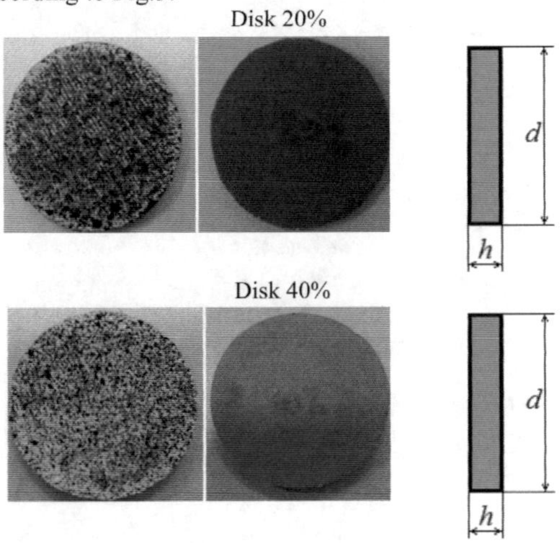

Fig. 3. Disks made of PLA.

The experimental installation is shown in Fig.4 and consists of the following elements: the disk made of PLA which is placed on the copper cylinder which is heated by a Mark VII (MG VISHAI) soldering station. The temperature of the cylinder is monitored using a Pt100 type heat – resistant, connected to the Spider 8 data acquisition system (HBM) and the data is viewed, in real time, using the CatmanEasy software (HBM) according to Fig.5. [9], [10] PLA disk temperature is monitored, in real time, using a T400 (FLIR) thermal imaging camera, as shown in Fig.6. The Δh expansion of the PLA disk in the transverse plane is measured with the Q400 digital correlation system from Dantec Dynamics (www.dantecdynamics.com), that includes two CCD cameras, 1/8", 1624x1234 pixel resolution, frame rate up to 30 Hz, control electronic, lenses with 17 mm focal length and patented cold light system HILIS (High Intensity LED Illumination System) for very homogenous illumination of the specimen. [7], [11]

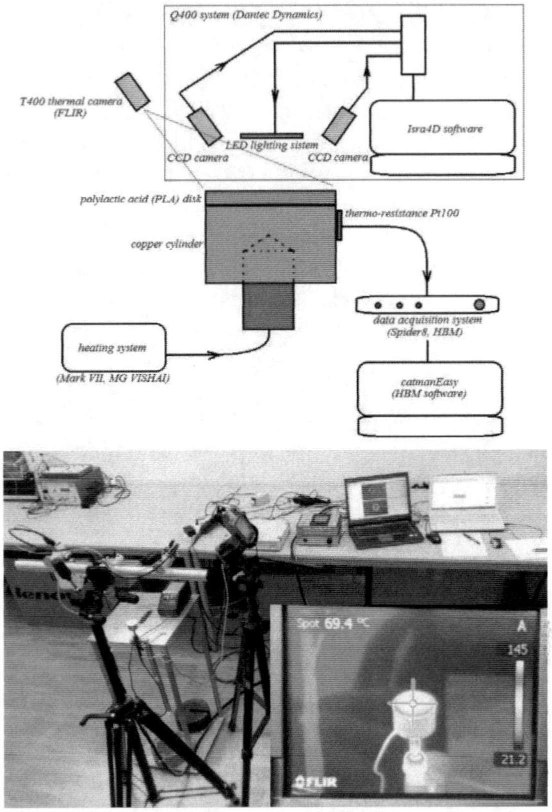

Fig. 4. The experimental installation.

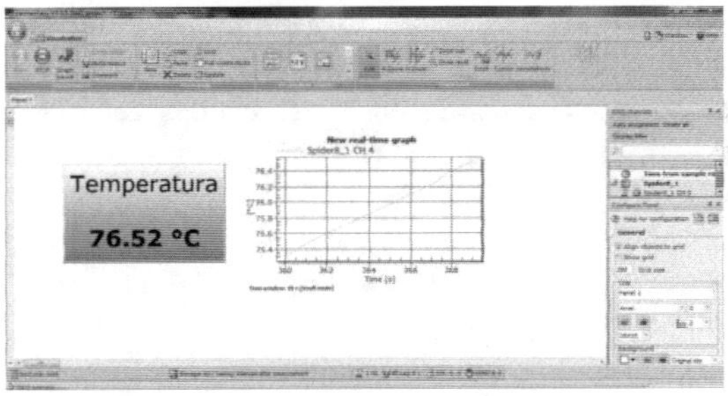

Fig. 5. Real – time monitoring of cylinder temperature with CatmanEasy, Spider8 and Pt100.

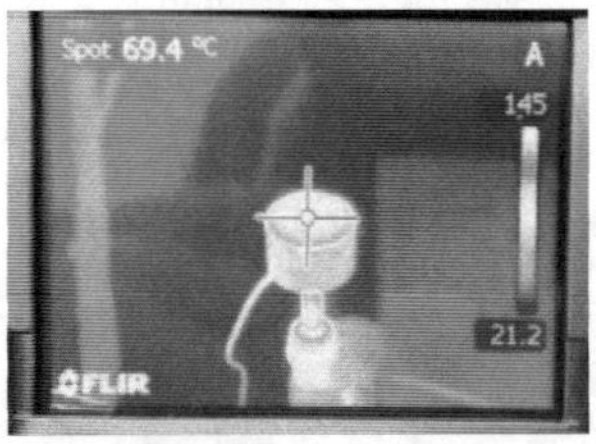

Fig. 6. Real – time monitoring of PLA disk with T400 thermal imaging camera.

Within the Q400 digital image correlation system, intrinsec parameters are determined by lens focal length, lens principal point, and radial and tangential lens distortion. To determine the intrinsec parameters, a calibration plate (Fig.7) is used, which is rotated and translated manually in front of the CCD cameras. The Isra4D software automatically records the nodal points of the calibration plate in a sequence of images and calculates intrinsec and extrinsec parameters.

Thus, the intrinsec CCD camera 1 parameters are: focal length (x; y) 3820±30; 3800±30, principal point (x; y) 610±20; 441±14, radial distortion (r^2; r^4) -0.14±0.03; -5.9±1.6, tangential distorsion (t_x; t_y) -0.0023±0.0009; 0.002±0.002 and for CCD camera 2 are: focal length (x; y) 3830±30; 3810±30, principal point (x; y) 700±40; 453±18, radial distortion (r^2; r^4) -0.26±0.03; -1.4±0.9, tangential distorsion (t_x; t_y) -0.0012±0.0013; -0.003±0.002.

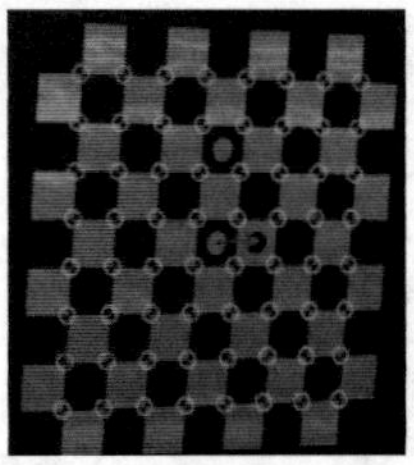

Fig. 7. Determining the intrinsec and extrinsec parameters of CCD cameras using the calibration plate.

Extrinsec parameters are represented by the translation vector and the rotation matrix. Thus, the extrinsec parameters for CCD camera 1 are: rotation vector (x; y; z) -3.020±0.004; -0.0679±0.0016; -0.769±0.008, translation vector (x; y; z) 5±2; 1.9±1.2; 323±3 and for CCD camera 2 are: rotation vector (x; y; z) -3.029±0.004; -0.014±0.002; -0.810±0.013, translation vector (x; y; z) -4±3; 11.6±1.6; 340±3.

In parallel with the determination of the deformations of the PLA disk in the transverse plane, the deformations of the copper cylinder (having a diameter d=37.51 [mm] and a height h=24.9 [mm]) will be measured according to Fig.8.

Once the displacement Δh are determined, depending on the temperature gradient Δt the CTE can be determined by the following relation:

$$\alpha = \frac{\Delta h}{h \cdot \Delta t}\left[\frac{mm}{mm \cdot {}^{o}C}\right]$$

For each disk (20% and 40% filling degree) or cylinder, three sets of measurements carried out were generally noted with Set1, Set2 and Set3. Parallel to the study of the CTE for PLA disks, the same coefficient will be obtained for the copper cylinder, which is the support and heating element of the disks.

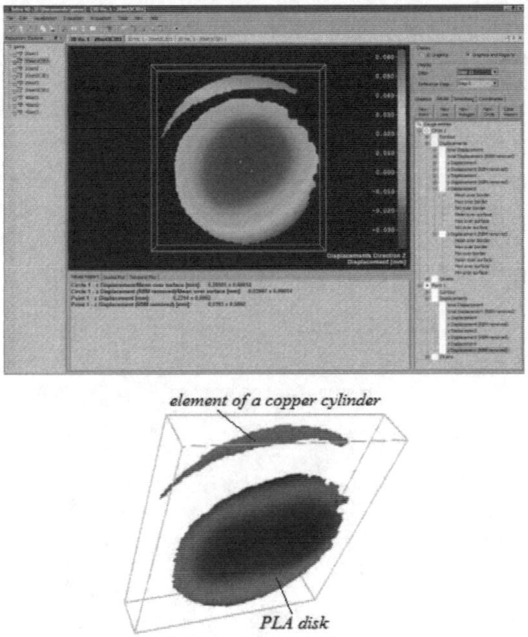

Fig. 8. Determination of the displacement in the transverse plane of the PLA disk and of the copper cylinder with the aid of digital image correlation method.

3 Results

In Fig.9 is plotted the temperature variation diagram of the PLA disk with 20% degree of filling according to time monitored with the thermal imaging camera T400 and in Fig.10 is plotted the temperature variation of the copper cylinder, according to time, monitored with Pt100 thermoresistor transducer.

In Fig.11 is plotted the temperature variation diagram of the PLA disk with 20% degree of filling according to time monitored with the thermal imaging camera T400 and in Fig.12 is plotted the temperature variation of the copper cylinder, according to time, monitored with Pt100 thermoresistor transducer.

In Fig.13 is plotted of the deformation Δh in the transverse plane for the PLA disk with 20% degree of filling according to the temperature gradient Δt monitored with the thermal imaging camera T400 and Fig.14 shows the variation of the deformation diagram Δh in the transverse plane for the copper cylinder according to the temperature gradient Δt monitored by the Pt thermoresistor transducer.

In Fig.15 is plotted of the deformation Δh in the transverse plane for the PLA disk with 40% degree of filling according to the temperature gradient Δt monitored with the thermal imaging camera T400 and Fig.16 shows the variation of the deformation diagram Δh in the transverse plane for the copper cylinder according to the temperature gradient Δt monitored by the Pt100 thermoresistor transducer.

https://doi.org/10.1051/e3sconf/20183201007

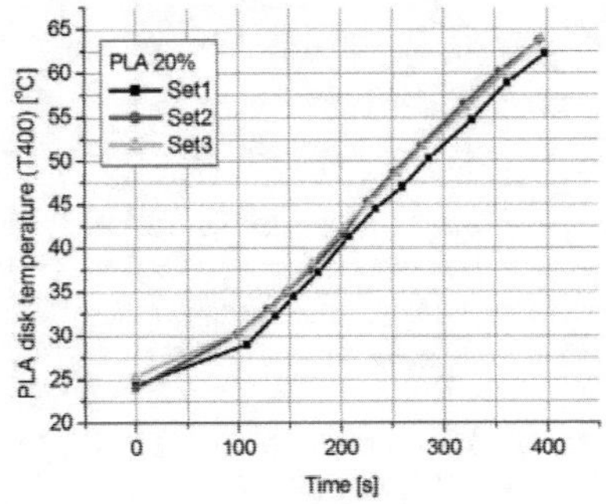

Fig. 9. Δt [°C] vs Δt [s] for PLA disk (20%).

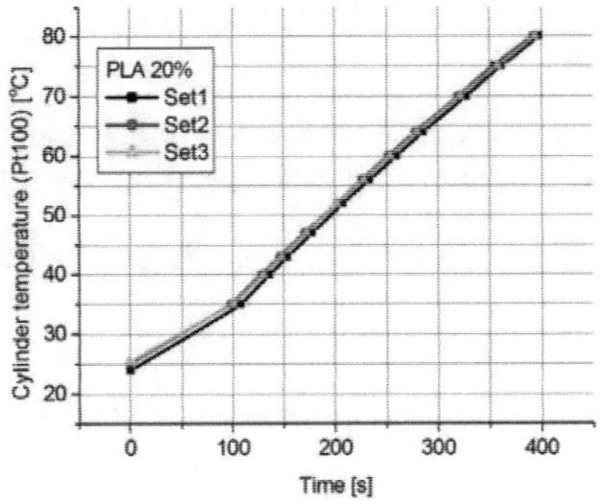

Fig. 10. Δt [°C] vs Δt [s] for copper cylinder.

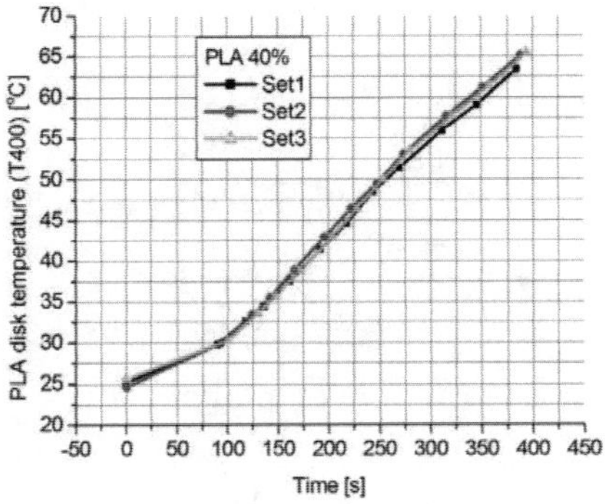

Fig. 11. Δt [°C] vs Δt [s] for PLA disk (40%).

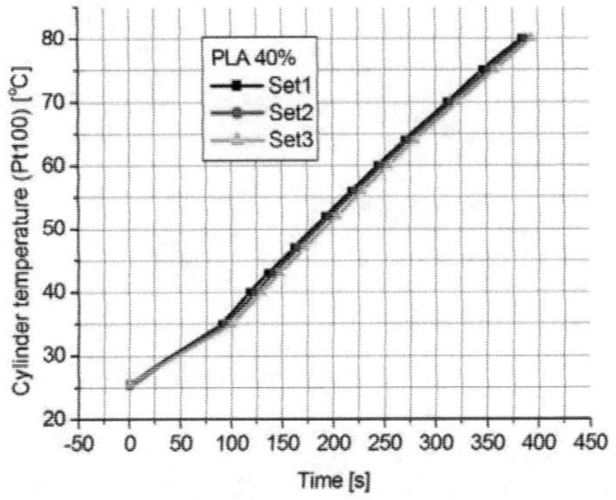

Fig. 12. Δt [°C] vs Δt [s] for copper cylinder.

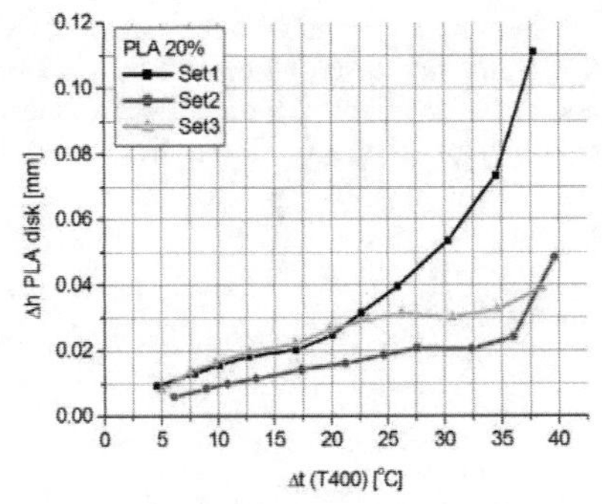

Fig. 13. Δh [mm] vs Δt [°C] for PLA disk (20%).

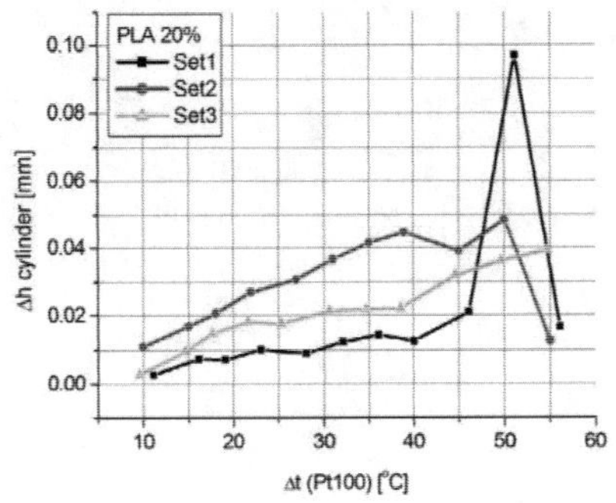

Fig. 14. Δh [mm] vs Δt [°C] for copper cylinder.

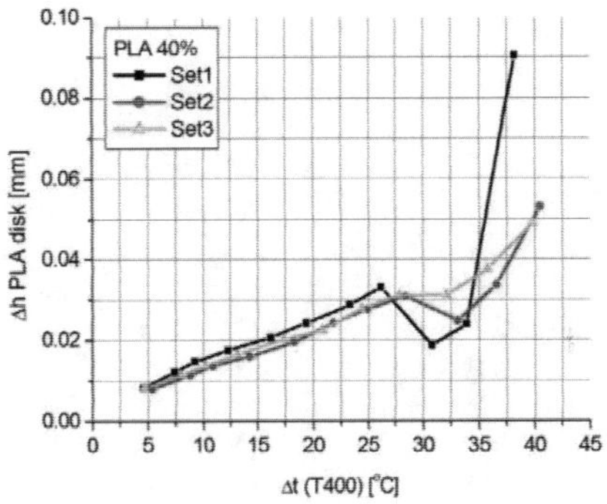

Fig. 15. Δh [mm] vs Δt [°C] for PLA disk (40%).

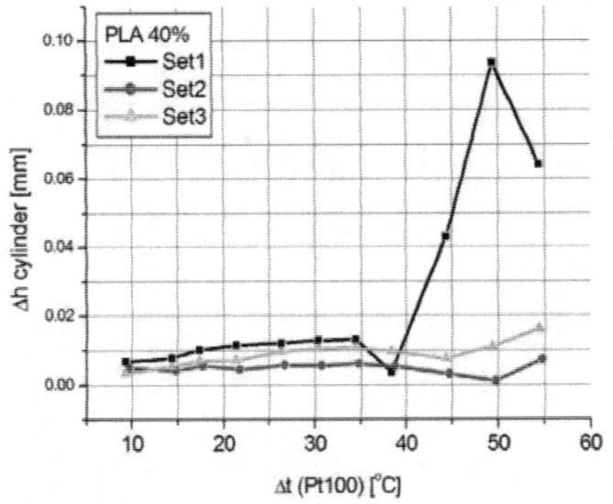

Fig. 16. Δh [mm] vs Δt [°C] for copper cylinder.

https://doi.org/10.1051/e3sconf/20183201007

Fig.17 shows the variation diagram of the CTE for the PLA disk having 20% degree of filling according to the temperature gradient Δt and Fig.18 shows the variation diagram of the CTE for the cooper cylinder according to the temperature gradient Δt.

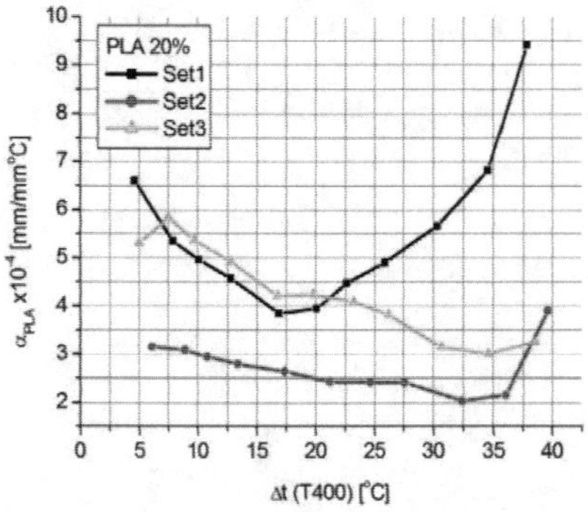

Fig. 17. α [mm/mm·°C] vs Δt [°C] for PLA disk (20%).

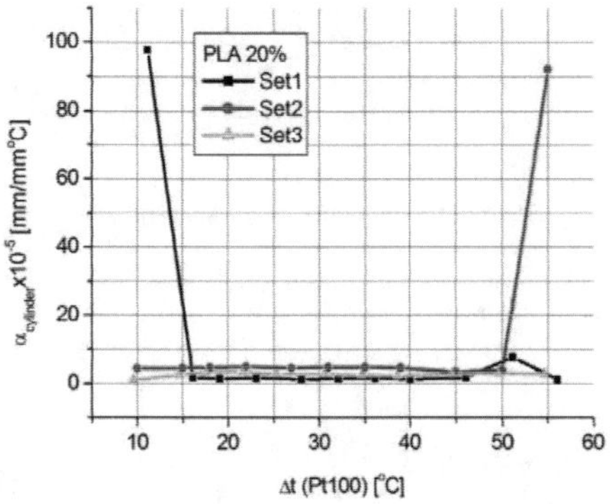

Fig. 18. α [mm/mm·°C] vs Δt [°C] for copper cylinder.

Fig.19 shows the variation diagram of the CTE for the PLA disk having 40% degree of filling according to the temperature gradient Δt and Fig.20 shows the variation diagram of the CTE for the cooper cylinder according to the temperature gradient Δt.

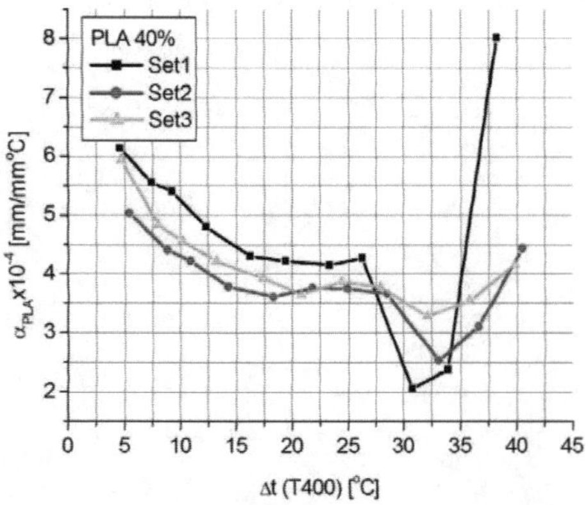

Fig. 19. α [mm/mm·°C] vs Δt [°C] for PLA disk (40%).

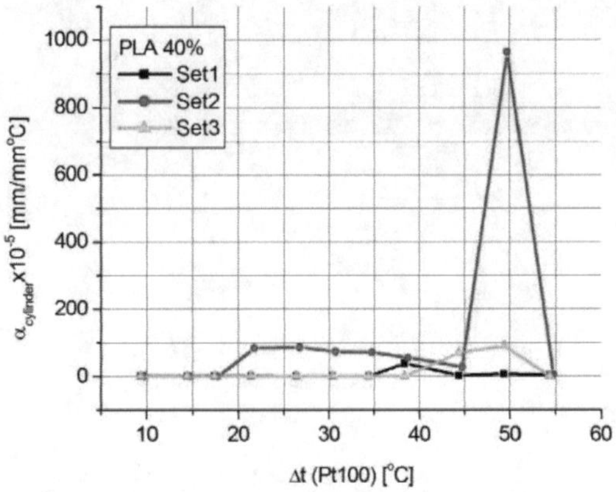

Fig. 20. α [mm/mm·°C] vs Δt [°C] for copper cylinder.

In Fig.21 is drawn the average line of the six curves representing the CTE for PLA as well as the representation of the linear function for this. In Fig.22 is drawn the average line of the six curves representing the CTE for copper as well as the representation of the linear function for this.

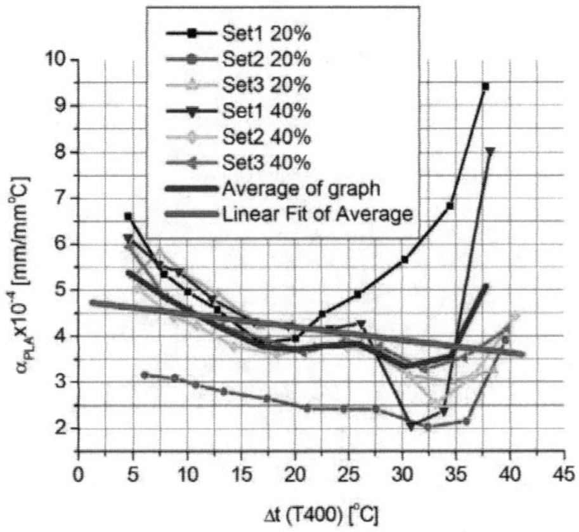

Fig. 21. α [mm/mm·°C] vs Δt [°C] for PLA disk (40%).

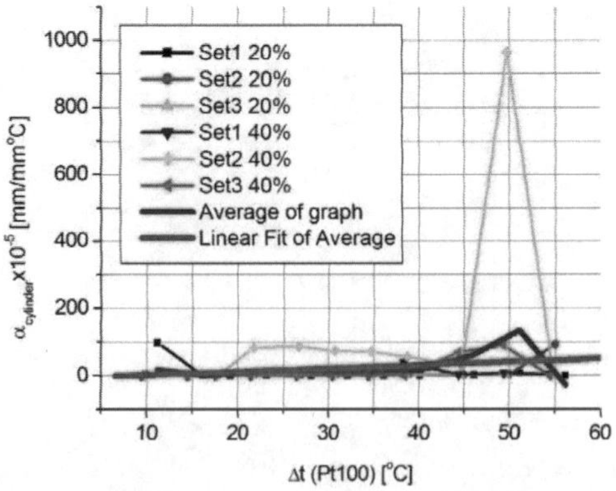

Fig. 22. α [mm/mm·°C] vs Δt [°C] for copper cylinder.

From the analysis of the obtained results and the diagrams drawn in the present paper, the following observations can be made:

For the PLA disk with 20% degree of filling the CTE has the following mean values: for Set1 $\alpha_{PLA}=4.29\cdot10^{-4}$ [mm/mm·°C], for Set2 $\alpha_{PLA}=2.74\cdot10^{-4}$ [mm/mm·°C], for Set3 $\alpha_{PLA}=5.5\cdot10^{-4}$ [mm/mm·°C]. The mean value is $\alpha_{PLA}=4.17\cdot10^{-4}$ [mm/mm·°C].

For the PLA disk with 40% degree of filling the CTE has the following mean values: for Set1 $\alpha_{PLA}=4.15\cdot10^{-4}$ [mm/mm·°C], for Set2 $\alpha_{PLA}=3.84\cdot10^{-4}$ [mm/mm·°C], for Set3 $\alpha_{PLA}=4.66\cdot10^{-4}$ [mm/mm·°C]. The mean value is $\alpha_{PLA}=4.55\cdot10^{-4}$ [mm/mm·°C].

The average CTE for the two disks is $\alpha_{mPLA}=4.36\cdot10^{-4}$ [mm/mm·°C].

Comparing this result with those reported in the literature ([3] $\alpha_{PLA}=0.7\cdot10^{-4}$ [1/°C]; [12] $\alpha_{PLA}=0.85\cdot10^{-4}$ [1/°C]; [13], [14], [15] $\alpha_{PLA}=7.4\cdot10^{-4}$ [1/°C]) a good concordance can be observed.

For the copper cylinder, when the PLA disk with a 20% degree of filling is placed on it, the CTE has the following mean values: for Set1 $\alpha_{Cu}=2.69\cdot10^{-5}$ [mm/mm·°C], for Set2 $\alpha_{Cu}=4.458\cdot10^{-5}$ [mm/mm·°C], for Set3 $\alpha_{Cu}=1.52\cdot10^{-5}$ [mm/mm·°C]. The mean value is $\alpha_{Cu}=2.88\cdot10^{-5}$ [mm/mm·°C].

For the copper cylinder, when the PLA disk with a 40% degree of filling is placed on it, the CTE has the following mean values: for Set1 $\alpha_{Cu}=1.35\cdot10^{-5}$ [mm/mm·°C], for Set2 $\alpha_{Cu}=1.33\cdot10^{-5}$ [mm/mm·°C], for Set3 $\alpha_{Cu}=2.2\cdot10^{-5}$ [mm/mm·°C]. The mean value is $\alpha_{Cu}=1.62\cdot10^{-5}$ [mm/mm·°C].

The average CTE for copper cylinder is $\alpha_{mCu}=2.25\cdot10^{-5}$ [mm/mm·°C].

Comparing this result with those reported in the literature ([16] $\alpha_{Cu}=(1.01\text{-}1.661)\cdot10^{-5}$ [1/°C]; [17], [18] $\alpha_{Cu}=2.2\cdot10^{-4}$ [1/°C]) a good concordance can be observed with the indication that the maximum and minimum values were taken into account from the results obtained after the measurements.

3 Conclusions

The present study describes how to determine CTE using a non – contact optical method, namely the 3D – DIC technique. This allows real – time and full – field analysis of the object to determine displacements and strain depending on a temperature gradient.

Although the displacements in the transverse plane of the two disks made of PLA and the copper cylinder are very low, the 3D – DIC method is of good accuracy provided that a few disturbing factors (vibrates, shocks, etc.) are avoided.

In this case, the digital image correlation method is also shown to be a good tool for evaluating the linear thermal expansion coefficient in a very short time span.

4 Acknowledgements

This study was made possible by the technical support given by PhD. Eng. Fechete Lucian, lecturer at the Automotive Engineering and Transportations Department, Faculty of Mechanics, Technical University of Cluj-Napoca.

References

1. *** Vishay Micro-Measurements, *Measurement of thermalexpansion characteristics of metals using strain gauges*, Technical note tn-513-1, Retrieved from www.vishaymg.com (2007)
2. http://ro.wikipedia.org/wiki/Acid-polilactic
3. L.T. Lim, Auras, R., Rubino, M., *Processing technologies for poly (lactic acid)*, Progress in Polymer Science, Volume 33, Issue 8, Pages 820-852 (2008)
4. http://www.icechim - pd.ro/ro/modif/modificare _biopolact_prezentare.html
5. http://despretot.info/wp-content/uploads /2012/08/ Imprimanta-3d- functionare.png
6. http://www.zspotmedia.ro/blog/printare-3d/
7. M.C. Dudescu, A.I. Botean, M. Hardau, *Thermal expansion coefficient determination of polymeric materials using digital image correlation*, Revista Materiale plastice, MPLAAM 50(1) (1-68), ISSN 0025/5289, Vol. 50, nr. 1, pag. 55-59, Bucureşti (2013)
8. P. Bing, X. Hui-Min, H. Tao, A. Asundi, *Measurement of coefficient of thermal expansion of films using digital image correlation method*, Polymer Testing, vol. 28, p.75-83 (2009)
9. K. Hoffmann, *An introduction to measurements using strain gages*, Hottinger Baldwin Messtechnik GmbH, Darmstadt (1989)
10. ***Spider8, Operating manual, Retrieved from www.hbm.com
11. C. Herbst, K. Splitthof, *Q400 Application Note - T-Q-400-Basics-3DCORR-002a-EN*, Dantec Dynamics GmbH, Germany, Retrieved from: www.dantecdynamics.com (2006)
12. https://omnexus.specialchem.com/polymer – properties / properties/coefficient-of-linear-thermal-expansion
13. https://unprinter.wordpress.com/2013/10/07/thermal-expansion-in-polylactic-acid/
14. F. Gao, *Advances in Polymer Nanocomposites*, 1st Edition (2012)
15. A.K. Mohanty, M. Misra, L.T. Drzal, *Natural Fibers, Biopolymers, and Biocomposites*, CRC Press (2005)
16. G.K. White, *Thermal expansion of reference materials: copper, silica and silicon*, J. Phys. D: Appl. Phys., vol.6, no.17 (1973)
17. K.G. Budinski, M.K. Budinski, *Engineering Materials. Properties and Selection*, Ninth Edition, Pearson
18. E. Avallone, T. Baumeister, A. Sadegh, *Marks' Standard Handbook for Mechanical Engineers*, 11th Edition

The necessity of HVAC system for the registered architectural cultural heritage building

Cătălin George Popovici[1], *Sebastian Valeriu* Hudişteanu[1], and *Nelu-Cristian* Chereche᪷[1,*]

[1]"Gheorghe Asachi" Technical University of Iaşi, Faculty of Civil Engineering and Building Services, Romania

Abstract. This study is intended to highlight the role of the ventilation and air conditioning system for a theatre. It was chosen as a case study the "Vasile Alecsandri" National Theatre of Jassy. The paper also sought to make a comparison in three distinct scenarios for HVAC Main Hall system - ventilation and air conditioning system of the Main Hall doesn't work; only the ventilation system of the Main Hall works and ventilation and air conditioning system of the Main Hall works. For analysing the comfort parameters, the ANSYS-Fluent software was used to build a 2D model of the building and simulation of HVAC system functionality during winter season, in all three scenarios. For the studied scenarios, the external conditions of Jassy and the indoor conditions of the theatre, when the entire spectacle hall is occupied were considered. The main aspects evaluated for each case were the air temperature, air velocity and relative humidity. The results are presented comparatively as plots and spectra of the interest parameters.

1 Introduction

The functionality of the HVAC system and other building services is a very important target for all type of buildings and even more for locations with high density of people, such as theatres.

An important and prestigious objective of Jassy is "Vasile Alecsandri" National Theatre, one of the most representative historical monuments of the city, rehabilitated and consolidated until 2012.

The "Vasile Alecsandri" National Theatre of Jassy was built between 1894 and 1896 on the site of the old town hall that was demolished [1, 2]. The project is the work of the Viennese architects Fellner and Helmer and the construction was carried out by the Construction Society of Bucharest. "Vasile Alecsandri" National Theater of Jassy is considered to be the oldest and most beautiful place of its kind in Romania.

HVAC system is one of the most important building services system of this famous theatre. Initial HVAC project was designed in 1896 by Koerting brothers in Paris and they only solved the problems of heating using natural circulation air combined with mechanical ventilation using a large pressured room. The existing installations have never been dimensioned for cooling [1, 2].

The new project solution of the engineering thermo-technical system included the system of insuring higher microclimatic conditions for the pleasant stay of visitors as well as actors and technical staff. Special care was given to the auditorium and to the stage. The previous solutions of heating and ventilation system with the new project were raised to a different comfort level with HVAC system. Under these circumstances and according to scientific literature specifications, the HVAC solution implemented for "Vasile Alecsandri" National Theatre of Jassy consists in four HVAC integrated systems serving: Main Hall, stage, foyer and adjacent areas [2]. It is a multi-zone solution which divides the building into 3 areas whose utilization is never simultaneous: Zone 1 - Main hall - stall, lodges and balcony; Zone 2 - stage and orchestra pit; Zone 3 - foyer and adjacent areas [6].

The HVAC technical data for the building are the following [2]:

- Total surface - 11.300 m^2;
- Capacity - 740 places;
- Total heating power - 800 kW;
- Boilers – 2 units;
- Total cooling power - 350 kW;
- Air chiller - 1 unit;
- Total supply airflow - 33000 m^3/h;
- Total exhaust airflow - 38000 m^3/h;
- Air handling unit - 6 units;
- Exhaust fan - 8 units.

2 Case description

The subject of this study is the Main Hall of "Vasile Alecsandri" National Theatre of Jassy – the space where the entire audience is present during spectacles. This represents a voluminous space with 10.7 m height, 18 m width and 21 m length.

The ventilation of the Main Hall is realized by using three independent systems for introducing the fresh air in different zones: stall, lodges and balcony. The exhaust air is evacuated through one system, placed under the

* Corresponding author: chereches@tuiasi.ro

https://doi.org/10.1051/e3sconf/20183201008

roof. The total airflow in the auditorium Main Hall is of 21,000 m^3/h. The ventilation is balanced, meaning that the introduced airflow is equal with the evacuated one.
The problem studied consist of the functionality of the HVAC system during the theatre show, in winter time, for three distinct scenarios:
First scenario - Ventilation and air conditioning system of the Main Hall doesn't work;
Second scenario – Only the ventilation system of the Main Hall works;
Third scenario - Ventilation and air conditioning system of the Main Hall works.

3 Numerical modelling

The comfort parameters were analysed numerically using the ANSYS-Fluent software for a 2D model of the building and the simulation of HVAC system functionality during winter time, in all three scenarios. A 2D model of the building was realized by using ANSYS-Fluent software, simulating the functionality of the HVAC system. For the studied scenarios, the external conditions of Jassy and the indoor conditions of the theatre, when the entire spectacle hall is occupied were considered. The main aspects evaluated for each case were the air temperature, air velocity and relative humidity.
The simulations are achieved in steady state regime, using turbulent flow and k-ε model [7, 8, 9]. Therefore, the temperature, velocity and relative humidity data inside the main hall of the theatre were obtained. Performing numerical simulations carried out by CFD tool, ANSYS-Fluent, the differential equations of heat transfer and fluid mechanics were solved [3]:
Momentum equation:

$$\frac{\delta}{\delta t}(\rho\vec{v}) + \nabla\cdot(\rho\vec{v}\vec{v}) = -\nabla p + \nabla(\bar{\bar{\tau}}) + \rho\vec{g} + \vec{F} \quad (1)$$

Energy conservation:

$$\frac{\delta}{\delta t}(\rho E) + \nabla\cdot(\vec{v}(\rho E + p)) = -\nabla\cdot(\sum_j h_j J_j) + S_h \quad (2)$$

Conservation of mass:

$$\frac{\delta p}{\delta t} + \nabla\cdot(\rho\vec{v}) = S_m \quad (3)$$

The geometry and mesh of the model, Fig. 1, were realized by using ANSYS-Design-Modeller and ANSYS-Meshing. The geometry of the building is a simplified one, for the cross section only, assuming a 2D model. It was determined the mass flow for each grille, taking into account that Fluent uses an auxiliary virtual dimension of 1 m long [3]. The mesh was created with refinements near walls, Fig. 2.b, resulting in 5596 nodes, 10418 2D interior faces and 5322 mixed cells.

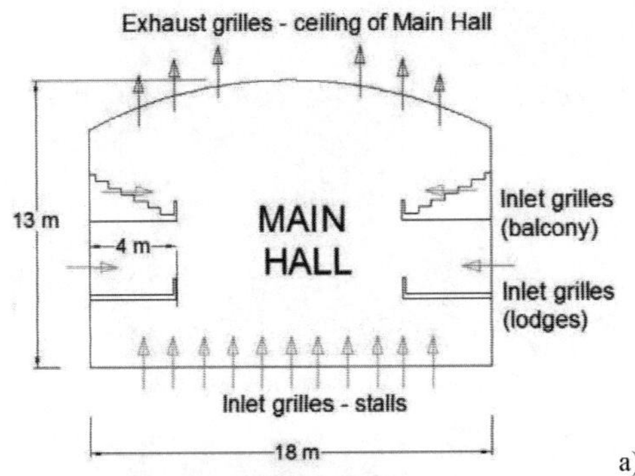

a)

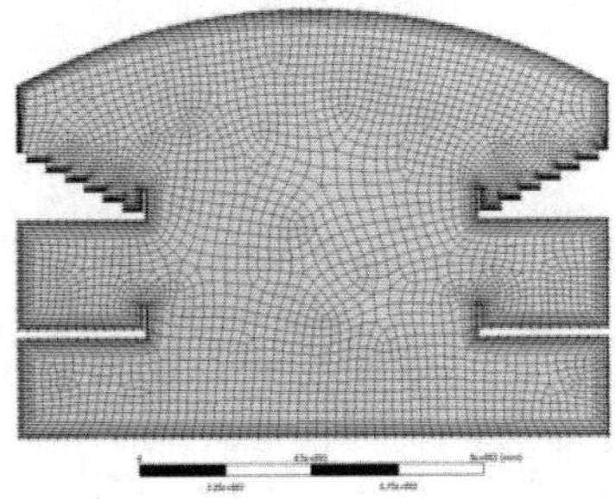

b)

Fig. 1. Cross section of the Main Hall: a) geometry of the 2D model; b) meshing.

The input data are the following:
- inlet sections, the conditioned fresh air temperature of 20 °C;
- outlet sections are set as outflow zones;
- exterior air temperature T_{ext} = -18 °C;
- convective heat transfer coefficient α = 24 W/m^2K;
- flow regime - turbulent (*k-ε* RNG);
- modelling humidity - species model;
- two volumetric species: air and water-vapours;
- moisture for a sitting person: 70 g H_2O/h [4, 5, 10].

4 Results

The results are presented comparatively for the three studied scenarios. The qualitative data are revealed by contours of velocity, temperature and humidity, while the quantitative ones can be observed in the plots of the same parameters.

4.1. First Scenario - Ventilation and air conditioning system of the Main Hall doesn't work

All the air handling units are out of order or doesn't work in this first case. There are 740 peoples inside the Main Hall during the show, without fresh air, heating or cooling. Fig. 2 highlights the first Scenario temperature distribution.
More accurate data is presented in Fig. 2.a, which offers information about temperature distribution at different height in cross section. The elevation planes were selected as h = 1 m, 4 m, 7.5 m, 9 m and 10.7 m, being most relevant from the point of view of human comfort in occupational zones.

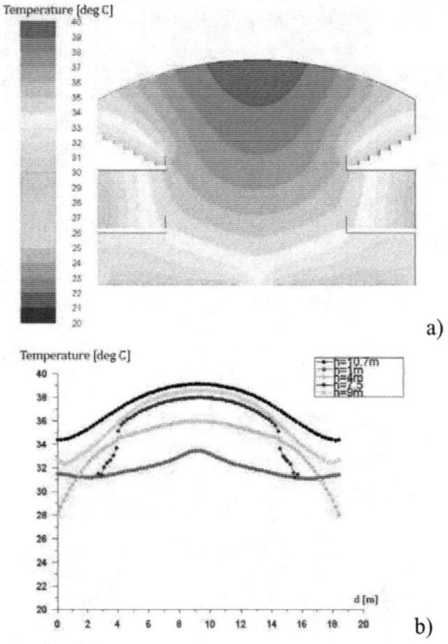

Fig. 2. The first scenario indoor temperature distribution: a) Temperature spectra; b) Temperatures at different heights.

Indoor air temperatures reach values of 28 - 39 °C and of 28 - 34 °C for stall, lodges and balcony areas. The thermal discomfort is maximum. The indoor air is almost impossible to breathe.
Fig. 3 shows the velocities spectra and velocity of the indoor air at different heights. All the values are near zero. Very quiet but too hot.

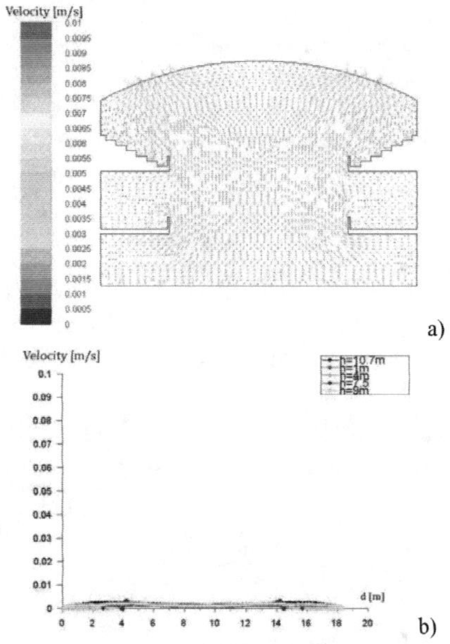

Fig. 3. The first scenario velocity distribution: a) Velocity spectra; b) Velocity at different heights.

The values of indoor relative humidity lie between 65 - 75 %, Fig 4. The maximum values are recorded at the top of the balcony and the risk of condensation is the greatest.

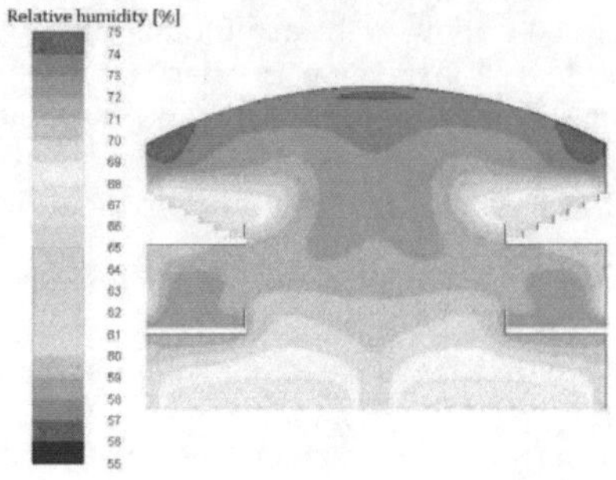

Fig. 4. The first scenario - Indoor relative humidity spectra.

4.2. Second scenario – Only the ventilation system of the Main Hall works.

Only the ventilation system works in the second scenario. The Main Hall capacity is full with 740 peoples inside and only the fresh air is provided. The cooling system doesn't work.
Fig. 5 shows the second Scenario temperature distribution.

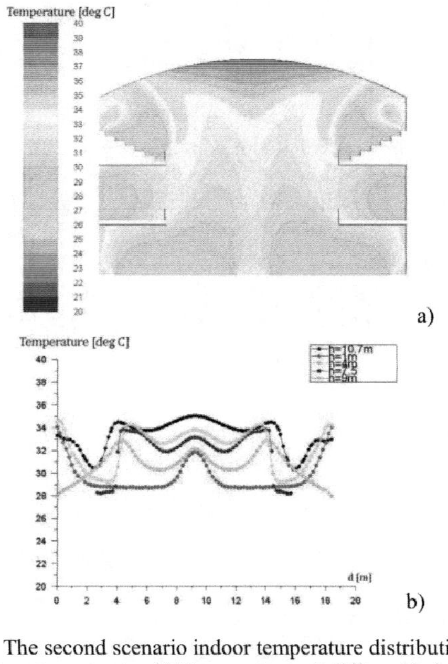

Fig. 5. The second scenario indoor temperature distribution: a) Temperature spectra; b) Temperatures at different heights.

The range of the indoor air temperatures values are of 27 - 35 °C and of 27 - 31 °C for stall, lodges and balcony areas. Thermal discomfort is felt.
The average velocity is around 0.3 m/s, Fig. 6, and doesn't affect the comfort of the occupants.

https://doi.org/10.1051/e3sconf/20183201008

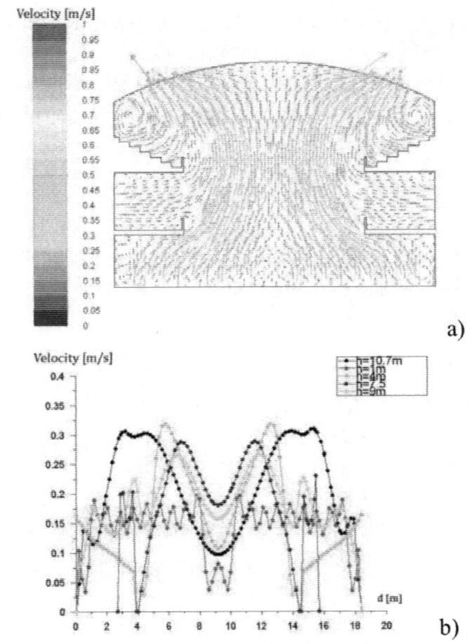

Fig. 6. The second scenario indoor velocities distribution: a) Velocity spectra; b) Velocity at different heights.

The indoor relative humidity value is between of 65 – 70 %, Fig 7. The maximum values are recorded at the top of the balcony.

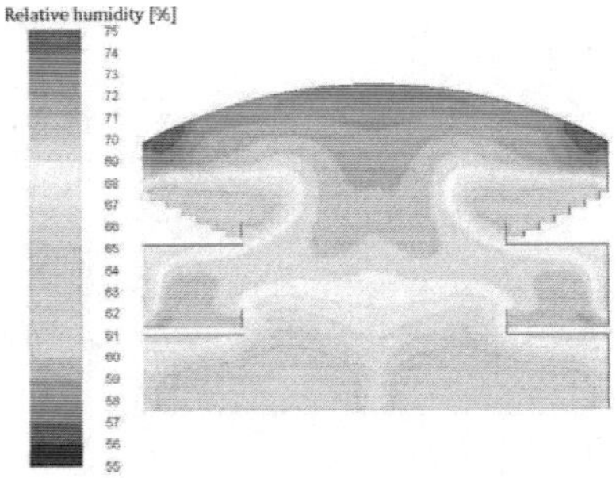

Fig. 7. The second scenario - Indoor relative humidity spectra.

Third scenario - Ventilation and air conditioning system of the Main Hall works.

All the air handling units are in use, so ventilation and air conditioning system of the Main Hall works in this third case. There are 740 peoples inside the Main Hall during the theatre show, enjoying fresh air, heating and cooling. Humidity control works too.
Fig. 8 shows the third scenario temperature distribution.

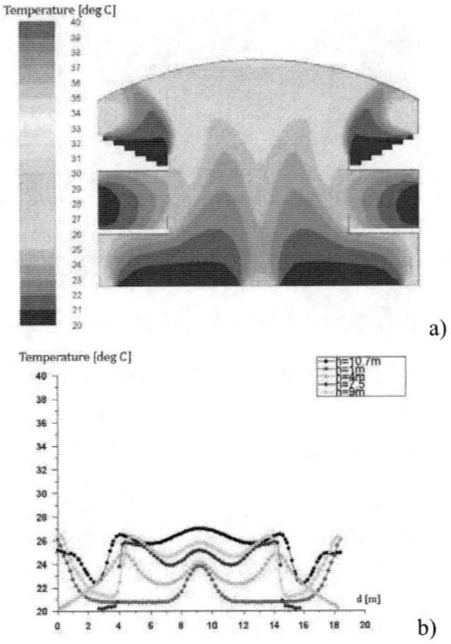

Fig. 8. The third scenario indoor temperature distribution: a) Temperature spectra; b) Temperatures at different heights.

The range indoor air temperatures values are of 22 - 27 °C and of 22 - 25 °C for stall, lodges and balcony areas. All the thermal comfort parameters are optimal.
The average velocity in around of 0.3 m/s, Fig. 9, and doesn't affect the comfort of the occupants.

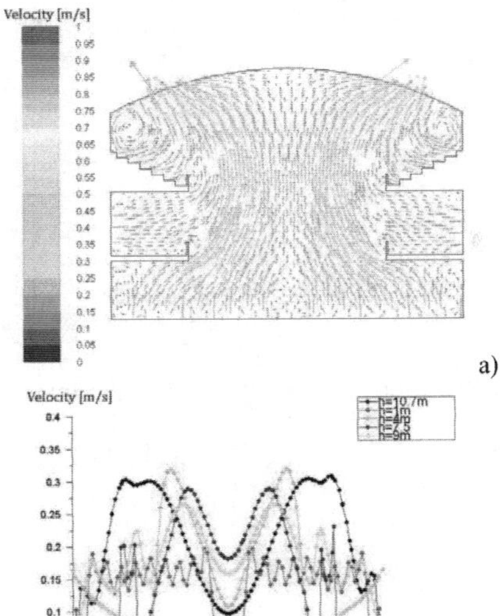

Fig. 9. The third scenario indoor velocities distribution: a) Velocity spectra; b) Velocity at different heights.

Fig. 10 shows that the indoor relative humidity value is between of 55 - 60 %.

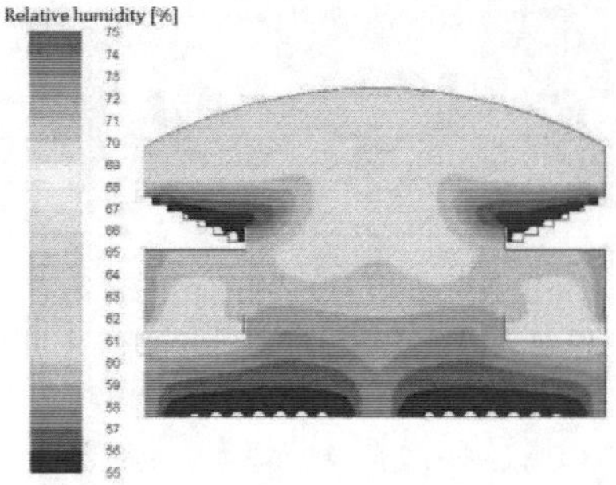

Fig. 10. The third scenario – Indoor relative humidity spectra.

4 Conclusions

The Main Hall HVAC system of the real building "Vasile Alecsandri" National Theatre of Jassy was studied by using a 2D model, with specific software ANSYS-Fluent.

The Main Hall HVAC system is subjected to an analysis of its functionality, by evaluating the supplied parameters when operating in three different scenarios as follows:

First scenario - Ventilation and air conditioning system' of the Main Hall doesn't work;

Second scenario – Only the ventilation system of the Main Hall works;

Third scenario - Ventilation and air conditioning system of the Main Hall works.

When the ventilation and air conditioning system of the Main Hall doesn't work, the indoor air is almost impossible to breathe. The indoor air temperatures are between of 28 - 35 °C, indoor air velocities are near zero and indoor relative humidity varies between 65 - 75 %. The thermal discomfort is maximum.

The ventilation system of Main Hall works in the second scenario and the thermal discomfort is felt. The indoor air temperatures reach values of 27 - 31 °C, indoor air velocities values are around 0.3 m/s and indoor relative humidity is lies between 65 - 70 %. The indoor air temperature and relative humidity values exceed the limits of thermal comfort.

The ventilation and air conditioning system of the Main Hall works in this third scenario and the thermal comfort parameters are optimal. The indoor air temperatures are between of 22 - 25 °C, indoor air velocities values around 0.3 m/s and indoor relative humidity varies between 65 - 70 %.

The recently implemented HVAC Main Hall system reaches its task and provides adequate comfort conditions.

After an overall outlook of the results, it can be stated with certainty that it is not possible to have a theatre show on Main Hall of "Vasile Alecsandri" National Theatre of Jassy without ventilation and air conditioning system. This is the main conclusion of the study and it doesn't conceive of a theatre without the operation of the ventilation and air conditioning system. In any season the humidity control is another need for registered architectural cultural heritage buildings to ensure the integrity of paintings and conserving the original architectural elements of the buildings.

This study confirms the necessity of HVAC system for the registered architectural cultural heritage building and theatres in a particular case.

Notations

p - Pressure (Pa);
α - convective heat transfer coefficient (W/m$^{2}\cdot$K);
v - Overall velocity vector (m/s);
t - Time (s);
ρ - Density (kg/m^3);
τ - Shear stress (Pa);
g - Gravitational acceleration (m/s^2);
F - Force vector (N);
E - Total energy (J);
h - Enthalpy (J/kg);
h_0 - Standard state enthalpy of formation (energy/mass, energy/mole);
J - Mass flux; diffusion flux (kg/m$^{2}\cdot$s);
S_h - Source of heat added;
S_m - Source of mass added to the continuous phase;
T_{ext} - temperature of external air (°C).

References

1. National Archives of Romania, 2007.
2. Project and Execution of the Construction, 2012.
3. ANSYS-Fluent 13.0, Theory Guide.
4. Peri A., Fernandes P. M., Vishwanadha C., Numerical simulation of air flow in a general ward of a hospital. IJRRAS 8, 400 - 444 (2011).
5. Balocco C., Petrone G., Cammarata G., Vitali P., Albertini R., Pasquarella C., Indoor Air Quality in a Real Operating Theatre under Effective Use Conditions. Biomedical Science and Engineering, 866 - 883 (2014).
6. Popovici CG, Hudişteanu SV., Numerical simulation of HVAC system functionality in a sociocultural building, Procedia Technology 22; 2016, p. 535–542.
7. Thiyagarajan G. M., Raj Kumar S., Velocity optimisation using computational fluid dynamics as a tool in leed rated building to reduce energy consumption. Journal of Chemical and Pharmaceutical Sciences, 142 - 145 (2015).
8. Uyttenhove W., De Paepe M., Janssens A., CFD-modelling of Temperature and Humidity Distribution in the St. Pieter's Church. IAE-Annex 41, Subtask 1, 2004.
9. Mahu R., Popescu F., Ion I. V., CFD Modeling Approach for HVAC Systems Analysis. Chemical Bulletin of "Politehnica" University of Timisoara, 69 - 73 (2012).
10. Bronström T., Fundamentals of indoor climate at the seminar in Riga regarding Indoor Climate in Churches – Problems and solutions, 2004.

Unforced Rayleigh instability of an immersed liquid jet

Claudiu Patrascu[1,*], *Alina* Chipaila[1],and *Corneliu* Balan[1]

[1] UPB, Hydraulics Department, REOROM Group, 313 Splaiul Independentei sect. 6, Bucharest, Romania

Abstract. Motivated by the occurrence of the injection of liquids in various technical processes, we study the capillary instability of a liquid jet surrounded by another liquid. The study focuses on the natural developing Rayleigh instability,hence without an imposed perturbation. We also point out the influence of viscosity on the main drop diameter, resulted after jet breakup, and on the breakup length itself. Modifications brought by a decrease of the capillary nozzle are also emphasized for a particular case.

1Introduction

The study of liquid jet breakup is motivated by practical applications such as ink-jet printing, DNA sampling or dosing consumer products [1-3]. Besides the applicative side of jet instability, one may find that liquid jets can be used to highlight the influence of physical macroscopic properties such as viscosity, density, interfacial tension or elasticity in interfacial phenomena.For example, the relaxation time of weakly elastic polymer solutions, can be determined via R.O.J.E.R, a device specifically created that makes use of jet dynamics and filament thinning in the vicinity of drop detachment [4,5]. The density contrast between an injected fluid and its surrounding medium is still a subject of interest, considering that when immersed, the force of gravity is reduced by a factor proportional to the density difference between the two fluids [6].The injection of a fluid in a certain industrial process might be subject to many perturbations, therefor the breakup process will be dependent of the superposition of allexcited modes. This causes a delay or a growth in the time needed for a drop to form and detach. The common approach to control jet breakup is to introduce a specific perturbation and excite a certain mode, but this approach results in a costly process. The study of an unforced breakup of a liquid jet may offer insight and a spawning ground for an alternative controlling technique.

The present study is structured as follows. We study the natural occurring capillary instability of a liquid jet surrounded by sunflower oil with no imposed perturbation. We emphasize the influence of viscosity and nozzle radius on the main and satellite drop diameter and on the breakup length.

2Theoretical aspects of jet breakup

A liquid emanating from a nozzle experiences instability which leads to the formation of droplets (Fig. 1-a). In the case of a jetting regime, the breakup process can lead to the formation of main drops and satellite droplets. Usually the diameter of the main drop, D_d is much greater than that of the satellite droplet, D_s. This is encountered whether the exterior medium is a gas or a liquid, provided that the liquids are immiscible. At a flow rate sufficiently high that ensures the convection of any perturbation that the interface might experience, the dynamics are governed by capillarity and the phenomenon is called Rayleigh instability. It represents a convective type of instability, where the growth rate is considered a complex number and the wave number real and positive. Suitable in this case is the temporal analysis applied to perturbations that grow uniformly in space. This was first carried out by Lord Rayleigh [7] which showed that the dispersion equation, relating the growth rate of disturbance and the wave number, exhibits a maximum, which lead him to conclude that the fastest growing mode is responsible for jet breakup. The effect of viscosity on the dispersion relation was emphasized by Chandrasekhar [8] and the effect of a surrounding medium by Tomotika [9] which has the following form:

$$i\underline{\omega} = \sigma[1-(kR_0)^2]f(kR_0,\beta)/2R_0\eta_2, \quad (1)$$

where$\underline{\omega}$ is the complex growth rate (quantifying the rate at which perturbations grow in time),$\sigma$is interfacial tension, k is the wave number, $R_0$the inner radius of the nozzle, $\eta_2$is the viscosity of the outer fluid and $f(kR_0,\beta)$a function of the nondimensional wave number and the viscosity ratio, respectively. Fig1-b shows the prediction for the growth rate as a function of the nondimensional wave number for three different injected liquids. Having determined the maximum of a particular curve, one can predict the breakup length by [10]:

$$L_b = cV/i\underline{\omega}_M . \quad (2)$$

In Eq. (2) V is the mean velocity of the liquid jet and c a pre-factor. The increase of concentration results in an increase of the viscosity of the injected liquid and subsequently of the viscosity ratio. This produces a

*Corresponding author: claudiu.t.patrascu@gmail.com

https://doi.org/10.1051/e3sconf/20183201009

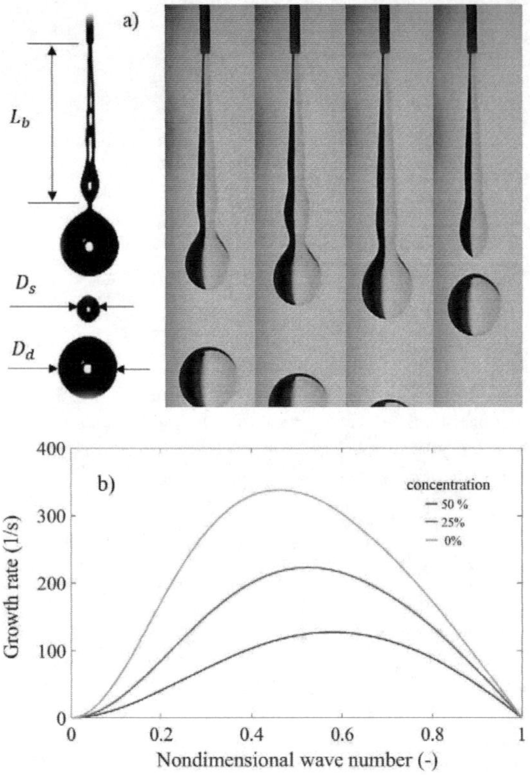

Fig. 1a) Schematic representation of the pursued parameters (left) and snapshots of the capillary instability of a liquid jet (right).The figure shows the breakup length, the main drop and the satellite droplet; b) Dispersion relation for three water and glycerin mixtures. The graph shows the dependence of the growth rate as a a function of the wave number and the influence of increasing the concentration of glycerin in water.

decrease in the maximum growth rate as predicted by Eq. (1). Decreasing the maximum value of the growth rate results in a decrease of the liquid intact length, if one expects the pre-factor to remain constant. Making the hypothesis that the fastest growing mode is responsible for breakup, we show that the pre-factor must change in order to capture the behaviour of the jet as viscosity is increased. The pre-factor c is dependent on the radius of the nozzle and on the value of the initial amplitudeA_0 introduced by the injection system:

$$c = \ln(R_0/A_0). \quad (3)$$

Reference regarding this pre-factor can be found in [3].

3 Experimental details

The experimental setup consists of a glass tank filled with sunflower oil in which a water – glycerine mixture is injected with the use of a blunt tip nozzle. Images of the phenomenon were taken by a Nikon 1 J5 camera with 3840×2160 resolution at 400 fps. For the injection process we used a Harvard 33 syringe pump. Physical properties

Table 1. Values of the material properties at 25° C for the fluids subject to analysis.

Sample	$\rho\ [kg/m^3]$	$\eta\ [mPas]$	$\sigma\ [N/m]$
water	1000	1	0.025
water + 25% glycerin	1076	2.4	0.023
water + 50% glycerin	1143	7	0.021
sunflower oil	925	55	-

of the fluids in question were determined as follows: the viscosity was determined using a rotational rheometer in a standard oscillatory test, the density by a mass per volume method and interfacial tension via a pendant drop method. The values of the above-mentioned material properties are listed in Table 1. We differentiate the injected fluidfrom the outer medium by sub-index 1 andsub-index 2,respectively. The viscosity and density of the injected fluid were increased by increasing the concentration of glycerine (in volume). The breakup of the liquid jet was not triggered in an imposed manner, but by a naturally occurring perturbation, caused by the injection process. Two characteristic parameters were measured, the breakup length, which represents the distant between the tip of the capillary nozzle and the point of drop detachment, and the diameters of the detached drops. For each parameter a series of 50 punctual measurements were performed.

4 Results and comments

A first comment can be made regarding the average value of the breakup length. Increasing the concentration of glycerine and therefor the viscosity and density of the injected fluid results in a decrease of the average value of

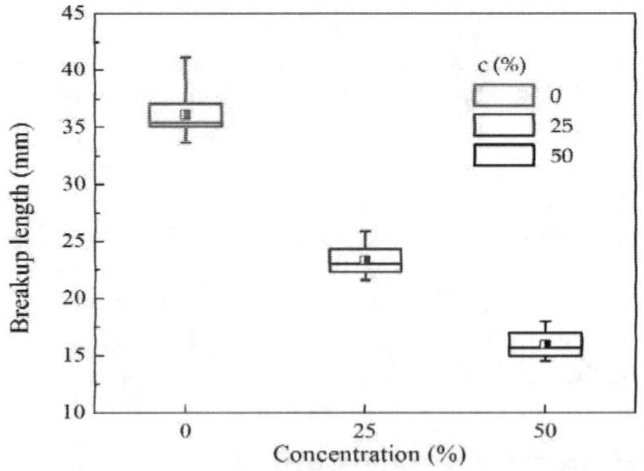

Fig. 2 Average valuesof the breakup length as a function of concentration of glycerin in water at a flow rate of 11 ml/min. The inner radius of the capillary tube is 0.3 mm, the outer fluid being sunflower oil in all three cases.

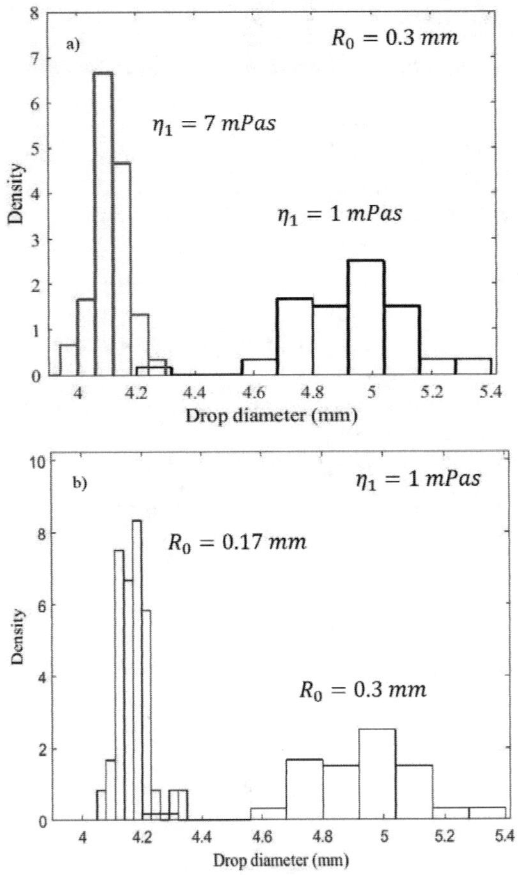

Fig. 3a) The influence of viscosity on the main drop diameter. A comparison between deionized water and a 50% glycerin in water mixture is presented, showing a decrease in both drop diameter and data dispersion; b) The influence of modifying the inner radius of the capillary tube on the main drop diameter in the case of deionized water. The same decrease in drop diameter and data dispersion is observed.

the breakup length (Fig 2). By inspection of Fig. 1-b and Eq. (2) we see that the maximum value presented by the growth rate decreases with viscosity, implying that the breakup length increases if one considers a constant value for the pre-factor. Under the hypothesis that the most unstable mode is responsible for breakup, the pre-factor must decrease, thus the initial amplitude must increase, although it is common knowledge that viscosity has a damping role. The injection system might also excite multiple modes, a subject not treated here, that could alter the linear predictions of Eq. (2).

Next, we look at the diameter of the main drop in two particular cases. First, the viscosity of the injected fluid is increased from 1mPas to 7 mPas (Fig. 3-a). The immediate effect of the latter is a decrease in the average value of the main drop diameter and also a decrease in data dispersion. The same effect is encountered if one modifies the inner radius of the nozzle. Both data dispersion and average value of the main diameter decrease (Fig. 3-b). A decrease in drop diameter with viscosity is seen in cases where the jetting regime is replaced by a dripping regime. Partially, this is due to the fact that interfacial tension decreases and density increases,causing the drop diameter to decrease.

For a deionized water in sunflower oil system, the satellite droplets can be very small in size compared to the main droplets. The data is dispersed, but the phenomenon still retains its periodicity when speaking of drop generation. The main drop is first generated followed by one or two smaller in size. When increasing viscosity of the injected fluid, the ratio between two consecutive drop diameters decreases(Fig. 4). Usually the smaller drop is referred to as a satellite drop, but in this case the latter is comparable in size with the main drop. This suggests that increasing the viscosity even further would totally eliminate the occurrence of satellite droplets. Further investigations are required to establish if this is indeed the case.

It is worth mentioning that above a certain critical concentration of glycerine in water the breakup length starts to increase, despite the variation that one would expect by prospecting Fig. 2. For example, a water glycerine mixture with the same viscosity as the surrounding medium can exhibit a breakup length greater than that observed for water(Fig. 5). The breakup length starts to grow when the concentration of glycerine in water is above 65%, which in terms of Ohnesorge number is approximately 0.2. We report that the occurrence of satellite droplets may be supresses if the viscosity of the

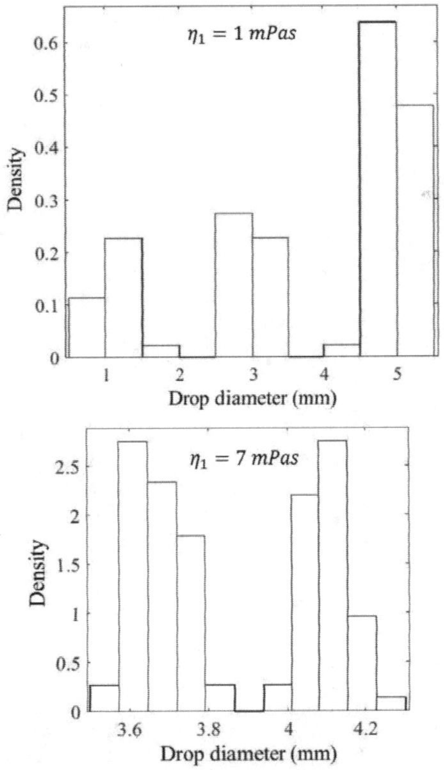

Fig. 4 The influence of viscosity on drop diameter, taking into account both the main drop diameter and the satellite droplets.

A comparison between deionized water (upper graph) and a 50% glycerin in water mixture with a viscosity of 7 $mPas$.

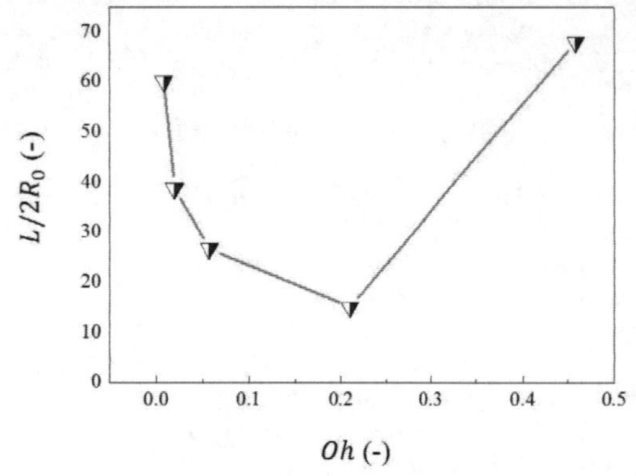

Fig. 5 The dependence of the nondimensional breakup length on Ohnesorge number, $Oh = \eta_1/\sqrt{(\rho_1 R_0 \sigma)}$. The graph shows a critical point above which the breakup length starts to grow when the viscosity of the injected fluid is increased.

injected fluid is between $7\,mPas$ and $25\,mPas$, the latter representing the viscosity for 65% glycerine in water.

5 Conclusions

In this article, we study the unforced Rayleigh instability of a liquid jet surrounded by another viscous fluid, the resulting system being in an immiscible state. A viscous outer fluid has a direct effect on the dispersion relation relating the growth rate and the wave number, decreasing the value of the most unstable growth rate. From an experimental point of view the second advantage of an external medium is a decrease in the time scales associated with the phenomenon. This offers a way to probe drop detachment and pinch-off without the use of expensive cameras.

Referring to jet breakup, increasing the viscosity of the inner fluid has a decreasing effect on the breakup length. This tendency is not monotonic, above Oh numbers greater than 0.2, the breakup length starts increasing. The second effect of increasing viscosity is a decrease in the main drop diameter and an increase of the satellite drop diameter.The main drop diameter also decreases when the inner radius of the nozzle is decreased. We observe that just by increasing the concentration of glycerine in water the satellite droplets could potentially be eliminated.

This work was partially supported by the grant of the Romanian National Authority for Scientific Research, CNCS, UEFISCDI, PHANTOM - Passive flow control for heat and mass transfer enhancement of impinging jets PN-III-P4-ID-PCE-2016-0758 and by the grant of the Romanian space agency ROSA, QUEST - Advanced air diffusion system of the crew quarters for the ISS and deep space habitation systems, STAR-CDI-C3-2016-577.

References

1. S. Lin,*Breakup of liquid sheets and jets,* Cambridge University Press, (2003)
2. C. Clasen, P.,Phillips, L. Palangetic, J. Vermant, AIChE J.**00**, (2011)
3. N.Ashgriz, *Handbook of atomization and sprays: theory and applications,*Springer Science & Business Media, (2011)
4. B. Keshavarz et al.,J.N.N.F.M.**222,** (2015)
5. E. Greiciunas, J.R**61**467, (2017)
6. A. Myshkis, *Low-gravity fluid mechanics*,Springer-Verlag, (2012)
7. L. Rayleigh,Proc. Math. Soc, **1.1** (1878)
8. J.Eggers, E. Villermeux, *Physics of liquidjets*, Reports on progress in physics**71.3**, (2008)
9. S. Tomotika,Proc. R. Soc. Lond. A, **150** (1935)
10. A. Kalaaji, B. Lopez, P. Attane, A. SoucemarianadinPhys. Fluids **15** 9, (2008)

https://doi.org/10.1051/e3sconf/20183201010

A state of the art regarding urban air quality prediction models

Cristiana Croitoru[1,*] *and Ilinca Nastase*

[1]CAMBI Research Center, Technical University of Civil Engineering Bucharest, 021414 Bucharest, Romania

Abstract. Urban pollution represents an increasing risk to residents of urban regions, particularly in large, over-industrialized cities knowing that the traffic is responsible for more than 25% of air gaseous pollutants and dust particles. Air quality modelling plays an important role in addressing air pollution control and management approaches by providing guidelines for better and more efficient air quality forecasting, along with smart monitoring sensor networks. The advances in technology regarding simulations, forecasting and monitoring are part of the new smart cities which offers a healthy environment for their occupants.

1 Introduction

1.1 Importance of indoor and exterior pollution

Urban air pollution represents a growing risk to citizens, particularly in large, over-industrialized cities knowing that the traffic is responsible for more than 25% of air gaseous pollutants and dust particles. In Europe, an average of 80% of the population lives in urban areas [1]. The report of OECD [2] it is stated that air pollution is one of the most serious environmental risks. According to the World Health Organization (WHO) fine particulate matter is associated with a broad spectrum of acute and chronic illness, such as lung cancer, chronic obstructive pulmonary disease (COPD) and cardiovascular diseases. Worldwide, it is estimated to cause about 25% of lung cancer deaths, 8% of COPD deaths, and about 15% of ischaemic heart disease and stroke [3]. The economic costs of the derived effects were quantified in 2010 at 1.5 trillion euros [4].
The air pollutants can be divided into two groups based on the origin of the chemical composition and their physical states. According to WHO, the main emission sources of these pollutants are the use of fossil resources, the transportation vehicles, the power plants, and the production of oil and gas [5]. The primary type of air pollutants is concerning those that enter directly into the atmosphere, such as sulfur dioxide (SO_2), the nitrogen oxides (NO_2), and the hydrocarbons (HC). The secondary type of contaminants is concerning those that can be generated within the atmosphere by photochemical reactions of hydrolysis or oxidation, such as the ozone (O_3). Aerosols are considered an important source of pollution in urban areas [6, 7]. Aerosols (particulate matter-PM) include dust, gas emission, smoke, fog and fibers. PM are a complex mixture of chemical species, originating from a variety of sources, both anthropogenic (traffic, industrial activities and combustion processes in general) and natural (dust storms, forest fires, pollen, volcanic eruptions). They may be emitted as primary pollutants or formed as a secondary product through atmospheric processes from gaseous precursors. According to the latest report of the European Environmental Agency (EEA), in 2017 a large number of European cities displayed values above the admitted concentrations of the air quality standards for PM10 and PM2.5, with the majority of critical concentrations observed in Eastern Europe and the Balkan region [6]. For PM10 particles, the concentrations above the EU daily limit value were registered at 19 % of the reporting stations in 20 of the 28 EU Member States [8]. A comprehensive study [9] reviews the basic evidence on the health effects of particulate matter. It addresses the major health risks in order to alert policy-makers so that more stringent strategies can be implemented to reduce air pollution and its health effects.
On the other hand, ozone can increase the incidence of asthma episodes, cause shortness of breath, aggravate lung diseases and can cause permanent lungs damage through long-term exposure. High ozone levels are linked to increases in hospitalizations, emergency room visits and premature death [10].
Considering the indoor pollution, besides the actual health problems, some other symptoms can be observed at the occupants affected, which cannot be linked directly to a specific pollutant. Starting with the late 1970s, it was noted that nonspecific symptoms were reported by tenants in newly constructed buildings, but the term was established by the World Health Organization (WHO) in 1982 [11]. In 1984 the WHO issued a report suggesting that up to 30% of new and remodeled buildings may have excessive complaints about indoor air quality [12]. As stated in an article [13] from the prestigious medical journal, The Lancet, SBS is an increasingly frequent problem since it can be

* Corresponding author: cristiana.croitoru@utcb.ro

uncomfortable, even disabling, and whole workplaces can be turned into non-functional.
Even though levels of particle matter and ground-level ozone pollution are significantly lower than in the past, levels still have a negative impact on human health. Both pollutants are the result of emissions from various sources and can travel long distances, affecting large regions.

1.2 Importance of pollution prediction

Air quality modelling plays an important role in addressing air pollution control and management approaches by providing guidelines for better and more efficient air quality forecasting[14]. As air pollution is a complex blend of contaminant components with considerable impact on human health, predicting air pollution concentrations appears as a priority for improving life quality [15]. It is demonstrated that the knowing how the pollution evolve helps in a better urban management with regards to inhabitants, allowing obtaining risk maps, especially when air quality indices are a tool for alertness, particularly important for vulnerable populations. The purpose of pollution modelling is to predict certain levels for certain situations in order to provide their citizens an opportunity to modify their personal habits to minimize exposure [16]. While substitute measures, such as distance to source of pollution, have been associated to large health effects [17], the problem of lowering the impact still exists. It is still needed to know how a certain concentration could evolve in certain environmental conditions and what effects will produce. The methods to evaluate the pollution dispersion utilize geographic information systems (GIS) to combine available geographic data with short-term monitoring information to develop exposure models capable of identifying small-area variations in pollution. Results from these models can then be overlapped on other geo-referenced data, like health risks, in order to emphasise the exposure of people at their homes, work etc.

2 Air pollution modelling

Air pollution modelling is using mathematical approaches to explain or predict the way pollutants behave in a certain environment or time scale, even we talk about indoor or outdoor. Numerical models can be repeated indeterminately with strict parameters at relatively low costs comparing to physical modelling. Modelling helps evaluate the correlation between the pollutants sources and their effects on ambient air quality. Simulations can be used to run scenarios, to test concepts and to acknowledge environmental impact under different emission sources and levels, climate and conditions. Different methods and techniques exist, but the purpose is permanently the same – evaluate the pollutant impact over a given dimension (space, time) using an existing set of measured data. Although there are no clear differences among diverse categories, models might be classified after [18] into groups according to their physical (e.g. reduced-scale) or mathematical principles (e.g. box, Gaussian, regression, CFD) or their level of complexity (e.g. screening, semi-empirical, numerical).

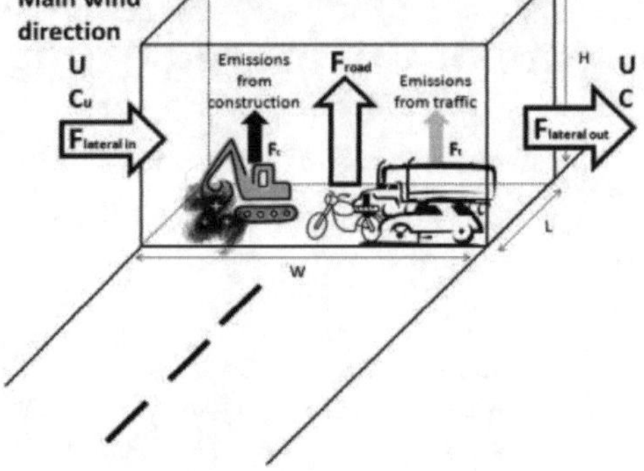

Fig. 1. Example of schematic diagram of the box model used to calculate PM10 emissions from the road construction phase of the road widening [19]

Chemical reactions between the pollutants can be integrated in these models, predicting also the effects generated by a certain critical situation with regards to secondary species which are not directly emitted as such, but form when other pollutants (primary pollutants) react in the atmosphere. Examples of a secondary pollutant include ozone, which is formed when hydrocarbons (HC) and nitrogen oxides (NOx) combine in the presence of sunlight. NO2 is formed as NO, reacts with oxygen in the air. Acid rain is formed when sulfur dioxide or nitrogen oxides react with water[20].
Moreover, urban air pollution simulation encounter problems with complex turbulent airflows like local-scale topography (e.g. canyons, mountains, tall buildings, urban streets etc.), airborne particles and gases in the atmosphere, air circulation due to thermal effects or due to other sources etc. These problems can be resolved by specific simulations and the Computer Fluid Dynamics (CFD) methods and tools have become extensively used. CFD modelling is a specific term used to describe the quantitative and qualitative evaluation of systems involving fluid flow, heat transfer and associated phenomena (e.g. chemical reactions, diffusion etc.) using computer-based numerical methods. Among the first CFD studies related to pollution, we remark Svensson's[21], who was one of the firsts who started to use commercial CFD tools for various air-pollution problems. However, for outdoor air pollution other specific problems might appear and should be treated carefully. A substantial problem is the discrepancy of input and output boundary conditions for meteorological profiles taking into account the Coriolis forces, surface layer structure etc.[22]. Atmospheric pollution domain, CFD models are generally used when pollutant dispersion is highly influenced by the geometry of an urban area. Accurately describing the topography of a certain urban zone is very important. Chu et al.[23] even used geographic information system (GIS) software to obtain detailed coordinates and dimensions of the buildings in the interest area. Neofytou et al. [24]

 https://doi.org/10.1051/e3sconf/20183201010

studied, using CFD methods, the formation of the vortices in the streets canyons due to the wind direction and the speed and the influence of the characteristics of the flow fields on the concentration distributions. Haghighat et al.[25] studied the impact of non-uniform urban surface temperature on pollution dispersion in urban areas, introducing also a radiation and a soil model to simulate the effect of heat storage. Other urban characteristics can be implemented to the model: in [26] Jeanjean et al. studied the effect of trees to disperse road traffic emissions at a city scale. The authors found that trees have a local beneficial impact on road traffic emissions by increasing turbulence and reducing ambient concentrations of road traffic emissions by 7% at pedestrian height. We may see that conditions imposed in models can vary on different aspects, but in the same time, a complex model will require important computation resources. The turbulence closure techniques used for the CFD models, when considering air pollution at urban level, two main types of turbulence models are used: Reynolds-averaged Navier-Stokes (RANS) and Large-Eddy Simulation (LES) [14] as the atmospheric turbulent flow in and above street canyons involves turbulent eddies on a variety of scales [27].

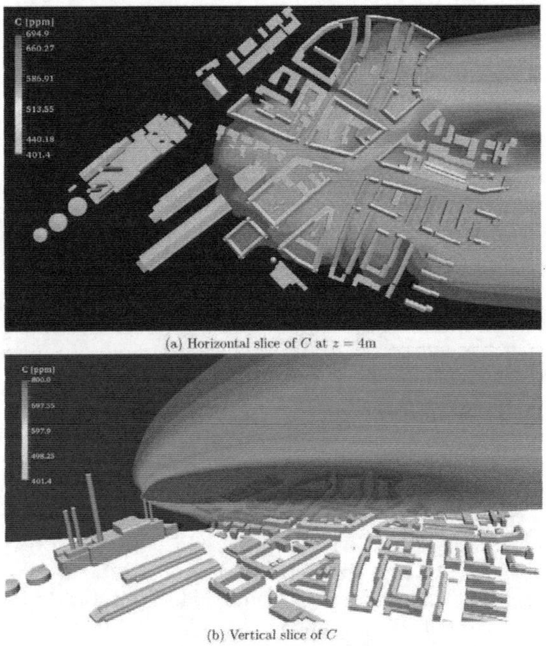

Fig. 4. Example of CFD modelling: images of the CO_2 concentration[28]

An interesting alternative to these models are the neural network models [29, 30]. Artificial intelligence mechanisms are more and more used since they learn from real measured data and predict a certain parameter, e.g. artificial neural networks, Bayesian neural networks or adaptive neuro-fuzzy algorithms.

Most current air quality predicting models use direct, simple approaches like box models, Gaussian models or linear statistical models. Those models are easy to implement and allow a fast calculation and low computation resources. However, in general, they do not define the relations and non-linear relationship which govern the transport and behaviour of pollutants in the atmosphere[31]. To overcome these problems, artificial intelligence models have become popular in air quality forecasting and other atmospheric domains.

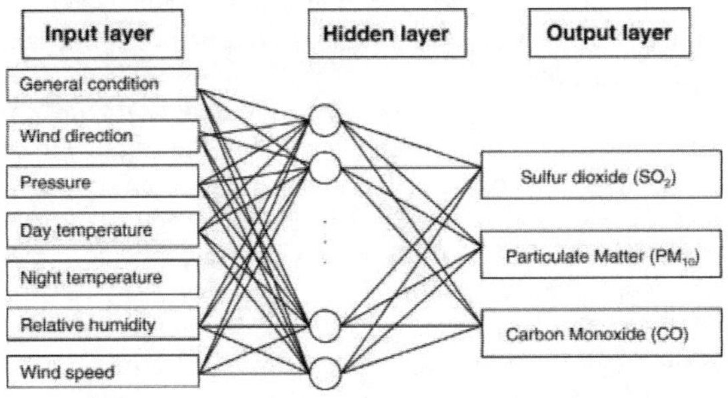

Fig. 3. Example of a neural network model[32]

Air quality management organisms worldwide use specific models which rely on mathematical and numerical techniques to simulate the physical and chemical processes of pollutants as they disperse and react in the atmosphere. The most commonly used air quality (AQ) models include: Dispersion Modelling - estimates the concentration of pollutants at specified ground-level receptors surrounding an emissions source; Photochemical Modelling - used in regulatory or policy assessments by estimating pollutant concentrations and deposition of both inert and chemically reactive pollutants over large spatial scales; Receptor Modelling - use the chemical and physical characteristics of gases and particles measured at source and receptor to quantify source contributions to receptor concentrations[33]. Different types of such models are analysed in a model evaluation study regarding the ozone level in Europe and Norths America [34], offering an opportunity to investigate methodologies for generating skilful ensembles of regional air quality models outputs. .

Air pollution is mostly addressed by national projects focusing mostly to tackle the health effects of ultrafine particles and air pollution due to combustion of fossil fuels [35]. On the other hand, most atmospheric models used in air quality management don't have a user-friendly interface that can synthesize the model results produced from different air pollutant emission inputs, and do not offer adequate data visualization for supporting policy making. To improve the efficiency of data synthesis, it is necessary to develop a tool to translate the massive amount of model data to policy-relevant visualization [36].

3 Air pollution monitoring

Field measurements can provide immediate data on pollutant concentration, air flow and pollutant dispersion, depending to the limitations of measurement techniques. Models which require learning data or those which require certain specifications are relying on the measured parameters. The restrictions encountered are often related to challenges to data interpretation, uncontrollable meteorological conditions, low spatial coverage, or typically high costs.

Pollution monitoring is a first step in resolving this problem, but we often remain at monitoring phase. Current sector-specific initiatives and priorities are mostly elaborated in isolation from each other – they lack consistency and sometimes counter-act each x should allow a cohesive solution for urban air pollution problems. Ideally, such policy decision support tools should be efficient, easy-to-use, and capable of estimating air quality changes in the quality of environment in real time for introducing given emission reduction scenarios for a particular city and/or relevant control actions. The literature survey indicates that pollution monitoring at urban scale has been approached in several projects, even with the dedicated purpose for policies implementation [37]. However, communities still need the link between the monitoring process and action taking.

Monitoring of urban pollutants is usually done through a network of sensors placed in specific areas. It also comprises the predicting step and data interpretation. The monitoring systems depend on the type of pollutant(s) or the target (local monitoring, district, city area etc.). Monitoring stations should be generally located near populated regions where pollutant exposure is the highest. Critical pollutants or with high risk, like ozone, have specific monitoring platforms, using certain types of technologies[38]. At global level it is also important to know the reference levels of pollutants in order to consider the urban levels. In a comprehensive review, Martin et al. [39] mentions the capabilities for satellite remote sensing of certain species(aerosols, tropospheric O_3, tropospheric NO_2, CO, HCHO, and SO_2) in the boundary layer, along with physical processes affecting their accuracy and precision. The purpose is to achieve higher spatial resolution to resolve urban scales, facilitate validation and reduce cloud contamination that increases remote sensing error.

At the urban level, the monitoring networks usually transmit the information and different alert levels are issued. Different indices have been issued so far for air quality, from which we mention for example the Air Quality Health Index (AQHI) which is a public information tool designed in Canada to evaluate the impact of air quality on health.

Fig. 5. Air Quality Index (AQI)calculated by EPA

A similar index is proposed by US Environmental Protection Agency: The Air Quality Index-AQI is an index for reporting daily air quality also with impact on citizens health. European Environment Agency is proposing the European Air Quality Index with values for different cities across Europe, but less data for South-eastern Europe.

Large, polluted cities have become more and more sensitive to air quality since pollution levels reach critical values. A heavy 16-day pollution episode occurred in Beijing was studied in [40] to reveal the combined effects of traffic restrictions and meteorology on urban air quality. Results indicated that additive functions of traffic restrictions, suitable relative humidity and temperature are more effective on concentration reduction only for some pollutants. More solutions are necessary to determine efficient ways for pollution control.

Urban networks are now equipped with wireless sensors, which transmit the data to dedicated servers where real time decision making takes place. The advances in technology have permitted the installation at large scale, monitoring large urban areas. Common architectures of information flow in wireless sensor networks is presented in figure 6.

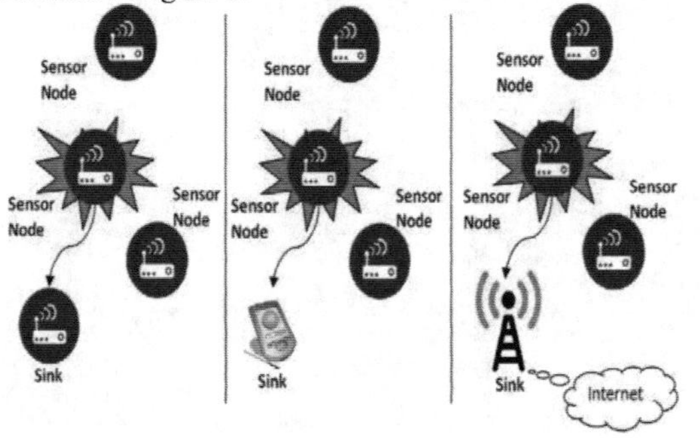

Fig. 6. Prospective architectural topologies of wireless sensor networks[41]

Recent studies reveal the importance of local-level pollution data, including human personal exposure to air pollutants. Connecting an array of wireless sensors to a server it is a simple and robust alternative of former urban sensor networks. The data can be transmitted via a General Packet Radio Service Modem (GPRS-Modem) and a Global Positioning System Module (GPS-Module), a simple and low cost solution.[42].

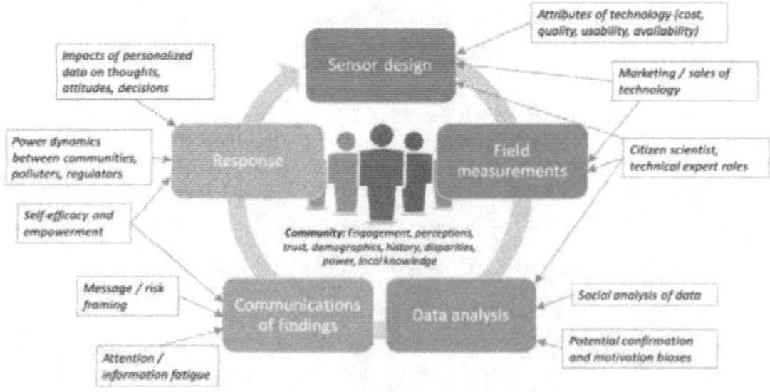

Fig. 7. Social sciences can contribute to the understanding of all elements of sensor design and use[43]

Moreover, utilizing low-cost air quality sensors and the ordinary smartphones, end-users (simple citizens) can be more involved and are able to extract, analyse and share the local air pollution information[44, 45].
This the concept of "participatory sensing", which has a promising trend, especially when influenced by social context. In participatory sensing, individuals and groups engage in the data collection actively and the helps the public administration to make proper decisions[43, 46].

4 Urban pollution- regulatory aspects and lack of consistency

Pollution monitoring is a first step in resolving this problem, but we often remain at monitoring phase. Current sector-specific initiatives and priorities are mostly elaborated in segregation from each other – they lack of consistency and sometimes counter-act each other [47]. Large amount of collected data has not been integrated into suitable software applications which should allow a consistent solution for urban air pollution management. Ideally, such policy decision support tools should be efficient, easy-to-use and capable of estimating air quality changes in real time. The purpose is to introduce emission reduction scenarios previously considered for a particular region and/or relevant verification actions. In addition, the tools should offer intuitive policy-related visualizations [36].
Studies worldwide are showing that current policies regarding air pollution control are mostly ineffective or applied in an unproductive manner. The air quality is a major issue for both, developed and developing countries. The World Health Organization [3] draws the attention about this global issue, especially in high urbanized regions with increased levels of traffic. Overall, in developing countries, the management practices to improve urban air quality are generally limited and the budget allocated for urban air quality management is also insufficient. However, exceptional cases can become a good practices example, like Tokyo, where air quality measurement capacity and management enabling capabilities are excellent [48]. Other significative countries are evaluated in [1] concerning urban air quality management system/process, from which we mention USA and European Union. USA has introduced the Clean Air Act (CAA) since 1963, setting responsibilities and relationships among federal, state, tribal, and local agencies. In Europe, the air quality management practices were introduced in 1980 with directive S0/779/EEC. However, the actual implementation of the newest Directive 2008/50/EC is left to the discretion of each state member. For example, a decrease in emission levels of NO_x and $PM_{2.5}$ from transport sector from 1990 to 2005 has resulted in a reduction in health impacts in the range of only 2.5% in Bulgaria to 25% in Luxembourg and Switzerland[49].
Dedicated platforms which are connected to local authorities are utterly needed so that the dedicated institution take actions, resulting a healthier urban environment.

5 Conclusions

The problem of urban air quality remains a sensitive subject, considering measuring techniques, modelling, forecasting, management and population health risks. New technologies of pollution monitoring, which have been simplified and can be accessible to the simple citizens, give the opportunity to raise awareness about clean environment, human health and civic implication.

This work was supported by the grants of the Romanian National Authority for Scientific Research, CNCSIS – UEFISCDI, SMARTSENSE - Real time SMART application for urban air quality management respecting the SENSitivE categories of population, PN-III-P2-2.1-PED-2016-1285.

References

1. Gulia, S., et al., Urban air quality management- A review. Atmospheric Pollution Research, 2015. **6**(2): p. 286-304.
2. OECD, The Economic Consequences of Outdoor Air Pollution. OECD Publishing.
3. WHO, Ambient air pollution: A global assessment of exposure and burden of disease, W.H. Organization, Editor. 2016.
4. WHO, Reducing global healh risks through mitigation of short-lived climate pollutants. Scoping report for policymakers, W.H. Organization, Editor. 2015.
5. WHO, Global surveillance, prevention and control of chronic respiratory diseases. A comprehensive approach, ed. W.H. Organization. 2007.
6. Diapouli, E., et al., Evolution of air pollution source contributions over one decade, derived by PM10 and PM2.5 source apportionment in two metropolitan urban areas in Greece. Atmospheric Environment, 2017. **164**(Supplement C): p. 416-430.
7. Daher, N., et al., Characterization, sources and redox activity of fine and coarse particulate matter in Milan, Italy. Atmospheric Environment, 2012. **49**(Supplement C): p. 130-141.
8. Agency, E.E., Air quality in Europe — 2017 report. 2017.
9. Kim, K.-H., E. Kabir, and S. Kabir, A review on the human health impact of airborne particulate matter. Environment international, 2015. **74**: p. 136-143.
10. EPA, Clean Air Act Overview- Air Pollution: Current and Future Challenges, U.S.E.P. Agency, Editor. 1990.
11. ORGANIZATION, W.H., Indoor air pollutants:exposure and health effects, in Reports and Studies, EURO, Editor. 1982, WHO: Copenhagen.

12. ORGANIZATION, W.H., Indoor air quality research, in Reports and Studies, EURO, Editor. 1984, WHO: Copenhagen.
13. Redlich, C.A., J. Sparer, and M.R. Cullen, Sick-building syndrome. The Lancet, 1997. **349**(9057): p. 1013-6.
14. Zhong, J., X.-M. Cai, and W.J. Bloss, Coupling dynamics and chemistry in the air pollution modelling of street canyons: A review. Environmental Pollution, 2016. **214**: p. 690-704.
15. Peng, H., Air quality prediction by machine learning methods. 2015.
16. Adams, M.D. and P.S. Kanaroglou, Mapping real-time air pollution health risk for environmental management: Combining mobile and stationary air pollution monitoring with neural network models. Journal of Environmental Management, 2016. **168**: p. 133-141.
17. Hoek, G., et al., Estimation of long-term average exposure to outdoor air pollution for a cohort study on mortality. J Expo Anal Environ Epidemiol, 2001. **11**(6): p. 459-69.
18. Vardoulakis, S., et al., Modelling air quality in street canyons: a review. Atmospheric environment, 2003. **37**(2): p. 155-182.
19. Font, A., et al., Degradation in urban air quality from construction activity and increased traffic arising from a road widening scheme. Science of The Total Environment, 2014. **497-498**: p. 123-132.
20. Atkinson, R., Atmospheric chemistry of VOCs and NO x. Atmospheric environment, 2000. **34**(12): p. 2063-2101.
21. Svensson, U., PHOENICS in environmental flows. A review of applications at SMHI. Lecture Notes in Engineering, 1986. **18**: p. 87-96.
22. Baklanov, A., Application of CFD Methods for Modelling in Air Pollution Problems: Possibilities and Gaps. Environmental Monitoring and Assessment, 2000. **65**(1): p. 181-189.
23. Chu, A.K.M., R.C.W. Kwok, and K.N. Yu, Study of pollution dispersion in urban areas using Computational Fluid Dynamics (CFD) and Geographic Information System (GIS). Environmental Modelling & Software, 2005. **20**(3): p. 273-277.
24. Neofytou, P., et al., Computational Fluid Dynamics Modelling of the Pollution Dispersion and Comparison with Measurements in a Street Canyon in Helsinki. Environmental Modeling & Assessment, 2008. **13**(3): p. 439-448.
25. Haghighat, F. and P.A. Mirzaei, Impact of non-uniform urban surface temperature on pollution dispersion in urban areas. Building Simulation, 2011. **4**(3): p. 227.
26. Jeanjean, A.P.R., et al., A CFD study on the effectiveness of trees to disperse road traffic emissions at a city scale. Atmospheric Environment, 2015. **120**: p. 1-14.
27. McNabola, A., B.M. Broderick, and L.W. Gill, A numerical investigation of the impact of low boundary walls on pedestrian exposure to air pollutants in urban street canyons. Science of The Total Environment, 2009. **407**(2): p. 760-769.
28. Toja-Silva, F., et al., CFD simulation of CO2 dispersion from urban thermal power plant: Analysis of turbulent Schmidt number and comparison with Gaussian plume model and measurements. Journal of Wind Engineering and Industrial Aerodynamics, 2017. **169**: p. 177-193.
29. Lal, B. and S.S. Tripathy, Prediction of dust concentration in open cast coal mine using artificial neural network. Atmospheric Pollution Research, 2012. **3**(2): p. 211-218.
30. Nejadkoorki, F. and S. Baroutian, Forecasting Extreme PM10 Concentrations Using Artificial Neural Networks. International Journal of Environmental Research, 2012. **6**(1): p. 277-284.
31. Luecken, D.J., W.T. Hutzell, and G.L. Gipson, Development and analysis of air quality modeling simulations for hazardous air pollutants. Atmospheric Environment, 2006. **40**(26): p. 5087-5096.
32. Kurt, A., et al., An online air pollution forecasting system using neural networks. Environment International, 2008. **34**(5): p. 592-598.
33. Wilson, R., Air Pollution, the Automobile, and Public Health. Environment: Science and Policy for Sustainable Development, 1989. **31**(4): p. 25-27.
34. Solazzo, E., et al., Model evaluation and ensemble modelling of surface-level ozone in Europe and North America in the context of AQMEII. Atmospheric Environment, 2012. **53**: p. 60-74.
35. Cartier, Y., T. Benmarhnia, and A. Brousselle, Tool for assessing health and equity impacts of interventions modifying air quality in urban environments. Evaluation and Program Planning, 2015. **53**: p. 1-9.
36. Prasad, K., A.K. Gorai, and P. Goyal, Development of ANFIS models for air quality forecasting and input optimization for reducing the computational cost and time. Atmospheric Environment, 2016. **128**: p. 246-262.
37. Borrego, C., et al., Urban scale air quality modelling using detailed traffic emissions estimates. Atmospheric Environment, 2016. **131**: p. 341-351.
38. Bowman, K.W., Toward the next generation of air quality monitoring: Ozone. Atmospheric Environment, 2013. **80**: p. 571-583.
39. Martin, R.V., Satellite remote sensing of surface air quality. Atmospheric Environment, 2008. **42**(34): p. 7823-7843.

40. Hu, D., et al., Urban air quality, meteorology and traffic linkages: Evidence from a sixteen-day particulate matter pollution event in December 2015, Beijing. Journal of environmental sciences (China), 2017. **59**: p. 30-38.
41. Rashid, B. and M.H. Rehmani, Applications of wireless sensor networks for urban areas: A survey. Journal of Network and Computer Applications, 2016. **60**: p. 192-219.
42. Al-Ali, A.R., I. Zualkernan, and F. Aloul, A Mobile GPRS-Sensors Array for Air Pollution Monitoring. IEEE Sensors Journal, 2010. **10**(10): p. 1666-1671.
43. Hubbell, B.J., et al., Understanding social and behavioral drivers and impacts of air quality sensor use. Science of The Total Environment, 2018. **621**: p. 886-894.
44. Hasenfratz, D., et al., Participatory air pollution monitoring using smartphones. Mobile Sensing, 2012. **1**: p. 1-5.
45. Sammarco, M., et al., Using geosocial search for urban air pollution monitoring. Pervasive and Mobile Computing, 2017. **35**: p. 15-31.
46. Ahmed, A.A.N., et al., A Participatory Sensing Framework for Environment Pollution Monitoring and Management. arXiv preprint arXiv:1701.06429, 2017.
47. Sá, E., et al., Climate change and pollutant emissions impacts on air quality in 2050 over Portugal. Atmospheric Environment, 2016. **131**: p. 209-224.
48. Santosa, S.J., T. Okuda, and S. Tanaka, Air Pollution and Urban Air Quality Management in Indonesia. CLEAN – Soil, Air, Water, 2008. **36**(5-6): p. 466-475.
49. Pulles, M., Impact of selected policy measures on Europe's air quality. 2010.

Fire modeling in a nonventilated corridor

Marius Dorin Lulea[1,*], *Vlad* Iordache[2], and *Ilinca* Năstase[2]

[1] PHD at Technical University of Civil Engineering of Bucharest, Romania

[2] Research Center CAMBI, Technical University of Civil Engineering of Bucharest, Romania

Abstract. The main objective of this study was to determine the effect of fire in a nonventilated corridor. A real-scale model of a corridor has been modeled in Fire Dynamics Simulator(F.D.S.) in order to determine the evolution of indoor temperatures, the visibility and the oxygen quantities during a fire. The start time of a sprinkler has also been determined. The use of sprinklers in buildings has become a necessity and a requirement imposed by technical norms. The provision of this type of installation has become a common feature in buildings with a high fire risk, with two main effects: fire extinction and protection of structural and partition elements from high temperatures[15]. The ultimate goal is to ensure optimal conditions for saving the building users, intervention teams and maintaining the stability of the building. Low temperatures and good visibility on the escape routes during a fire are the basic conditions to ensure the optimal evacuation of users.

1 Introduction

The main purpose of fire safety processes is to save the users of the buildings. Beside this main purpose, the following aspects have a particular importance: ensuring fire-fighting interventions conditions, preventing the collapse of buildings, limiting the spread of fire inside the buildings or to other buildings and, ultimately, limiting financial losses[1].

Under these conditions it is imperative to protect the escape routes[2] by maintaining low temperatures, good visibility and low concentration of toxic gases resulting from the fire. For a period of time hallways, stairwells and corridors must be protected area in case of fire, required to ensure the evacuation safely of users. The norms[2] requires that the time to evacuate users to be as short as possible. For example, for one-way evacuation, it should be limited to 20 seconds- 6m (buildings made from combustible materials) and up to 70 seconds - 21m (non-combustible materials with fire performance).

Fire protection of a building can be achieved by the application of passive protection measures or active protection measures[3]. Passive fire protection of buildings refers, in general, to structural and architectural compliance. This category includes fire requirements for: columns, beams, floors, interior and exterior walls, roof and covering[2]. Active fire protection refers to the use of automated systems that do not require human intervention and which also affect the fire, the source of the fire and its products. This category includes fire extinguishers with sprinklers[4], mechanical ventilation systems for fire protection, inert gas installations, fog extinguishing systems, powder extinguishing systems and others.

The protection of the escape routes can be done by using fire-resistant partition elements(walls, floors, doors), smoke and hot gas automatic extraction(natural or mechanical evacuation of smoke), or/and sprinkler equipment.

Sprinklers produce beneficial effects due to the following mechanisms[5]: reduce air temperatures and material temperatures, decrease heat load of materials(the combustion) by increasing humidity, makes a non-combustible water film on the surface of the materials, decreases the oxygen concentration by increasing the amount of water vapor resulting from contact of droplets with hot surfaces and others.

There are many design cases where the experts are asking themselves: what is the maximum acceptable length of an unventilated or unprotected corridor without significantly disrupting users' evacuation? How long time is available for the users to evacuate in safety condition?

The present study is oriented towards to determine the time during which an unventilated and unprotected corridor, affected by fire, becomes unsafe for evacuation. The study also aims to determine the time at which the first sprinkler will be initiated in the corridor.

* Corresponding author: LuleaMariusDorin@gmail.com

There are three methods currently used for modeling fires:

- simplified mathematical models[6];
- zonal models[7];
- field models[8].

The most common and simplest method at this time is the one given by the Eurocode[6] and which provides a function time- temperature of indoor air(fig. 1). The method is simple to use but presents the disadvantage that it leads to an overvaluation of the consequences of the fire and can leads to unreasonable consumptions of materials. The model proposes equal temperature in all building volumes, and the representation is not realistic[8].

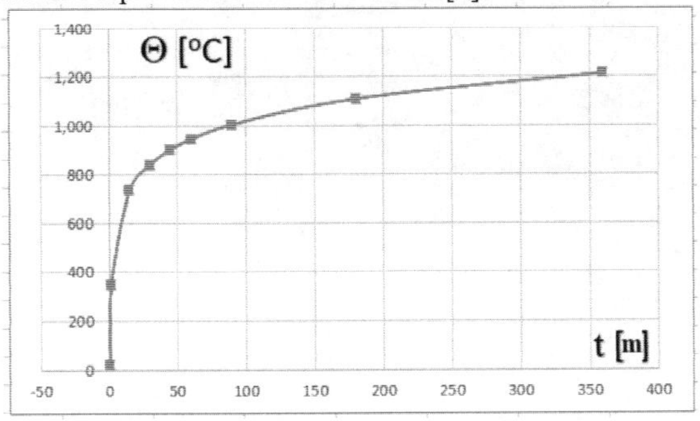

Fig. 1. The Time-Temperature variation curve SR EN1991-1-2

Global direction is to shift from the simplified models to field models that provide a more realistic representation of the fire[9]. This is happening with the development of computing technologies. In the case of important constructions, realistic modeling of the fire is required because it leads to an efficient consumption of materials and helps the designer to establish correct and justified design solutions. Simplified modeling, for example on the basis of the standard Eurocode temperature curve, is a covering and often leads to ineffective solutions.

In order to reach our proposed goals, we made a modeling in F.D.S. of a fire in a nonventilated corridor. In the present study it was analyzed the time evolution of the next parameters: indoor air temperature, visibility and oxygen concentration at different characteristic points.

2 Modeling Fire Dynamics Simulator(F.D.S.) and Case Study

This research involves modeling of a fire in a nonventilated corridor using F.D.S. This is a computational fluid dynamics (CFD) model of fire-driven fluid flow. The software solves numerically a form of the Navier-Stokes equations appropriate for low-speed, thermally-driven flow(Ma<0,30), with an emphasis on smoke and heat transport from fires.

The Fire Dynamics Simulator and Smokeview are the products of an international collaborative effort led by the National Institute of Standards and Technology (NIST) and VTT Technical Research Centre of Finland. The F.D.S. currently takes the form of a programming tools and it provides to users unlimited and explicit access to the algorithm or other data.

F.D.S. includes certain restrictions and simplifications for modeling fluid motion and make algorithms to respond effectively purpose: to study the effect of fire.

It is important, in the design of a building, to model the fire by controlling the parameters that lead to the role of fire safety systems:

- the distribution of temperature in time and space;
- the distribution of smoke in time and space;
- the distribution of combustion gases in time and space.

To determine the equations solutions, F.D.S. uses the following models:

Method to solve the Navier-Stokes equations- use finite volume method to solve the numerical problem;

Turbulent flow- Large Eddy Simulation(fig. 2), neglecting small turbulence which require important calculation resources, but having a low effect on the final result;

Geometry- F.D.S. uses a 3D model with rectiliniar grid with the possibility of multiple mesh(fig. 3);

Fire source- specify the heat release rate per unit area(HRRUA);

Burning- the burning process releases heat, gases and smoke; the combustion model assumes that each cell initiates, maintains or stops burning when a certain temperature is reached and there is a certain ratio of gas to oxygen (more than 14%). By default, F.D.S. uses the mixed fraction model with two parameters- the first parameter represents the unburn fuel fraction and the second parameter fraction of the fuel that has been burned. We can also choose a two-phase model (phase 1: obtaining CO and phase 2 CO2 production) with three parameters, involving a burned fuel fraction, the burned fuel fraction in phase 1 and the burned fuel fraction in phase 2 . To make the simulations tractable, the author limits the number of fuels to one, and the number of reactions to just one or two.

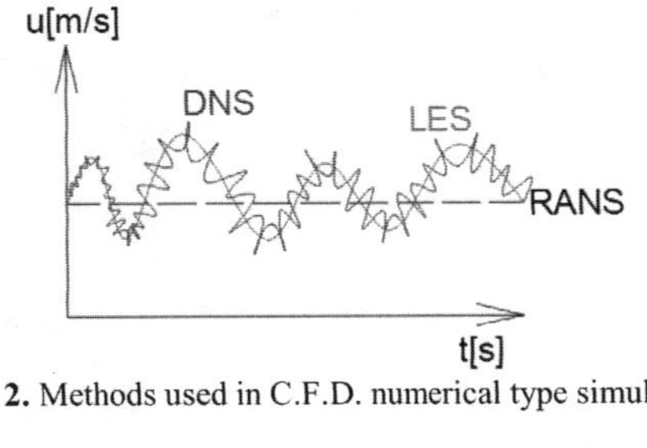

Fig. 2. Methods used in C.F.D. numerical type simulation

The F.D.S. modeling consist of divide the domain into thousands of cells. For each cell is known at a certain time various parameters such as: temperatures, velocity, visibility, gas concentrations, quantity of material (unconsumed combustible, burned fuel or noncombustible material)[9]. Each cell changes its parameters over time according to the characteristics of

 https://doi.org/10.1051/e3sconf/20183201011

the other cells from space and the external forces acting on the given domain.

The current study involved the burning inside a closed corridor with the dimensions of 7,80m x 1,80m x 2,80m(fig. 4). The corridor has no natural or mechanical ventilation.

The tridimensional rectilinear mesh is composed of regular cells with the following dimensions: 0,10m x 0,10m x 0,10m. This means that the domain analyzed consists of 176 000 cells, each coming with its own physical and chemical equations interrelated(fig. 3).

The corridor is included in an analysis space with the following dimensions: 11,00m x 4,00m x 4,00m(fig. 4).

Two slots(F1, F2) were provided at the lower part of the corridor: 1,00m x 0,10m- simulates air infiltration in adjoining rooms.

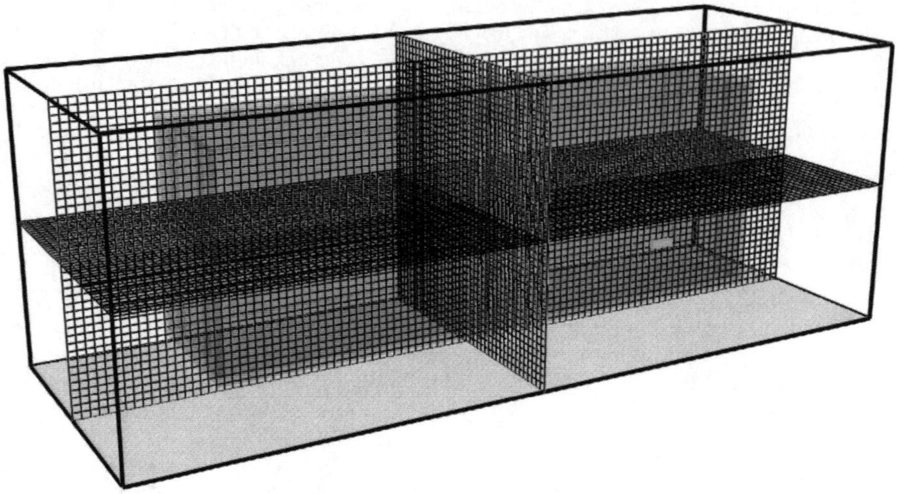

Fig. 3. Domain mesh- 10cm x 10cm

The fire source was provided at one end of the corridor. We have specified the fire by a given Heat Release Rate per Unit Area(HRRPUA): 450kW/m^2. The burner has the dimensions of 0,30m x 0,40m(fig.5).

$$HRR = HRRPUA \times S_{burner} \quad [kW] \qquad (1)$$

$$HRR = 54kW \ [kW]$$

The fuel used is "wood", characterized by a Lower Heating Value(LHV) of 17,700 [MJ/kg]. The quantity of soot released in the air during the fire is considered equal to 0,02 kg soot/ kg combustible.

For comparation, the maximum HRR reach about 4,0MW for a single car fire[12].

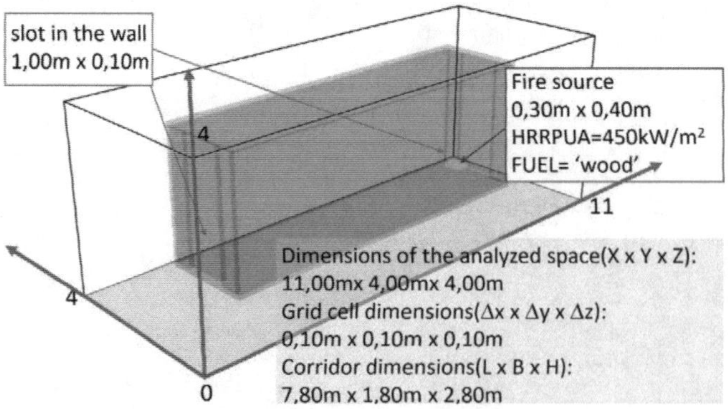

Fig. 4. Fire space characteristics

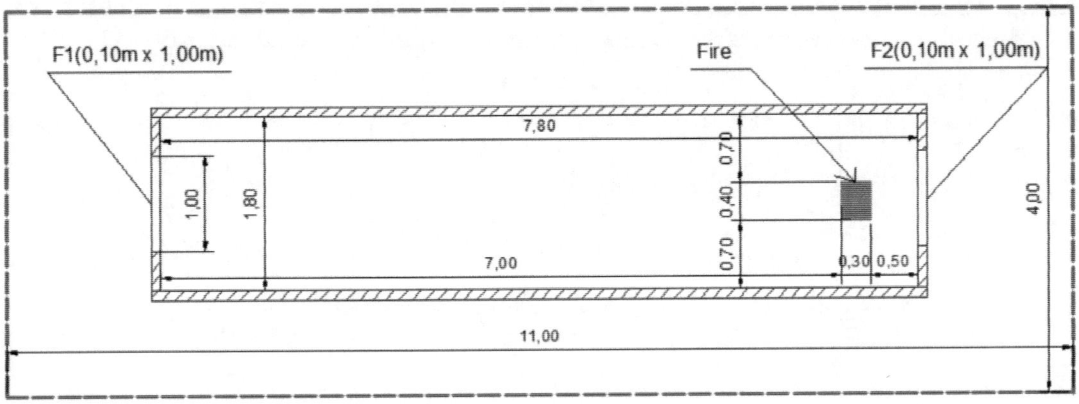

Fig. 5. Model plan

In the corridor there were mounted 15 temperature sensors(S1- S15) at 30cm from the ceiling(z=2,50m). The distance between sensors is 50cm(fig. 6). The sensors have been installed at the same level as usually sprinklers are.

The parameters that characterize the fire were followed for a period of 360 seconds.

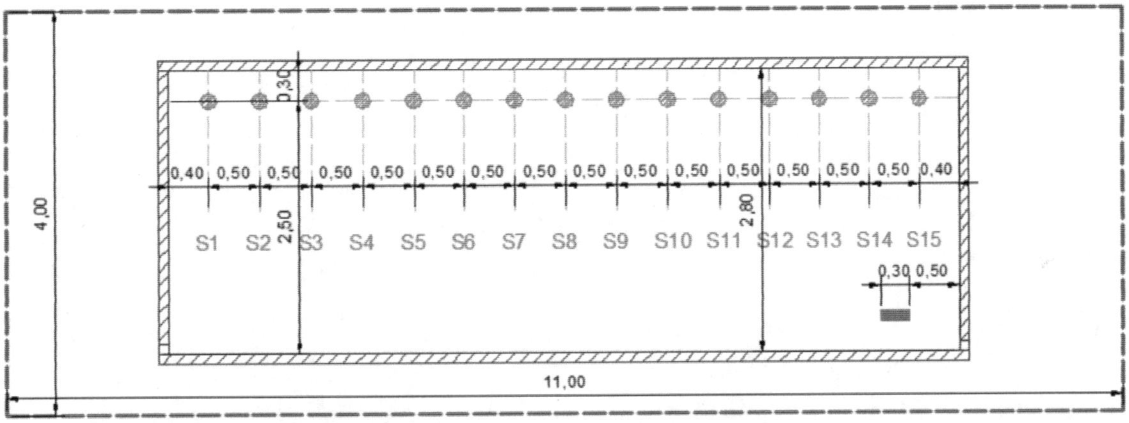

Fig. 6. Position of temperature sensors- section

Another important parameter monitored is the quantity of oxygen(concentration) in the air. An oxygen concentration value greater than 18% is required for a good evacuation and values below 16% are dangerous for humans. If the oxygen concentration is below 14% the burning will cease.

Two sensors for measuring oxygen concentration were placed on 1.50m from the floor: O1, O2(fig. 7). Other researchers also used the position z = 1,50m to determine the parameters that characterize the fire [13].

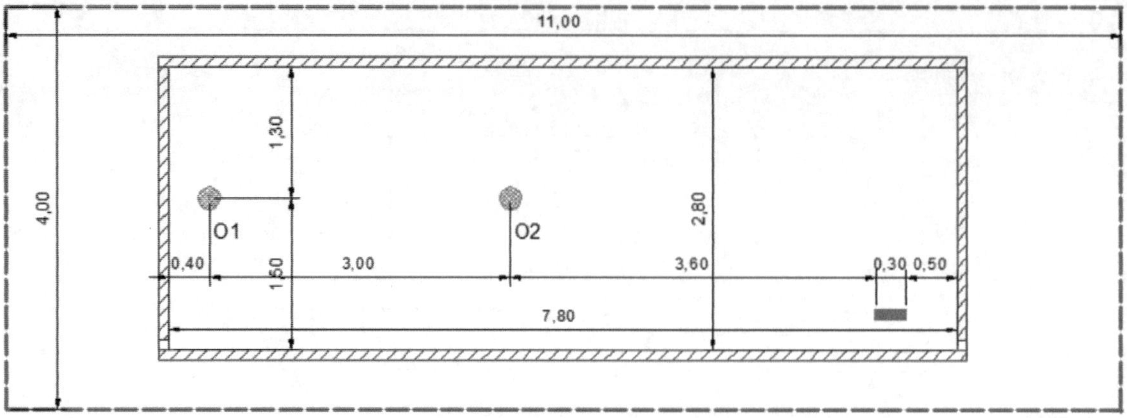

Fig. 7. Position of oxygen sensors- section

3 Results

In the current study, during the analysis period, the following output parameters were followed: the visibility, the oxygen concentration and the temperature.

3.1. Visibility

An important parameter which was followed is the visibility in the corridor. For each iteration the smoke density is known.

Visibility is the distance that an observer can identify an object relative to the background and obscuration is the amount light intensity is reduced as it passes through smoke[11].

$$I = I_0 e^{-K_m \rho_s L} \quad (3)$$

K_m- mass extinction coefficient [m^2/ kg],

L- the unit of length over wich the light is attenuated [m],

ρ_s- the density of soot [kg/m^3],

I- the intensity of monochromatic light [nm],

I_o- the incident intensity [nm].

The visibility was measured in O_1 point. It can be noticed that in less than 30 seconds the amount of smoke is significant but the visibility at 1.70m of floor is ensured through all the corridor; after 60 seconds the smoke will cover the corridor all its height but the visibility at 1.50m of flooring drops below 2.00m; after 120 seconds the smoke will significantly affect all the surrounding areas but the visibility at 1.50m of flooring drops below 1.00m(fig. 8).

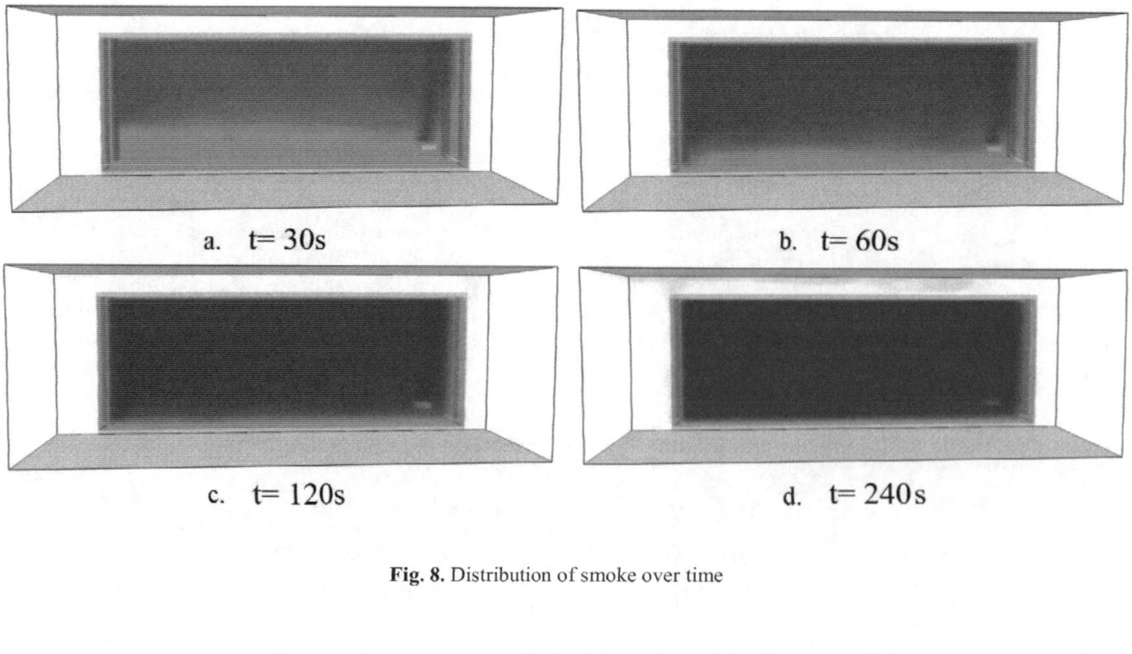

a. t= 30s b. t= 60s

c. t= 120s d. t= 240s

Fig. 8. Distribution of smoke over time

https://doi.org/10.1051/e3sconf/20183201011

3.2. Oxygen concentration

The oxygen concentration in the air is an important parameter that has been measured. The ouput values indicate a linear decrease over the analysis period(table 1)

Table 1. Oxygen concentration($S1_{O2}$, $S2_{O2}$)

Time	O_1	O_2
[s]	%	%
0	0,2078	0,2078
30	0,2035	0,2043
60	0,19990	0,2008
120	0,1931	0,1940
240	0,1811	0,1823
360	0,1694	0,1699

Fig. 9.a. and Fig. 9.b. show that in six minutes the quantity of O_2, decreased over about 20% for both sensors. A linear variation is observed, correlated with the fact that HRRPUA has a constant value, so there is a constant consumption rate of the quantity of O_2 in space. In fact, a constant value of HRRPUA does not exist in reality, but it is important to notice that the amount of O_2 consumed is linearly proportional to HRRPUA, if there is a predominant type of fuel.

A continuous decrease of the oxygen quantity is observed, as expected; at t = 240 seconds the quantity decreases from 0.2078(t=0s) to 0.1811. However, leaks from the wall bring fresh air into the corridor from the closer rooms, so that the rate of decline is not so accelerated over the studied period. The quantity of oxygen is inappropriate but it is over the limit to ensure correct evacuation. After six minutes it can be noticed that the values fall below 0.1700, which can significantly affect evacuation.

The time distribution of oxygen concentration, in the median axis of the corridor, can be seen in fig. 10.

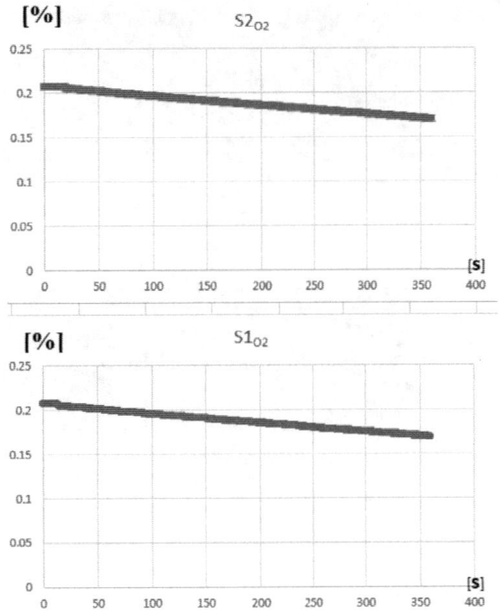

Fig. 9.a. The evolution of O_2

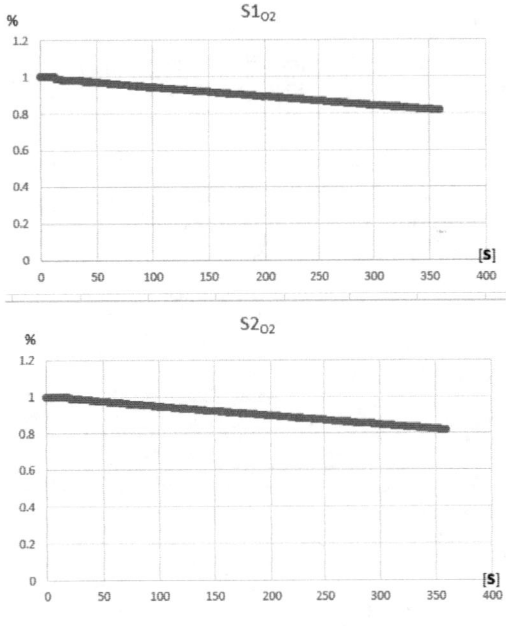

Fig. 9.b. The percentage evolution of O_2

Fig. 10

* Corresponding author: LuleaMariusDorin@gmail.com

 https://doi.org/10.1051/e3sconf/20183201011

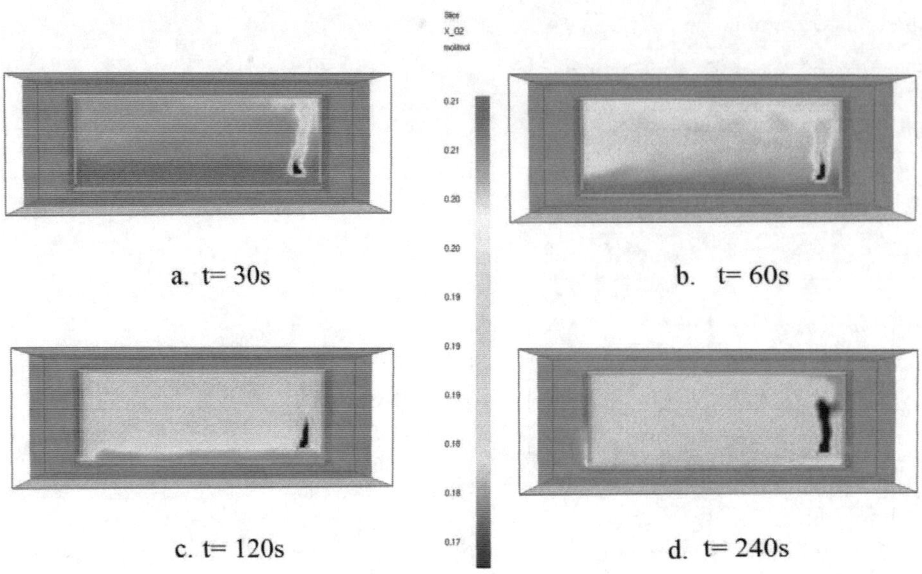

Fig. 10. The Oxygen concentration at y=0,90m.

3.3. Temperature

The temperature parameter was measured in the 15 sensors located at the ceiling. Table 2 shows the values at different times for three of them.

Table 2. Temperature(S2, S6, S10)

Time	S2	S6	S10
[s]	^oC	^oC	^oC
0	20,00	20,00	20,00
30	50,48	52,16	60,94
60	73,24	77,82	80,41
120	88,41	95,87	99,51
240	87,87	99,70	100,04
360	90,65	95,82	97,80

Sensors S2, S6, S10 measured the temperature at z = 2.50m, in the corridor's median area.

For sensor S10(fig. 6) the temperature value of 57°C is reached at time t = 22s and 68°C at t = 34s. These is the usual sprinkler firing temperatures in civil buildings.

In fig. 11 we can see that there is a stratification of hot air at the top of the corridor. At 60 seconds after the fire starts, the temperature difference between the top and bottom of the corridor is significant (the value is double). The bottom third of corridor is useful for evacuation, while the upper part has temperatures close to those that cause irrecoverable injuries.

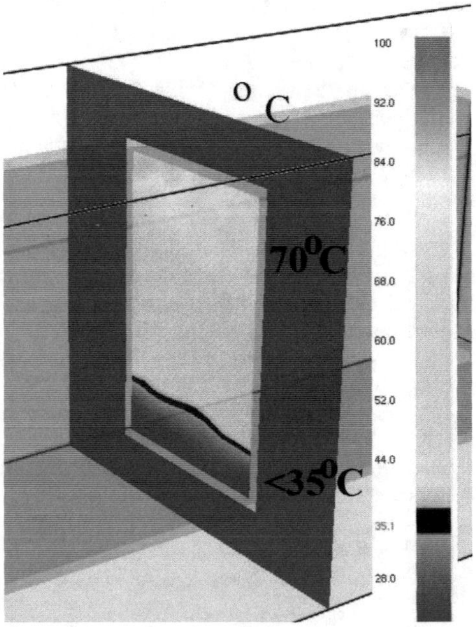

Fig. 11. Temperature distribution at z=3,60m, t=60s

We can see that the situation is almost similar after 120 seconds: the temperature at the bottom of the room is about 60°C, not perfect, but suitable to ensure evacuation. At the top of the corridor, the temperature exceeds 100°C, making this area improper for evacuating people. We conclude that it is important for users to know that evacuation using only the inferior third of the corridor or burning space can save lives.

* Corresponding author: LuleaMariusDorin@gmail.com

https://doi.org/10.1051/e3sconf/20183201011

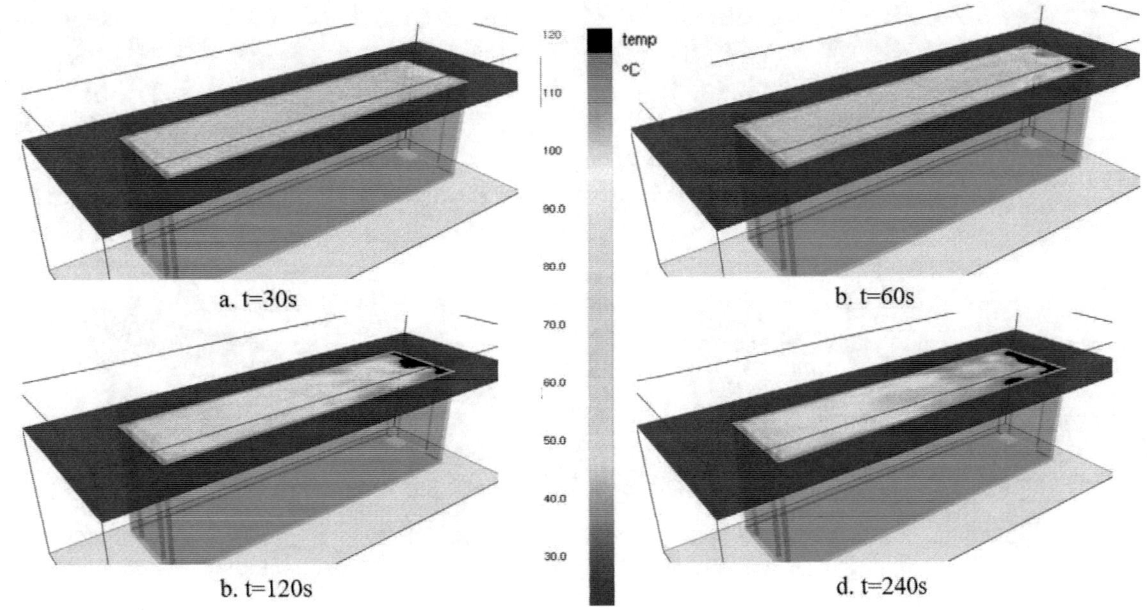

Fig. 12 Temperature distribution, z=2,50m

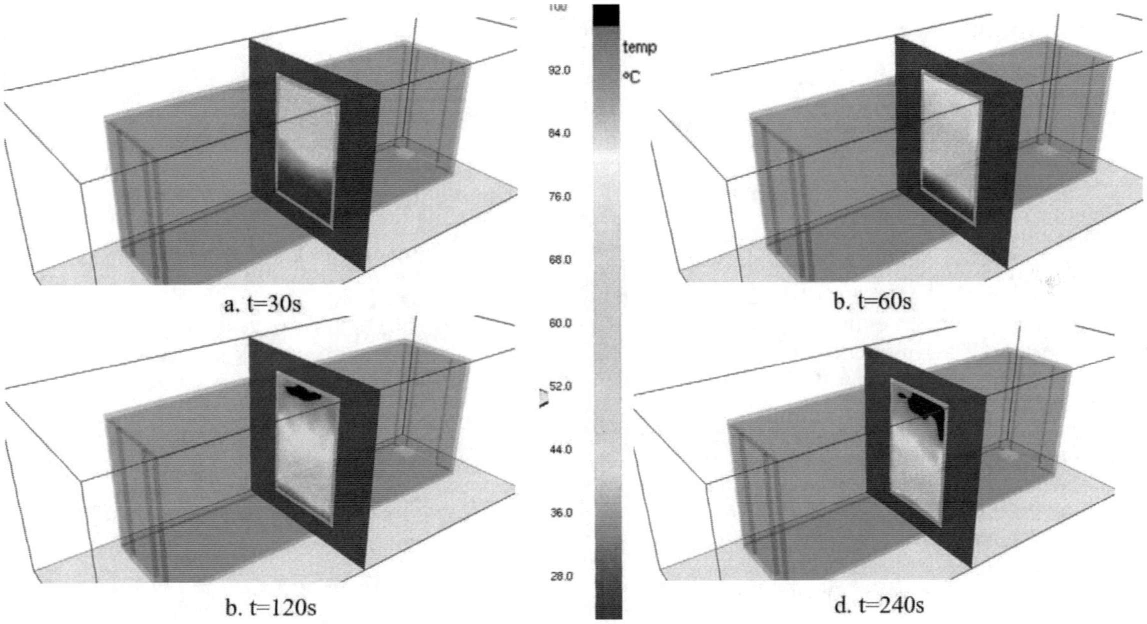

Fig. 13. Temperature distribution, x=3,90m

* Corresponding author: LuleaMariusDorin@gmail.com

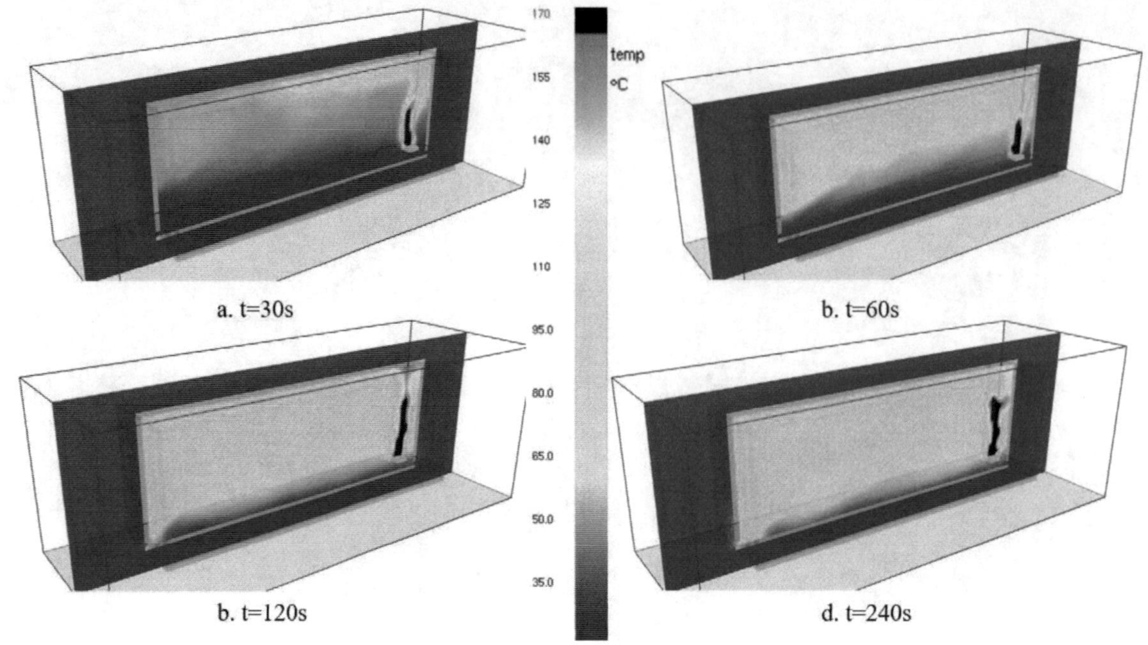

Fig. 14. Temperature distribution, y=0,90m

4 Conclusions

A modeling of a fire in an nonventilated corridor was performed with F.D.S.. The time evolution of the following parameters have been recorded: visibility, oxygen concentration and indoor temperatures. All the tracking parameters have an important role in determining the escape conditions.

In the first 4 minutes from the start of the fire in the corridor, the changes in the concentration of oxygen is within acceptable limits; after 6 minutes it drops below 0.1700, which affects the ability of people to react and the efficiency of the evacuation.

The stratification of smoke and hot gas at the top of the corridor was observed; for exemple: at t=60s, temperature S6=77,82^{O}C(z=2,50m), but at z=0,50cm the temperature is less than 35°C.

In the first 60 seconds from the start of the fire in the corridor, smoke is significant; more then 80% of the space is completely flooded with smoke, and the visibility measured at 1.50m is less than 2.00m.

The temperature values at the same level are sensitively the same. The temperature rises rapidly in space; at 0,30m from the ceiling the temperature reach 100°C in about 120 seconds.

During 30 seconds the sprinklers will start approximately, all at the same time, because the temperature at 0,30m from the ceiling reach 68°C.

Studying these aspects we can draw the conclusion that once a fire has started in the unventilated corridor, during less than a minute the corridor becomes inappropriate for evacuation. Therefore, an early alert of the users about the existence of a fire is required and it is necessary to take measures to prevent the fire to expands inside the corridor. The model shows that every second counts.

The short-term during which smoke floods the space shows that it is necessary for the corridors to be equipped with devices for the natural-organized smoke evacuation or mechanical smoke evacuation, preferably with manual and automatic start.

Another important conclusion is that the provision of sprinklers in such spaces produces positive effects because their action leads to limiting and even extinguishing the fire. The difference between the 57°C sprinklers and 68°C sprinklers is not so important because the trigger difference is only a few seconds. When ventilation is not possible for various reasons, the provision of sprinklers in the corridor may be an alternative protective measure, cumulative with passive fire protection measures(walls, floors, doors with fire resistance characteristics). It is known that the use of sprinklers together with mechanical ventilation systems causes a delay in triggering of the first category[14].

Acknowledgement

Thanks to *dr. Kevin McGrattan*(from NIST) and *dr. Lukas Arnold*(from Julich Supercomputing Centre), who have organized Summer School of Fire Dynamics Modeling 2017 and helped us to understand the phenomenon behind the F.D.S. algorithm, the simplifications and limitations of the software.

References

[1] Legea 307:2006, privind Apărarea împotriva incendiilor;

[2] P118-99 Normativ de siguranță la foc a construcțiilor

[3] R. Baldwin, P.H. Thomas, Passive and Active Fire Protection- The optimum Combination, Department of Environment and Fire Offices' Committee, 1974

[4] P118/2: 2013 Normativ privind securitatea la incendiu a construcțiilor. Partea a II-a- Instalații de stingere;

[5] NFPA 13: 2017, Standard for the Installation of Sprinkler Systems;

[6] SR EN 1991-1-2: 2013 Eurocod 1: Acțiuni asupra structurilor. Partea 1-2: Acțiuni generale. Acțiuni asupra structurilor supuse incendiului;

[7] Pietreanu Costel, Strugariu Robert, Petcana Cătălin, Inspectoratul General pentru Situaţii de Urgenţă

[8] Fire Dynamics Simulator. Technical Reference Guide. Volume 1: Mathematical Model, 2017

[9] K. McGrattan, S. Hostikka, R. McDermott, J. Floyd, C. Weinschenk, K. Overholt, Fire Dynamics Simulator. User's Guide, 2017

[10] G.W. Mulholland, C. Croarkin, *Specific Extinction Coefficient of Flame Generated Smoke*, Fire and Materials, 2000

[11] SR EN 12845:2016 Instalații fixe de luptă împotriva incendiului. Sisteme automate de stingere tip sprinkler. Dimensionare, instalare, întreținere;

[12] X.G. Zhang, Y.C. Guo, C.K. Chan, W.Y. Lin, *Numerical simulation on fire spread and smoke movement in an underground car park*, Elsevier, 2006

[13] Qing Li, Si-cheng Li, Zi-heng Wang, *Research on smoke exhaustic effect at different installation height of mechanical exhaustic port in ring corridor of high-rise building*, Elsevier, 2015

[14] P.A.(Tony) Enright, Case Studies in Fire Safety, *Impact of fan ventilation systems on sprinkler activation*, Elsevier, 2014

[15] J. Outinen, J. Samec, Z. Sokol, *Research on fire protection methods and case study "Futurum"*, Elsevier, 2012

Numerical simulation and comparison of two ventilation methods for a restaurant – displacement vs mixed flow ventilation

George Chitaru[1, *], *Charles*Berville[1] and *Angel* Dogeanu[1]

[1]UTCB Faculty of Building Services, 021414 PacheProtopopescu Blvd., Bucharest, Romania

Abstract. This paper presents a comparison between a displacement ventilation method and a mixed flow ventilation method using computational fluid dynamics (CFD) approach. The paper analyses different aspects of the two systems, like the draft effect in certain areas, the air temperatureand velocity distribution in the occupied zone. The results highlighted that the displacement ventilation system presents an advantage for the current scenario, due to the increased buoyancy driven flows caused by the interior heat sources. For the displacement ventilation case the draft effect was less prone to appear in the occupied zone but the high heat emissions from the interior sources have increased the temperature gradient in the occupied zone. Both systems have been studied in similar conditions, concentrating only on the flow patterns for each case.

1 INTRODUCTION

Since the energy crisis in the 1970s the criteria for designing the HVAC systems has constantly changed[1].Due to the petroleum shortage, certain countries that were affectedhave implemented quite drastic measures to reduce the energy consumption of the buildings[2]. These actions have had a negative impact on the comfort of the occupants due to the reduced functionality and size of the heating and ventilation systems[3].Nowadays, there is a high interest in properly designing the HVAC systems so that a desired quality of indoor air is achieved with minimal energy consumption[4]. Most often, this balance is hard to be obtained as a higher comfort level usually demands higher energy costs[5], but certain systems are more suitable than others for specific situations[6]. Therefore, an engineer's choice can have agreat impact not only on the energy consumption of the building but also on the wellbeing of the occupants.

Taking into consideration the difficulties that appear when choosing the type of the HVAC system, the current paper comprises a numerical only study that focuses on the comparison between two common ventilation systems: a displacement ventilation system and a mixing ventilation system. Both these systems will be used to provide the necessary cooling load and fresh air during a summer scenario for a restaurant type building. The performance of each of these systems will be evaluated in the same environmental conditions (interior volume, number of occupants, cooling loads, occupant placement, etc.)based on the ventilation efficiency and the comfort level of the occupants. The comfort level is determined by certain factors like the temperature distribution/gradient in the occupancy zone, the draft rate near the tables zoneand by the velocity fields near the occupant area. The inclusion of these parameters allows to achieve a holistic view of the efficiency of each ventilation procedure which, correlated with the costs of each system, can offer important information regarding the viability of the solution before putting into practice. Since the study is focused on determining the velocity and the temperature distribution in an open space area, with a non-uniform geometry, it was suitable to use a computational fluid dynamics program along with the finite volume method approach in order tocorrectly evaluate each case. The numerical simulations have been carried out with the academic version of the ANSYS AIM 18 software;therefore,certain simplifications to the model were necessary due to resource limitations.

2 PHYSICAL MODEL

2.1 Building's presentation

The current study concerns a self-service restaurant with the following characteristics:

* Corresponding author: chitarumadalin@gmail.com

- Activity: Self-service restaurant
- Location: Constanta, Romania
- Occupancy: 302 [pers.] (75 [pers.] *)
- Area: 1253 m² (262 m² *)
- Ceiling height: H=3.2m

For the building in question the heating, cooling and humidity loads have been previously determined using the Romanian Standard I5-2010 [7]for the corresponding climate region:

- Cooling loads (summer): 83 [kW] (20 [kW]*)
- Heating loads (winter): 18 [kW] (4.5 [kW]*)
- Humidity loads during summer: 14 [g/s] (3.5 [g/s] *)
- Humidity loads during winter: 5 [g/s] (1.2[g/s] *)

**The values presented between ( ) are the values used for the virtual model, which accounts for only a quarter of the real building.*

Fig. 1 Model of the real building

2.2 Air parameters

In order to properly size the ventilationunits for the virtual model, it was necessary to determine the external and internal air parameters and to also calculate therequired supply airflow needed to obtain an adequateair change rate.
The calculation procedure has been taken from the Romanian Standard I5-2010, as it follows:
The set point for the internal temperature has been considered atθ_c =26 [°C] with a relative humidity of 50%.
To calculate de required airflow rate of fresh air we used the relations below:

- Recommended specific flow per person: $q_{pers} = 25\ [m^3/(h*pers)]$
- Recommended specific flow per unit area: $q_{surf} = 2.52\ [m^3/(h*m^2)]$
- Total floor area: $S_{floor} = 262\ [m^2]$
- Fresh air flow required:

$$D_{FA} = 75[pers]*q_{pers} + q_{surf} * S_{floor} \quad (1)$$

$$D_{AF} = 2535\ [m^3/h]$$

2.3 Heat gains

Due to the activity inside the building, it was considered that,a large amount of the heat and humidity gains will come from the people and the food that is served. Moreover, significant heat gains would come from the exterior walls due to thedirect and diffuse solar radiation that enters through the glazing on the sides.
For the heat gains per person, for a restaurant, the Standard I5-2010 recommends a higher value that includes the heat gains from the warm food that is served:

- Heat gains from people and food: $Q_{(pers+nourr)}=12\ [kW]$
- Heat gains from glazed walls: $Q_{(par\ vitres)} =7.5\ [kW]$
- Heat gains from the inner walls: $Q_{(par\ int)} =0.5\ [kW]$

2.4 Ventilation systems

As it was said in the introduction, in this current study two common ventilation systems will be compared: a displacement ventilation system and a mixing ventilation system. In order to have a better understanding of these two systems and the choices concerning the sizing, in the following section the calculations that were used will be presented. Considering that the conditions for the two systems are the same, the same air flow calculation will apply for both cases.

2.4.1 Sizing and calculation

First, by knowing the cooling and humidity loads for the summer season, the indoor air parameters and the introduced air parameters have been determined using the Mollier diagram.

Table 1 Air parameters determined with theMollier diagram

Parameters	t	x	φ	h
unit	[°C]	[g/kg]	[%]	[kj/kg]
I_v	26	11.5	50	53.5
C_v	20	9	60	42.5

Where:C_v is the blowing temperature and I_v is the indoor comfort temperature.

With these parameters, the total supply airflow required to maintain the indoor temperature at 26 [°C] can be calculated. For this calculation, two formulas are proposed,from which the most important value will be chosen for the design. The first one is based on the enthalpy of the air and the other one on the humidity of the air:

$$D_{intake} = Q_v / \Delta h\ \ [kg/s] \quad (2)$$

$$D_{intake} = G_v / \Delta x\ \ [kg/s] \quad (3)$$

Where: Q_v is the total heating loads [kW], $\Delta h$is the enthalpy difference [kj/kg], G_v is the humidity gain [g/s] and Δx [g/kg] humidity content difference.

The maximum of the two airflow results has been chosen, which in our case corresponds to the necessary value to eliminate the excess humidity content. With the calculated air flow, the air change rate can be determined and then compared to the recommended values of the Standard I5-2010 to verify the choice.

In this case, the standard I5-2010 recommends a value between 5-10 [h^{-1}] for non-smoking rooms, and our result was 8.3 [h^{-1}]. Next, the value for the supply air flow rate has been calculated for each of the ventilation units.

2.4.2 Mixing ventilation

The mixing ventilation is the most widespread ventilation system technology and it is characterized by supplying air in such a mannerthat the entire room volume is fully mixed.
In the mixing ventilation case, the fresh air is introducedand extracted through the ceiling after it has been mixed with the interior air[8].
The dimensions used for the inlets and outlets are the following:

- Inlets: 0.5*0.43 [m]
- Outlets: 0.5 [m^2]

Since, for this case, the outlets and inlets are both placed on the ceiling it was taken into consideration that a sufficient gap between the intake and extraction grilles is necessary to avoid a potential negative interaction.

2.4.3 Displacement ventilation

Displacement ventilation systems are characterized by an introduction of the air at low velocity, which causes minimal induction and mixing. Displacement inlets introduce the air through the lower part of the room and outlets are located inthe higher part of the room. Displacement ventilation systems use the buoyancy forces generated by the interior heat sources like the occupants. This way, polluted air and internal heats gains are removed more efficiently [9].
Considering the dimensions of the room, it was decided that four displacement unitsshould be installed (one in each corner of the room) as it presented in Fig.2.
To properly choose the dimensions of the units, the inlet air flow on each unit must first be calculated, taking into account the thermal loads, the parameters of the air introduced and the parameters of the indoor air. As previously specified, the total airflow will remain the same as for the mixing ventilation.In addition, for displacement ventilation, the velocity of the introduced air must be at very low valuesin order to avoid areas of discomfort for the occupants. Therefore, a proper dimensioning of the units was needed.

Each inlet unit has a height of 2.1m, and the surface depends on the placement in the room:

- Inlet unit 1: $S_{(unit\ 1)} = 2.87$ [m^2]
- Inlet unit 2: $S_{(unit\ 2)} = 2.87$ [m^2]
- Inlet unit 3: $S_{(unit\ 3)} = 3.5$ [m^2]
- Inlet unit 4: $S_{(unit\ 4)} = 1.97$ [m^2]

The outlets are the same as for the mixing ventilation case: 0.5 [m^2].

3 MATHEMATICAL MODEL

The geometry of the virtual model has been realized with the integrated ANSYS AIM module of Space Claim. In order to decrease the simulation time, and the complexity of the model, it was considered best to use a simplified version of the real building, therefore the model only contains a quarter of the open restaurant area.

The virtual building model presents the following characteristics:

- Two exterior walls: 20m + 13m
- Two adjacent walls: 26.5m + 11.25m
- Ceiling height: 3.2m
- A ground floor: 262 m^2
- A roof top: 262 m^2

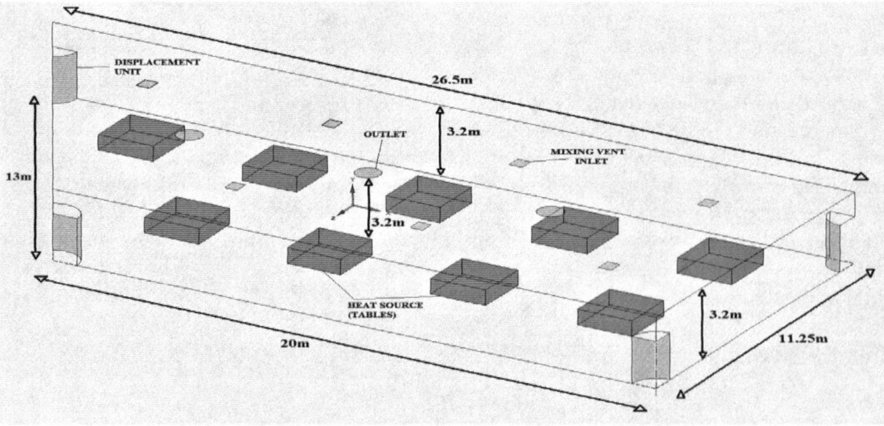

Fig. 2 Virtual model of the restaurant (1/4 of the real building)

The meshof the virtual model has been created using an automated tetrahedral mesh generation with element size constraints imposed on important boundary surfaces: inlets, outlets and heat sources (face element size set to 0.05m). The mesh resolution has been set on its highest setting, resulting in aminimum element size of 0.003 [m], a maximum face size of 0.29 [m] and a maximum element size of 0.58[m]. No inflation has been added near the walls as there is no interest in studying the boundary-layer flow which would have increased the complexity of the model. The resulting mesh, and its characteristics, for each of the two cases can be seen in Fig. 3 and Fig.4 below:

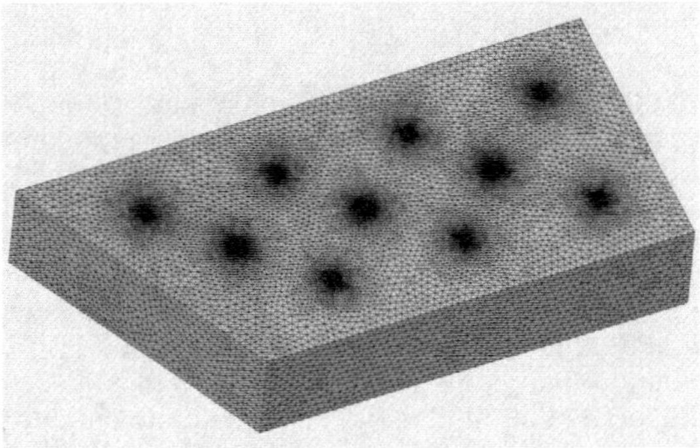

Fig. 3 Mixing ventilation mesh (2.686.615 Cells, 480086 Nodes, Average Element Quality 0.84, Max Aspect Ratio 13.2, Average Skewness 0.21)

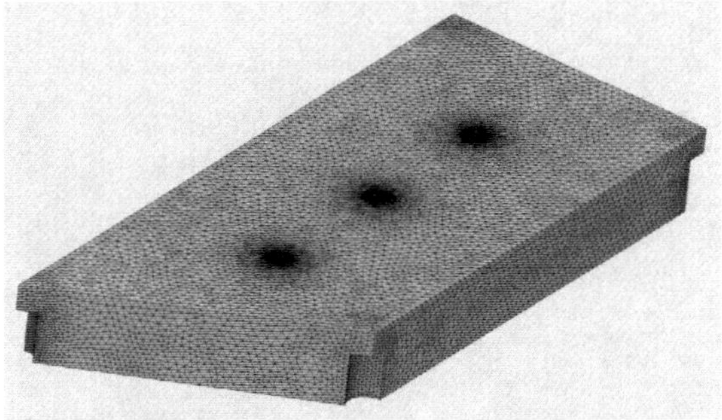

Fig. 4 Displacement ventilation mesh (2.698.948 Cells, 612979 Nodes, Average Element Quality 0.77, Max Aspect Ratio 32, Average Skewness 0.22)

Considering the assumptions made in the previous paragraphs, for this study, a steady-state numerical simulation was used to determine the velocity and temperature distribution for a given situation. The Reynolds-Averaged Navier Stokes (RANS) equations were used, along with the Standard k-εturbulence model. To account for the natural convection, the Boussinesq approximation has been used to determine the buoyancy driven flows inside the room (especially near and above the tables). Next, the boundary conditions have been imposed for each of themodel's elements. The building envelope has been considered as adiabatic since the approximation of the heat-flux through each element could affect the behavior of the model.

The configuration for the numerical solver is presented in Table 2 below:

Table 2. Mathematical Solver Configuration

Physics/Setting	*Model/ Value*
Calculation type	Static/Steady
Turbulence	RANS
	k-ε Standard model
Energy	Heat Transfer Model
Flow (Buoyancy)	Boussinesq approximation
Gas properties	Default air properties (ρ=1.183 [kg/m^3] ; c_p= 1006.30813404 [J/kg* K]; μ=1.84480821367e-05 [Pa*s]
Numerical controls (Discretization and Solution controls)	Green Gauss Cell Based Gradient Method
	Second Order Upwind Advection
	PRESTO Pressure Scheme
	Coupled Pressure Velocity Coupling
	0.5 High Order Term Relaxation Factor
Initial conditions	T=20 [°C]
	P=101325 [Pa]

Therefore, all the heat gains that are supposed to come from the exterior walls (roof and floor included), and from the adjacent interior walls, have been introduced as an internal heat source that is uniformly distributed throughout the whole interior volume (P_i=8000 W).
On the other hand, the interior heat gains from the occupants and the food that is served (heat gains from electric/electronic devices, household appliances, etc. are not taken into consideration) are equally divided between 9 parallelepipeds with dimensions of 1.8mx1.8mx0.7m (LxWxH), which results in an approximate heat flux of:

$$\Phi_{(food+occ)}=12000\ W/54.36\ m^2=220\ [W/m^2] \quad (3)$$

Each parallelepiped is considered as a table with 8 persons placed around it and for each person an additional heat flux due to warm food is taken into account.

The boundary conditions for each of the surfaces are presented in Table 3.

Table 3. Boundary conditions for each case

Boundary	Displacement	Mixing
Inlet(Velocity)	v=0.2 [m/s]	v=1.29 [m/s]
Outlet (Pressure)	$P= P_{atm}$	$P= P_{atm}$
Heat source term (interior volume)	P_i=8000 [W]	P_i=8000 [W]
Ext + Int Walls	Insulated (adiabatic)	Insulated (adiabatic)
Heat source term (Food+Occupants)	$P_{occ+food}$=12000 [W]	$P_{occ+food}$=12000 [W]
Ground floor	Insulated (adiabatic)	Insulated (adiabatic)

4 RESULTS

For both cases the temperature and velocity fields were analyzed for different height planes, each one corresponding to a human body zone which is prone to draft effects: 0.1m for the feet zone, 0.6m for the pelvic region, 1.1m for the head zone at seated position and 1.7m for the head zone at standing position[10]. The interest was to determine an approximate draft sensation rate in certain zones of the room where the combined effect of high velocity, low temperature and high turbulence could affect the occupant comfort. The relation used to calculate the draft effect[11] in these regions is:

$$DR = (34 - t_a)*(v - 0.05)*0.62)*(0.37* v* T_u+3.14) \quad (4)$$

Where: $t_a$is the average temperature of the air in the region, v is the average velocity of the air in the region and T_u is the local turbulence intensity

Table 4. Draft results for targeted areas

Plane height [m]	Displacement			Mixing		
	Average velocity* m/s	Average temperature* °C	Draught sensation** %	Average velocity* m/s	Average temperature* °C	Draught sensation** %
Feet 0.1m	0.25	21	5.08	0.35	24	5.86
Body 0.6m	0.2	22	3.51	0.3	25	4.39
Head (Seated) 1.1m	0.15	25	1.76	0.4	25.5	5.81
Head (Standing) 1.7m	0.07	26.5	0.29	0.45	26	6.26

*Values calculated beneath air diffusers for the mixing ventilation case and above or around the tables for the displacement ventilation case. The most uncomfortable zones have been taken into consideration to avoid using average values that take into accountregions of no occupancy (room corners, entrance, etc.)
** Value calculated with a Turbulent Intensity of 10%

Analyzing the above table, it can be said that the displacement ventilation system presents a considerable reduction in draft sensation compared to the mixing ventilation;therefore, itmight be more suited for buildings that require high performance in terms of indoor environmental quality. Even though the draft percentage is not very high in any of the cases, for the mixing ventilation system there are certain "hot spots" where the level of discomfort can increase dramatically.

As it can be seen in the figures presented below, for the mixing ventilation case, the high inlet velocity combined with the high turbulence of the flow might increase the draft sensation in the occupant area due to increased heat losses through body parts that are not covered by any clothes. It can be seen in the iso-surfaces in Fig. 5 that the flow arrives at the head level of seated occupants (1.1m) at a velocity of 0.3-0.4 m/s with a temperature as low as22°C, right beneath the diffuser. In comparison, the ISO 7730 recommends an average velocity of 0.1-0.15m/s in the occupant area. The situation is even worse if we analyze the differences for a person standing right beneath an air diffuser (velocity between 0.5-0.7 m/s at a temperature of 22-26°C).Of course, the direction and position of the diffusers relative to the occupants is an important factor[8]. The direction of the flow in this case has been considered normal to the boundary surface (flow in the direction of the gravity).

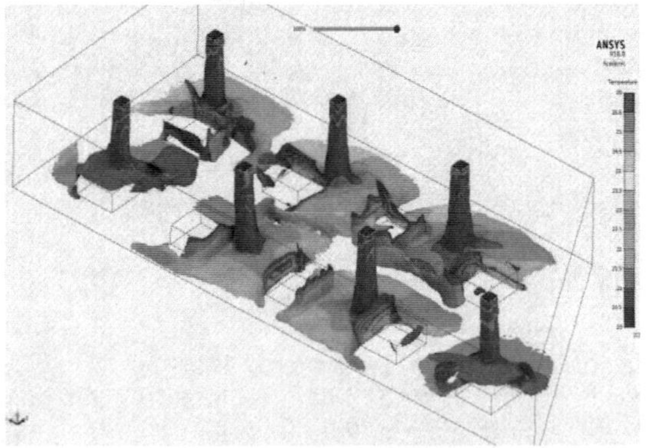

Fig. 5 Iso-surface Velocity-Temperature coupling for mixed ventilation method (Blue = 20°C – Red =26°C)

A better representation of the cold air courant that comes from the mixed ventilation inlets can be seen in Fig. 6. Here it can also be seen that the temperature distribution on the height of the room is more uniform (gradient of only 2°C in occupant zone). But, in the end,the overall uniformity highly depends on the positioning of the inlets and on the angle of introduction. Therefore, a more exhaustive analysis of this method should be done in order to have a thorough comparison.On the other hand, for the displacement ventilation case, the draft sensation is drastically reduced due to the low inlet velocity and lower turbulence intensity near the occupant area.On the other hand, for the displacement ventilation case, the draft sensation is drastically reduced due to the low inlet velocity and lower turbulence intensity near the occupant area.

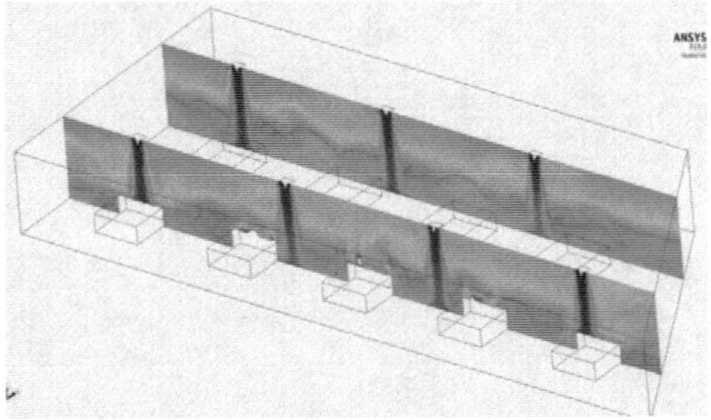

Fig. 6 Temperature contours beneath the mixed ventilation inlets (Blue = 20 °C, Red = 30 °C)

The inlet temperatures are also usually higher, but in this case the loads where high enough to go as low as 20 [°C] for the supply air.Due to this fact, higher discomfort can only be found at feet level as the velocity of the air is higher near the floor level and the temperature of the fluid is low.As it can be seen in the figures below, the iso-surfaces show a more evenly distributed temperature across the room, while the gradient in the occupant zone increases from 24-26°C to 21-26.5°C. In Fig. 7 it can be observed how the fresh air that enters the room, and ventilate the entire area where the occupants are eating and then how the buoyancy forces caused by the people and the warm food drive the contaminated air towards the ceiling where it is extracted.

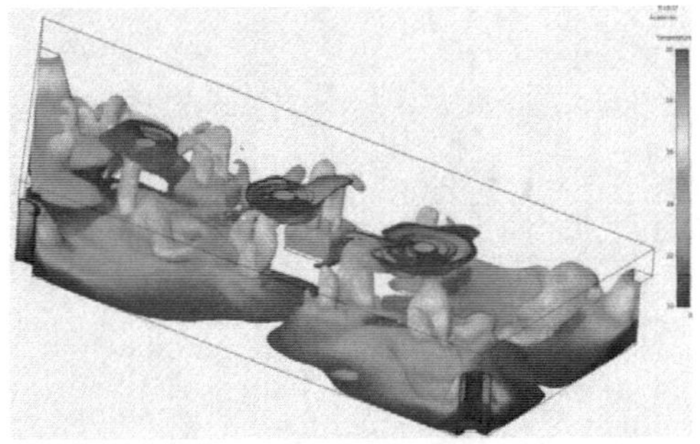

Fig. 7 Iso-surface Velocity-Temperature coupling for displacement ventilation method (Blue = 20°C – Red =30°C)

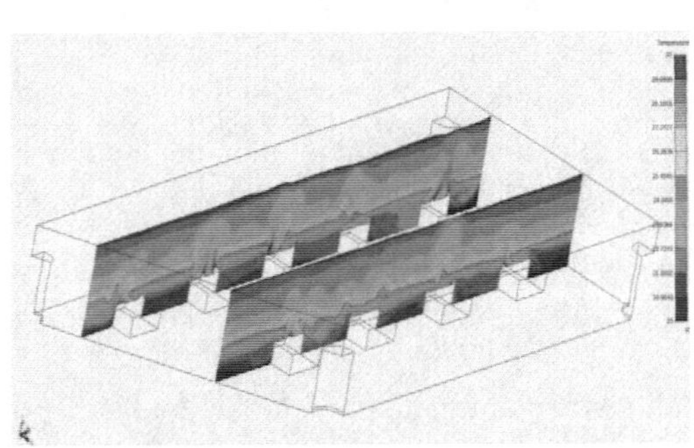

Fig. 8 Temperature contours above the tables for displacement ventilation (Blue = 20°C – Red =30°C)

In Fig. 8, temperature distribution across the length of the room is almost uniform, and how the temperature increases gradually with the height, arriving at almost 30°C near the ceiling.

5 CONCLUSION

In the conditions of this paper, the displacement ventilation system has shown some advantages over the mixed flow ventilation system, concerning the comfort level in the occupant area. Taking into account these results and the complementary advantages of a displacement system (reduced noise, energy consumption), it can be said that this type of ventilation is well suited for an open-space restaurant area.

The current study has only shown a possibility for the comparison of these two systems, therefore a more thorough numerical and experimental investigation is necessary in order to fully support these claims.

This work was supported by a Grant of the Romanian National Authority for Scientific Research, CNCS, UEFISCDI, Project code: PN-III-P2-2.1-BG-2016-0158.

REFERENCES

1. Courtney, T. *The energy crisis of the 1970's*. WSU Online Library, 2015.
2. W. Burleigh Seaver, A.H.P., *Decreasing fuel-oil consumption through feedback and social commendation.* Journal of applied behavior analysis, 1976. **9**(2): p. 147-152.
3. Papadopoulos, A.M., *Energy cost and its impact on regulating the buildings' energy behavior*, T. AUTh, Greece, Editor 2011.
4. EPA, U.S., *Indoor Environments Division, Energy Cost and IAQ Performance of Ventilation Systems and Controls* 2000.
5. Yun, G.Y., *Influences of perceived control on thermal comfort and energy use in buildings.* Energy and Buildings, 2017. **10**(1).
6. Cristina Becchio, S.P.C., Michele Vio, Giulia Crespi, Leonardo Prendin, Micaela Ranieri, DavideVidotto, *Toward NZEB by optimizing HVAC system configuration in different climates*, in *AiCARR 50th International Congress; Beyond NZEB Buildings*, Elsevier, Editor 2017, Energy Procedia. p. 115-126.
7. MDRAP, *I5 - National norm for design, execution and exploatation of HVAC systems*, 2010.
8. Limited, P.I., *Price Engineer's HVAC Handbook: A Comprehensive Guide to HVAC Fundamentals*, 2011.
9. Limited, P.I., *Engineering Guide: Displacement Ventilation*2016.
10. ASHRAE, *''Thermal environmental conditions for human occupancy,'' ANSI/ASHRAE Standard 55-2004, American Society of Heating, Refrigerating and Air-Conditioning Engineers, Atlanta, GA,* 2004.

11. *ISO 7730 - Ergonomics of the thermal environment—Analytical determination and interpretation of thermal comfort using calculation of the PMV and PPD indices and local thermal comfort criteria*. 2005.

E3S Web of Conferences **32**, 01013 (2018)
EENVIRO 2017
https://doi.org/10.1051/e3sconf/20183201013

Impinging jets – a short review on strategies for heat transfer enhancement

Ilinca Nastase[1,*] and *Florin* Bode[1,2]

[1]CAMBI, Technical University of Civil Engineering in Bucharest, Building Services Faculty, 020396, Bucharest, Romania
[2]Technical University of Cluj Napoca, Mechanical Engineering Faculty, Cluj-Napoca, Romania

Abstract. In industrial applications, heat and mass transfer can be considerably increased using impinging jets. A large number of flow phenomena will be generated by the impinging flow, such as: large scale structures, large curvature involving strong shear and normal stresses, stagnation in the wall boundary layers, heat transfer with the impinged wall, small scale turbulent mixing. All these phenomena are highly unsteady and even if nowadays a substantial number of studies in the literature are dedicated, the impinging jets are still not fully understood due to the highly unsteady nature and more over due to great difficulty of performing detailed numerical and experimental investigations.

1 Introduction

In industrial applications, heat and mass transfer can be considerably increased using impinging jets. Impinging jets can produce very high heat/mass transfer rates since they exhibit amongst the highest known levels of transfer capabilities for single phase flows, especially at low nozzle-to-plate distances [1, 2]. An oriented fluid flow versus a surface can transfer substantial amounts of mass and heat between that surface and the fluid. For example, compared with conventional convection cooling by confined flow parallel to the cooled surface, the use of jet impingement produces in increase in the heat transfer coefficients that are up to three times higher at a given maximum flow speed. Applications to engineering fields include turbine blade cooling [3], paper and fabric drying [4], furnace heating [5], tempering of glass and metal sheets [6, 7], food processing [8] and many others. Heat transfer enhanced by forced convection with impinging jets is known to yield high local and average heat transfer coefficients in the cooling of electronic components [9]. In this field, technological advance relies very much on the ability to dissipate extremely large heat fluxes [10]. Thus, interest in the present topic, from the standpoint of both theoretical and technical assessments, continues to be unquestioned.

The investigation tools, either experimental or numerical, have undergone tremendous developments which allow for a deeper comprehension of the involved phenomena. Since Martin [11] published in 1977 his pioneering review article on heat and mass transfer of impinging jets, several additional literature reviews have appeared [12-15]. In the case of the impinging jet, turbulence is generated by the flow itself and by possible external disturbances and varies significantly with the nozzle shape, the upstream conditions and the position within the jet [16]. The details of the nozzles, and of the upstream flow facility, is important in the design of heat transfer equipment in which turbulent jets impinge at relatively short nozzle-to-plate spacing H/D, but are secondary if impingement occurs beyond H/D > 8(Fig. 1).

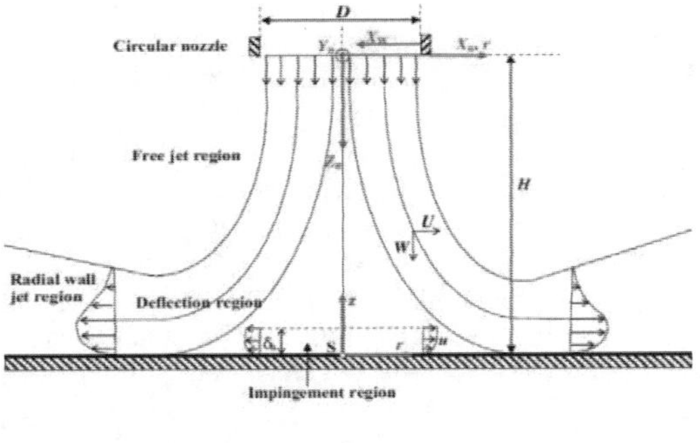

Fig. 1 Schematic description of impinging jet on a flat plate [17]

2 Some aspects related to the physics of the impinging jets

The commonly used designs of jet nozzle orifices are either circular nozzle, producing an axisymmetric velocity profile, or slitdiffusersproducing a wide thin jet mainly having a plane two-dimensional velocity profile.

The axisymmetric and plane free jets have been extensively studied in the past and their mean flow is described in reference works on free jets [18, 19]. Other review works are available on their impinging counterparts [20-22].

* Corresponding author: ilinca.nastase@utcb.ro

A schematic representation of an axisymmetric impinging jet is represented in Fig.1. A fluid flow with uniform velocity issued from a circular nozzle and impinging orthogonally on a wallis considered in this case. As shown in Fig. 1, the flow of a submerged impinging jet elapses through several different regions, some of which may not exist according to the distance between the nozzle originating the jet and the impinging plate H.

A jet emerges from a nozzle with characteristicmean and turbulent velocity and temperature profiles dependent upon the upstream flow conditions and nozzle shape[19]. For a pipe-shaped nozzle, the flow generally develops into the velocity profile typical of pipe flows with a certain turbulence, initialized upstream. For a flow generated by a differential pressure across a thin plate with an orifice, just downstream of this orifice a vena contractaphenomenon is produced, which is characterized by an initial flow having a flat velocity profile with low turbulence intensity. After exiting the nozzle, an impinging jet passes through a region where it is sufficiently far from the impingement surface and so behaves as a free submerged jet. Generally, the observation of the circular free jet emerging from a diffuser in an immobile environment, or itself animated with a velocity lower than the one the jet, shows a widening of the flow starting from the exit plane. It is common to distinguish three regions away from the jet outlet and on its axis: a potential core region, a settlement region, and a fully established region [19, 23-25]. In the region of the potential core at the confluence of the two fluids, an annular mixing zone of increasing thickness is generated, delimiting a so-called potential cone within which the flow has the same conditions as those of the jet at its outlet (velocity, temperature, density ...). A geometric definition of the potential cone, also called potential core, would be given by the place where the longitudinal mean velocity U would be 0.99U0, where U0 is the axial longitudinal velocity at the exit plane. Its length varies for a circular jet between 4 and 6 diameters of the nozzle [19, 23, 26].

The mixing layer around the annular potential cone is the seat of a strong shear. On either side of the potential core, the velocity decreases to join to the lower speeds of the environment. This connection is made by braking the velocity of the potential cone through the entrainement of the ambient fluid that mixes with the injected fluid[19]. In this zone, the average longitudinal components of velocity and of its turbulent component obey to similitude laws [18, 19, 27]. In addition, the turbulent energy is concentrated around the axis of the mixing layer where the radial gradients of the velocity are maximal. We find there vortices centered on the axis of the mixing layer whose size increases with the longitudinal distance X. These vortices initially have a spatio-temporal coherence [28].

In the proximity of the plate, the axial velocity of the flow decays quickly and the jet is forced to undergo a sudden change of direction in a region called stagnation region.In the neighbourhood of the impinging point S the flow spreads in radial directions parallel to the wall. A stagnation point with zero velocity exists on the wall in the intersection with the jet axis. The axial velocity component decreases and is transformed into an accelerated horizontal component. As a consequence, a high static pressure is created, over and above the wall, which transmits upstream the presence of the latter. Downstream the stagnation region, the flow spreads radially outward almost parallel to the impinged plate and develops into a semi-confined flow. Due to the jet finite width and the momentumexchange with quiescent surroundings and plate, the stagnation flow finally evolves into a wall jet flow (wall jet region).

As the plate gets closer to the nozzle exit, some of the abovereported regions are no longer present. First, the jet fully developedregion starts to disappear and then, the whole decaying jet regionmay not exist. If the plate lies within a distance of about two nozzlediameters, the high static pressure in the stagnation region hasa significant influence on the flow development up to the nozzle exit; therefore, also the usual quasi-conical shape of the core at small H/D values, thevelocity profile of the jet does not have sufficient place to developand the arrival velocity is essentially uniform.

2Passive and active strategies for heat transfer enhancement

Instability of the free shear layer downstream of the nozzle exit and turbulence development in the jet near field, are of fundamental importance in mass and heat transfer of impinging jet. Gardon and Akrifat [169], have studied the effect of turbulence on the heat transfer between two-dimensional jet and flat plate. It was shown that some seemingly anomalous heat-transfer phenomena can be explained as effects of the turbulence occurring in jets. Turbulence is generated by the jets themselves and by possible external disturbances and varies significantly with the nozzle shape, the upstream conditions and the position within the jet. The results presented by these authors suggest that details of the nozzles, and of the upstream flow facility, may be important in the design of heat transfer equipment in which turbulent jets impinge at relatively short nozzle-to-plate spacing H/d, but are secondary if impingement occurs beyond H/d = 8.

The great implications that vortical features can have on surface heat transfer rate and distribution motivated the flourishing of countless investigations devoted to passive and active strategies to enhance the heat flux between a jet and the impinged surface. Passive strategies are mainly based on the shape of the nozzle [29-31].

The geometries of the nozzles became progressively complex in the past decades in order to enhance the self-induction of the asymmetric coherent structures [14-18]. The earliest studies conducted on rectangular jets revealed significant information on the structure and the dynamics of this type of jet flow. An interesting phenomenon was particularly discovered, namely, the switching of the major and minor axis of the jet. Krothapalli et al. [22] proposed a linear dependence between the distance from the nozzle to the first cross-over location and the nozzle aspect ratio. Later, Quinn [26] showed the existence of four pairs of counter rotating streamwise vortices next to the exit of the rectangular jet. These structures play an important part in the momentum transport. This author also reported a distortion of the jet's transverse geometry from a rectangular shape in the initial region to an elliptic shape in the far field. This phenomenon would be explained by the distortion of the azimuthal roll-up structures in addition to the nonuniform self-induction in the shear layer.

Furthermore, Hussain and Husain [15] showed that elliptic jets, unlike circular or plane jets, are characterized by a variation of the vortical structures azimuthal curvature, leading to non-uniform self-induction and consequently to a complex three-dimensional distortion that results in axis switching. They also emphasized a shorter potential core compared to the reference circular jet and higher levels of turbulent kinetic energy in this region. In a continuous effort to increase the spreading and mixing performance of the jet flows, other various methods have been explored. Rectangular and circular tabbed nozzles displayed a particularly improved mixing efficiency compared to the same geometries without mechanical tab intrusions [18, 28-30]. Each tab produces a pair of counter rotating streamwise vortices modifying the turbulent structure and increasing mixing in the jet flow.Geometrical manipulation of diffuser nozzles has been subsequently oriented to more complex shapes [31, 32], and it has been suggested in recent work of Hu et al. [33-35], that the lobed nozzle is an extraordinary mixing device.

The previous observations have been exploited for imping jet applications. For instance, nozzles with chamfered outlets produce 20–30% increase of the heat transfer rate when compared to non-chamfered ones [32]. Variations from circular to elliptic nozzle exit section shapes are studied in [33], where, for an elliptic cross sectional area nozzle with aspect ratio equal to 4, a maximum stagnation heat transfer 15% larger than for the circular configuration is found and a higher enhancement is found elsewhere. Other passive strategies are related to the use of vortex and turbulence generators in the nozzle geometry. In particular, Gao et al. [34] show that, for a nozzle-to-plate distance of 4 diameters, triangular tabs placed around a circular orifice (chevron nozzle) lead to an enhancement higher than 25% with respect to the round configuration for the same nozzle-to-plate distance. With respect to the circular impinging jet, a cross-shaped lobed jet [16, 35] attain the highest average heat transfer rate in the considered range of the Reynolds number compared with other devices in the literature, with very high heat transfer peaks (up to 75% enhancement at $H/D = 2$).

Mixing and turbulence enhancement can be also promoted directly on the impinged surface with structure modification, for instance with fins having a triangular, square or rectangular cross section - an increase of the average Nusselt number value by 12–23% [36] can be produced. Surface roughness, in the form of cubes, produce heat transfer augmentation in the range of 8–28% [37] while a dimpled surface is reported to reduce the heat transfer coefficient as compared to a smooth on [38]. Slot jet impinging on square and triangular ribbed walls are analysed in [39, 40]. The reported increases may be due to combined effect of two factors, namely: the enhanced turbulence mixing caused by the rib distortion of the flow fields and the extension of the heat transfer surfaces (fin effect) provided by the ribs. In [41], vortex generators in the form of equilateral triangles are analysed and it is observed that the heat transfer coefficient between the impinging jet and the target plate is sensitive to the shape of the fin with an increase of the heat transfer coefficient up to 77%. The augmentation of the heat transfer for the surfaces vortex generators is higher than that of the finned surfaces.

The active control methods generally use the acoustic [42-46]or mechanical [47-52]excitation to the exit plane of the flow, which favorsunder certain conditions, the amplification of the instabilities and the appearance of large highly energetic vortex structures that are generating mixing enhancement.Liu and Sullivan [53] studied the forcing of the impinging jet with flow visualization as well as velocity and heat transfer spectra from hot wire and hot-film sensors. Both heat transfer enhancement and reduction are found and are related to the development of largescale vortical structures in the wall jet region. When the excitation frequency is close to the natural frequency of the impinging jet, the initiated intermittent vortex pairing produces a chaotic lump eddy which contains a great deal of the small scale random turbulence which enhances the local heat transfer. Once the forcing is nearthe sub-harmonic of the natural frequency, stable vortex pairing is promoted. The strong large scale well organized vortices, formed after the stable pairing, induce the unsteady separation of the wall boundary layer and, hence, lead to the local heat transfer reduction.

3Advanced experimental methods and perspectives

A large number of flow phenomena will be generated by the impinging flow, such as: large scale structures, large curvature involving strong shear and normal stresses, stagnation in the wall boundary layers, heat transfer with the impinged wall, small scale turbulent mixing. All these phenomena are highly unsteady and even if nowadays a substantial number of studies in the literature tried to illuminate this matter, the impinging jets are still not fully understood due to the highly unsteady nature and more over due to great difficulty of performing detailed numerical and experimental investigations [62].

There are several studies on the mass transfer between a plate and an impinging jet using limiting diffusion current technique. The study of Chin and Tsang[54] considered the mass transfer between an impinging jet and a circular disk electrode located in the stagnation region of the flow. They found that the electrode possesses a property of "uniform accessibility" to the diffusing species if the electrode radius is less than one nozzle diameter for the turbulent jet and half a nozzle diameter for the laminar jet. The authors have presented semi empirical correlations in both laminar and turbulent flow regimes for nozzle-to-plate distances from 0.2 to 6 nozzle diameters. The wall jet region was found to begin approximately at a radial distance of four nozzle diameters.

For the fluid flow measurement, a suitable technique is Particle Image Velocimetry (PIV). Due to three-dimensional nature of the envisaged configurations (special geometries of the nozzles, impacting surface modification) the recent developments in the volumetric optical techniques such as tomographic PIV provide a suitable approach for the fluid flow analysis. Volumetric PIV is based on the double-pulsed laser illumination of a fluid volume seeded with particles. The particles distribution is reconstructed from multiple camera views and then the volumetric flow field is obtained through volume cross-correlation which allow for the computation also of derivative quantities giving a detailed description of vortex and turbulence dynamics [34]. Due to the laser scattering at the wall surface, velocity measurements with PIV method very close to the impinging surface are neither usually possible nor reliable. In this case, the electrodiffusion method (ED) is the most reliable method for measuring the wall shear rate [35]. The ED method is based on the measurement of the limiting diffusion current on a working electrode (probe) [36]. This technique has been developed for measuring the average rates of mass transfer and their fluctuations on a wall. The transfer area consists of a small polarized electrode flush mounted in the wall. An electrochemical redox reaction whose rate is limited by mass transfer is employed. ED measurements of the wall shear were first applied by Vallis et al. [37].

Yapici et al. [55] measured the surface distributions of the local shear stress values for a fully developed turbulent velocity profile, round, submerged jet, impinging on a flat surface. The results showed that the normalized local shear stresses values decreased with increasing Reynolds number and had a tendency to converge at higher Reynolds numbers. Alekseenko and Markovich [56] measured wall shear stresses of circular jet at a Reynolds number of 41600 and for nozzle-to-plate distances from 2 to 8 nozzle diameters while Baleras et al. [57] studied the effect of the normal velocity on the segment currents of a three-segment circular probe in the vicinity of stagnation point at Reynolds number 640. No study on the wall shear stress and mass transfer for a lobed impinging jet using the electrodiffusion technique was proposed until our paper[16]. Since the lobed orifice jet is characterized by intensified vortical dynamics and turbulent kinetic energy in its initial region, it is believed that such a complex jet flow will also display enhanced mass transfer when considered in impinging configuration. In the following are presented results from our article [16]. In this study we compare, in terms of mass transfer rates, a cross-shaped orifice nozzle to a reference convergent circular nozzle having close equivalent diameter. The knowledge issued from previous analysis of the near field of a cross-shaped orifice jet in free air-air jet configuration [40, 46], is used in an attempt to gain insight into the differences observed for the two studied impinging jets. Indeed, different variations of the radial distribution of wall shear rates and of the mass transfer at the stagnant point were obtained with the lobed nozzle flow relatively to the flow generated by the classical circular convergent nozzle. The wall shear rate of the impinging jets on a flat plate was measured by electro-diffusion technique and mass transfer was calculated from these values in the impingement region.

The difficulty of the experimental investigation arise from fact that flow field is extremely sensitive to external perturbations and hence the impossibility of introducing probes for measuring various quantities of the flow, the flow field being strongly influenced by these probes. In this case it is recommended the usage of nonintrusive research methods. The effect of the impinging jet in the heat and mass transfer for the studied flow is strongly dependent of the wall shear rate [77], hence by measuring this quantity, we can measure the imp act of the impinging jet. Also, the wall shear rate is influenced by the behavior of the flow. Therefore, in order to have a clear image of the impinging jet, we must determine in the same time the flow field and the wall shear rate.

In a first attempt of combining flow velocity and wall shear rate measurements El Hassan et al. [41, 42] performed simultaneously time resolved PIV and ED measurements. They have shown that the instantaneous wall shear stress is strongly dependent on the vortex dynamics. The interaction between primary and secondary structures, and their pairing process with a tertiary preceding structure, affects the ejection process near the wall and subsequently the wall shear stress variation. Recently, the team coordinated by Meslem [43] performed non-synchronized Time Resolved Tomo PIV and ED measurements comparing three circular jets: a convergent nozzle jet, a hemispherical nozzle jet and an orifice plate jet. A link between the wall shear stress and the mass transfer was revealed by the authors. The wall shear rate and the mass transfer were found to be in a close relationship with the near field flow features, themselves affected by the nozzle geometry. The instantaneous PIV fields indicated the formation of secondary vortices in the region where a secondary peak in local mass transfer emerges. The level of this secondary peak was found to be sensitive to the nozzle shape. The limited spatial resolution imposed by the used TR cameras in Tomo PIV configuration did not allowed the investigation of other geometries of nozzles. The vector fields were obtained by performing multi-pass direct cross-correlation using a final interrogation volume of 64x64x64 voxels (0.28D x0.28D x 0.28D). The intricate threedimensionalvortical characteristics of lobed jets for instance would not be capable of being captured by this configuration.

Conclusions and perspectives

In industrial applications, heat and mass transfer can be considerably increased using impinging jets.A large number of flow phenomena will be generated by the impinging flow, such as: large scale structures, large curvature involving strong shear and normal stresses, stagnation in the wall boundary layers, heat transfer with the impinged wall, small scale turbulent mixing. All these phenomena are highly unsteady and even if nowadays a substantial number of studies in the literature are dedicated, the impinging jets are still not fully understood due to the highly unsteady nature and more over due to great difficulty of performing detailed numerical and experimental investigations. If we want to gain knowledge about these fundamental phenomena, to understand how to control them for each practical application, we have to find methods that are adapted to our goal. We believe that the first step in this way is to perform classical volumetric PIV measurements allowing a much higher spatial resolution to capture the three-dimensional characteristics of more complicated impinging jet configurations. These measurements have to be synchronized with ED measurements. In addition, 3D LDV measurements triggered by the analogical ED signal will represent a useful, complementary, high resolution tool, allowing investigating specific regions of interests in the flows. Further understanding is needed about the vortex dynamics and breakdown as well as on the influence of streamwise vorticity. These issues necessarily call for the improvement of inspection capabilities of certain experimental techniques such as ED, IR thermography and volumetric PIV.

Advanced passive strategies such as the use of the lobed nozzles or structured surfaces are not sufficiently studied to provide higher cooling/heating efficiency. Most of the above mentioned issues have to be investigated not only from the experimental point of view but also from the numerical one with further refined methods. The nowadays available computing power allows to perform such theoretical studies without any major difficulties and with much less effort than the experimental ones. Nevertheless, it is obvious that experiments have an indispensable role since the development and validations of theories, models and algorithms still rely on the quality of experiments and call for even more challenging measurement.

Acknowledgement

This work was supported by the grant of the Romanian National Authority for Scientific Research, CNCS, UEFISCDI, PHANTOM - Passive flow control for heat and mass transfer enhancement of impinging jets PN-III-P4-ID-PCE-2016-0758

References

1. Narayanan, V., J. Seyed-Yagoobi, and R.H. Page, An experimental study of fluid mechanics and heat transfer in an impinging slot jet flow. International Journal of Heat and Mass Transfer, 2004. 47: p. 1827–1845.
2. Baydar, E. and Y. Ozmen, An experimental and numerical investigation on a confined impinging air jet at high Reynolds numbers. Applied Thermal Engineering, 2005. 25: p. 409–421.
3. Han, B. and R.J. Goldstein, Jet-impingement heat transfer in gas turbine systems, in Annals of the New York Academy of Sciences. 2001. p. 147-161.
4. Polat, S., Heat and mass transfer in impingement drying. Drying Technology, 1993. 11(6): p. 1147-1176.
5. Viskanta, R., Heat transfer to impinging isothermal gas and flame jets. Experimental Thermal and Fluid Science, 1993. 6(2): p. 111-134.
6. Gardon, R. and J.C. Akfirat, Heat transfer characteristics of impinging two-dimensional air jets. J. Heat Transfer, 1966. 88: p. 101-108.
7. Gardon, R. and J. Cobonpue, 1962: p. 454-460.
8. Sarkar, A., et al., Fluid flow and heat transfer in air jet impingement in food processing. Journal of Food Science, 2004. 69(4): p. CRH113-CRH122.
9. Fabbri, M., S. Jiang, and V.K. Dhir, A comparative study of cooling of high power density electronics using sprays and microjets. Journal of Heat Transfer, 2005. 127(1): p. 38-48.
10. Anwarullah, M., V.V. Rao, and K.V. Sharma, Experimental investigation for enhancement of heat transfer from cooling of electronic components by circular air jet impingement. Heat and Mass Transfer, 2012. 48(9): p. 1627-1635.
11. Martin, H., Heat and Mass Transfer between Impinging Gas Jets and Solid Surfaces, in Advances in Heat Transfer. 1977. p. 1-60.
12. Carlomagno, G.M. and A. Ianiro, Thermo-fluid-dynamics of submerged jets impinging at short nozzle-to-plate distance: A review. Experimental Thermal and Fluid Science, 2014. 58( ): p. 15-35.
13. Hall, J.W. and D. Ewing, On the dynamics of the large-scale structures in round impinging jets. Journal of Fluid Mechanics, 2006. 555: p. 439-458.
14. Harmand, S., et al., Review of fluid flow and convective heat transfer within rotating disk cavities with impinging jet. International Journal of Thermal Sciences, 2013. 67: p. 1-30.
15. Polat, S., et al., Numerical flow and heat transfer under impinging jets: A review. Annual Review of Numerical Fluid Mechanics and Heat Transfer, 1989. 2: p. 157-197.
16. Kristiawan, M., et al., Wall shear rates and mass transfer in impinging jets: Comparison of circular convergent and cross-shaped orifice nozzles. International Journal of Heat and Mass Transfer, 2012. 55(1–3): p. 282-293.
17. Meslem, A., et al., Flow dynamics and mass transfer in impinging circular jet at low Reynolds number. Comparison of convergent and orifice nozzles. International Journal of Heat and Mass Transfer, 2013. 67: p. 25-45.
18. Abramovich, G.N., The theory of turbulent jets. 1963, Cambridge, Massachusets: MIT Press. 668.
19. Rajaratnam, N., Turbulent jets. 1976, Amsterdam, Netherlands: Elsevier Scientific Publishing Company.
20. Martin, H., Heat and mass transfer between impinging gas jets and solid surfaces. Advances Heat Transfer, 1977. 13: p. 1-60.
21. Jambunathan, K., et al., A review of heat transfer data for single circular jet impingement. International Journal of Heat and Fluid Flow, 1992. 13: p. 106-115.

22. Carlomagno, G.M. and A. Ianiro, Thermo-fluid-dynamics of submerged jets impinging at short nozzle-to-plate distance: A review. Experimental Thermal and Fluid Science, 2014. 58(Supplement C): p. 15-35.
23. Awbi, H.B., Ventilation of Buildings. 1991, London, U.K.: E&FN SPON. 313.
24. ASHRAE, ASHRAE Handbook Fundamentals. 1993, Atlanta: American Society of Heating, Refrigerating and Air Conditioning Engineering. Chap. 31.
25. Tuve, G.L., Air velocities in ventilating jets, ASHVE Research Report No.1476, in ASHVE Transactions, 58. 1953. p. 261-283.
26. Oosthuisen, P.H. An experimental study of low Reynolds number turbulent circular jet flow. in ASME Applied Mechanics, Bioengineering and Fluids Engineering Conference,. 1983. Houston, U.S.A.
27. Rajaratnam, N. and B.S. Pani. Turbulent compound annular shear layers. in Proceedings ASCE, J. Hydraulics Division. 1972.
28. Davies, P.A.O.L., M. Fischer, and M.J. Barrat, The caracteristics of the turbulence in the mixing region of a round jet. Journal of Fluid Mechanics, 1963. 15: p. 337-367.
29. Chaudhari, M., B. Puranik, and A. Agrawal, Effect of orifice shape in synthetic jet based impingement cooling. Experimental Thermal and Fluid Science, 2010. 34(2): p. 246-256.
30. Garimella, S.V. and B. Nenaydykh, Nozzle-geometry effects in liquid jet impingement heat transfer. International Journal of Heat and Mass Transfer, 1996. 39(14): p. 2915-2923.
31. Pan, Y., J. Stevens, and B.W. Webb, Effect of nozzle configuration on transport in the stagnation zone of axisymmetric, impinging free-surface liquid jets. Part 2. Local heat transfer. Journal of Heat Transfer, 1992. 114(4): p. 880-886.
32. Brignoni, L.A. and S.V. Garimella, Effects of nozzle-inlet chamfering on pressure drop and heat transfer in confined air jet impingement. International Journal of Heat and Mass Transfer, 2000. 43(7): p. 1133-1139.
33. Lee, J. and S.J. Lee, The effect of nozzle aspect ratio on stagnation region heat transfer characteristics of elliptic impinging jet. International Journal of Heat and Mass Transfer, 2000. 43(4): p. 555-575.
34. Gao, N., H. Sun, and D. Ewing, Heat transfer to impinging round jets with triangular tabs. International Journal of Heat and Mass Transfer, 2003. 46(14): p. 2557-2569.
35. Bode, F., A. Meslem, and C. Croitoru, Numerical simulation of a very low Reynolds cross-shaped jet. Mechanics, 2013. 19(5): p. 512-517.
36. Hansen, L.G. and B.W. Webb, Air jet impingement heat transfer from modified surfaces. International Journal of Heat and Mass Transfer, 1993. 36 p. 989-997.
37. Chakroun, W.M., A.A. Abdel-Rahman, and S.F. Al-Fahed, Heat transfer augmentation for air jet impinged on a rough surface. Applied Thermal Engineering, 1998. 18(12): p. 1225-1241.
38. Ekkad, S.V. and D. Kontrovitz, Jet impingement heat transfer on dimpled target surfaces. International Journal of Heat and Fluid Flow, 2002. 23(1): p. 22-28.
39. Gau, C. and C.C. Lee, Impingement cooling flow structure and heat transfer along rib-roughened walls. International Journal of Heat and Mass Transfer, 1992. 35(11): p. 3009-3020.
40. Gau, C. and I.C. Lee, Flow and impingement cooling heat transfer along triangular rib-roughened walls. International Journal of Heat and Mass Transfer, 2000. 43(24): p. 4405-4418.
41. Nakod, P.M., S.V. Prabhu, and R.P. Vedula, Heat transfer augmentation between impinging circular air jet and flat plate using finned surfaces and vortex generators. Experimental Thermal and Fluid Science, 2008. 32(5): p. 1168-1187.
42. Zaman, K.B.M.Q. and A.K.M.F.Hussain, Vortex pairing in a circular jet under controlled excitation. Part 1. General response. Journal of Fluid Mechanics, 1980. 101(3).
43. Hussain, F. and H.S. Husain, Elliptic jets. Part1. Characteristics of unexcited and excited jets. Journal of Fluid Mechanics, 1989. 208: p. 257-320.
44. Zaman, K.B.M.Q., Axis switching and spreading of an asymmetric jet: the role of coherent structure dynamics. Journal of Fluid Mechanics, 1996. 316(1): p. 1-27.
45. Lai, J.C.S., Turbulence suppression in an elliptic jet. International Journal of Heat and Fluid Flow, 1992. 13(1).
46. Lin, Y.T., et al., Investigation on the mass entrainement of an acoustically controlled elliptic jet. International Communications in Heat and Mass Transfer, 1998. 25(3).
47. Wiltse, J.M. and A. Glezer, Manipulation of free shear flows using piezoelectric actuators. Journal of Fluid Mechanics, 1993. 249(261-285).
48. Parekh, D.E., et al., Innovative jet flow control : Mixing enhancement experiments. AIAA Paper 96-0808, 1996.
49. Suzuki, H., N. Kasagi, and Y. Suzuki, Active control of an axisymmetric jet with distributed electromagnetic flaps actuators. Experiments in Fluids, 2004. 36(498-509): p. 1-43.
50. Davis, M.R., Variable control of jet decay. AIAA Journal, 1982. 20(5).
51. Denis, S., Contrôle du developpement des couches de mélange axisymétriques subsoniques par jets impactant. 2000, Université de Poitiers. p. 280.
52. Collin, E., Etude de l'injection radiale de fluide dans une couche de mélange annulaire supersonique. 2001, Université de Poitiers. p. 282.
53. Liu, T. and J.P. Sullivan, Heat transfer and flow structures in an excited circular impinging jet. International Journal of Heat and Mass Transfer, 1996. 39(17): p. 3695-3706.
54. Chin, D.T. and C.H. Tsang, Mass transfer to an impinging jet electrode. Journal of the Electrochemical Society, 1978.
55. Yapici, S., et al., Surface shear stress for a submerged jet impingement using electrochemical

https://doi.org/10.1051/e3sconf/20183201013

technique. Journal of Applied Electrochemistry, 1999. 29: p. 185-190.

56. Alekseenko, S.V. and D.M. Markovich, Electrodiffusion diagnostics of wall shear stresses in impinging jet. Journal of Applied Electrochemistry, 1994. 24: p. 626-631.

57. Baleras, F., et al., A three-segement electrodiffusion probe in axisymmetric flow with stagnation and separation. Journal of Applied Electrochemistry, 1994. 24: p. 676-684.

Real scale experimental study for performance evaluation of unidirectional air diffuser perforated panels

Laurențiu Tăcutu[1], *Ilinca* Nastase[1*], *Vlad* Iordache[1], *Tiberiu* Catalina[1] *and Cristiana Verona* Croitoru[1]

[1]*CAMBI, Technical University of Civil Engineering in Bucharest, Building Services Department, 66 Avenue PacheProtopopescu; 020396, Bucharest, Romania; * correspondent author*

Abstract. Nowadays, there is an increasing emphasis on indoor air quality due to technological evolution and the fact that people spend most of the time in enclosed spaces. Also, energy efficiency is another related factor that gains more and more attention. Improving air distribution in an enclosure can lead to achieve these goals. This improvement can be done by adjustingthe air terminals position, theredimensions or the air diffuser perforations. The paper presents the study of 8 types of panels with different perforations shapes. The systems were characterized by flow, pressure loss and noise. Usualand special geometries were chosen, all having the same flowsurface. The perforated panels were mounted in a unidirectional air flow (UAF)diffuser, also called a laminar air flow (LAF)diffuser, that is placed in a real scale operating room (OR) in our laboratory.The purpose of this study is to determine whether changing the shape in the perforated panels can improve the technical parameters of the diffuser.

1. Introduction

In modern days most of the buildings, whether they arecommercial, residential or industrial, have mechanical ventilation systems for improving air quality in their premises[1,2]. The systems, commonly referee to as HVAC systems, have the purpose toimprove indoor air quality and thermal comfort.To reach these goal, air parameters and air distribution are adjusted in accordance.Adjusting the air parametersrepresents an approach that is frequently called as an active method, while modifying the air distribution by changing the geometries in the perforated panelsis usuallyreferred as a passive method.Optimizing theair distributionin a space is achieved by adapting the air terminals units to the specifics of the space. This means to choose the optimal position for the air terminals and to select the right dimension and geometry for the diffuser. It is well recognised the fact that air distribution has an influence on thermal comfort, indoor air quality or even energy efficiency[3]. This also influences aerosol distribution, an important parameter in ORs and clean rooms.The paperpresents the study of 8 types of panels with different perforations shapes, mounted in an air terminal unit. For each panel, the following technical parameters were determined: flow, pressure loss, noise.

Analysing these technical parameters represents the first step in determining how the geometry of the perforations in panels affects the air distribution in a room.The perforated panelshave been mounted on a unidirectional air flow (UAF) [4, 5] or laminar air flow (LAF)[6]diffuser that isinstalled in the ventilation system of a real scaleOR in our laboratory.Future research on these panelswill take in to account their performance in terms of thermal comfort. The thermal comfort measurements will bemade in the same climaticchamber. The equivalent temperature, PMV and PPD indices will be determined with the help of two humanoid thermal manikins and a ComfortSense equipment.

2. Experimental method

These measurements were made in a climatic chamber that simulates a real scale OR, which has the dimensions of 3.5x3.5x2.5m (lxLxH), temperature control on each wall and air ventilation system with an UAF diffuser. There were 8 types of panels analysed. The geometries selected were common ones, but also some special ones (Fig.1).For an easier reference, the panels were numbered with letters (Fig. 1, Table 1).

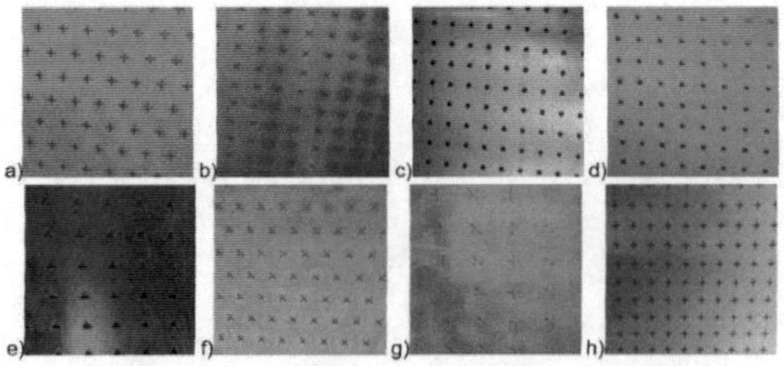

Figure 1. The geometriesof the panels

The special geometries were developed in previous studieswhere it was analysed, by numerical and experimental approaches, the flowing through such shapes[7-9]. Thepanels are made from analuminium sheet with the thickness of 1.5mm. A paneldimension is

* Corresponding author: ilinca.nastase@gmail.com

1255x635mm.For covering the whole surface of the diffuser, it requires 4panels, resulting in a total of 32 pieces. The free flow surface for each shape is approximately 19.625 mm²(S_{ff}) and the free flow areafor eachpanel is approximately 0.45m² (A_{ff}). All 4 panels, with the same perforations, were mounted at the same time and measurements were made on the whole surface of the UAF diffuser. In the middle of the UAF diffuser, on the entire length, is a lamp with the width of ≈120 mm. Inside, the plenumis organized as follow: an enclosure at the upper part in which air is supplied, panelswith round holesat the lower part of this enclosure for stabilizing the flow, a second enclosure that represents the space for mounting the HEPA (high efficiency particulate air) filters and afterthe perforated panels are mounted. For these measurements there were no HEPA filters mounted.

Table 1. Panel geometries and notations

No.	Panel type	Fig.1	Notation
1	„+” intercalated	a)	A
2	„x” in line	b)	B
3	„O”	c)	C
4	„□”	d)	D
5	„Δ”	e)	E
6	„x” intercalated	f)	F
7	„+” and „x”	g)	G
8	„+”	h)	H

The ventilation system consists in an axial fan, ducts and fittings, air flow regulators, plenumwith grillesand an automation system with a frequency converter mounted near the chamber.To have the same flow rate in the system,identical frequencies where selected when each type of panel was measured. The frequencies where selected to provide a minimum flow which the equipment can measure, not smaller than their error of measurement,and to allow at least 6 measurements across the entire range of the converter. The frequencies used for the measurements were 10.3Hz, 19.2Hz, 26.5Hz, 34.5Hz, 40.6Hz, 44.6Hz, 45.3Hz or max, on a Danfoss converter, type VLT HVAC Basic.

a) b)

Figure 2. a) Ventilation system; b) Frequency converter

The air flow was measured byusing a balometer from TSI, type EBT720/EBT721, with a hood of 610x610mm made out of textile. The hood has almost the same width as a panelbut only a half of its length. Two types of such balometers were used forverifying if there are any perturbations between the values obtained from measuring with only one equipment and measuring with both at the same time. There were no significant differences between the values obtained, and the onesobserved can be dueto measurement errors. Also, the balometers were positioned on different areas of the UAF diffusersurface. These measurements revealed that each panel had almost the same flow rate, regardless of his position in the UAF diffuser. This means that the aeraulic balancing of the ventilation system was done well. For minimizing air leaks between the panel and the frame of the diffuser, a sealing was made with black duct tape that behaves like a sponge.

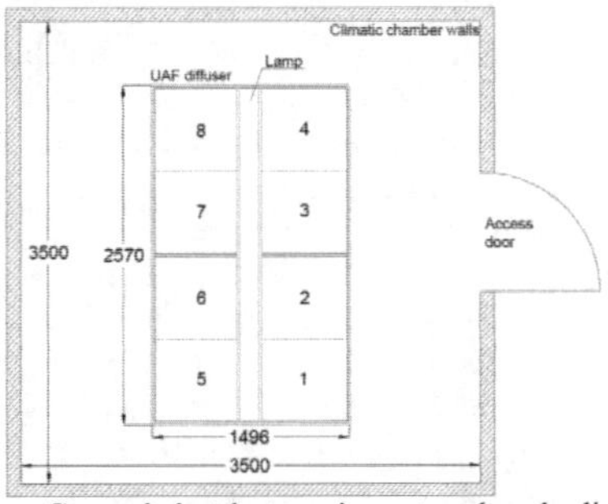

Figure 3. General sketch, top view, at real scale dimensions with the UAF diffuser and climatic chamber

Number notation (from 1 to 8) represents a measurement surface that can be covered by the hood of a balometer. The surface of a panel covers two notations (1 and 2). A total of 8 measurement surfaces for air flow have resulted on the surface of the UAF diffuser.
The pressure loss for each panel was measured by using a differential pressure manometer from Testo, type 521, in which two hoseswere connected.For measuring the pressure loss generated by each panel, one of the hose was mounted before the panel, inside of the plenum, in the space were the HEPA filters had to be mounted. The second hose was mounted after the panel, in the room. The door of the climatic chamber was closed during all measurements.

Figure 4. Equipment used, from left to right: a mounted hose and the manometer, sound meter, balometer

For sound measurements a portable sound meter from Bruel&Kajer, type 2250-S, was used. The sound measurements were made at a height that corresponds to medical staff, ≈1.8m, and at a height that corresponds to patient, ≈1m. For each height there was four measurement points, each point was placed in the middle of each panel. The acoustic measurements were made also without panels, to determine if they generate or attenuate the sound. The sound measurements were made predominantly in periods without any activity in the laboratory (night or morning), but there were also some that could be done only during the day.
The value obtained for each measurement is an average value over a 10÷20s interval. The measurements were made starting with the lowest frequency, determining the flow, followed by the pressure lossmeasurement, ending with acoustic measurements. Tests were made for

verifying if there are any differences between the values obtained when going ascending or descending on the frequency converter.No significant differences were observed between the values obtained and the ones observed can be due to reading or measurement errors.

3. Results and comments

Because no attenuationhas been implemented to eliminate the noise from the fan, noise levels resulted to be high. Also, due to this fact it will be wrong to assume that the technical characterization refers only to the panels, but rather they would representthe entire system (panel and ventilation system). Each system has beencharacterized, based on the values obtained, by pressure loss, noise and air flow. Charts for every system with these technical parameters were made. These charts are similar with the ones that can be found indiffusers datasheet. An example of such chart, with noise values for 1.8m and 1m height, can be seen in Fig. 4 and Fig. 5, for the ventilation system with panel type A.

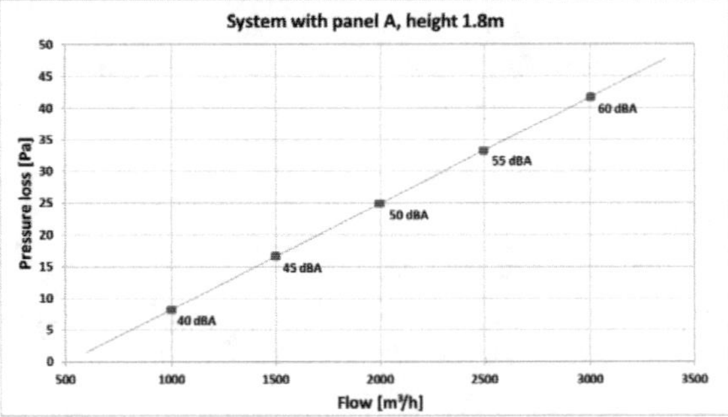

Figure 5. Ventilation system with panel type A, noise 1.8m

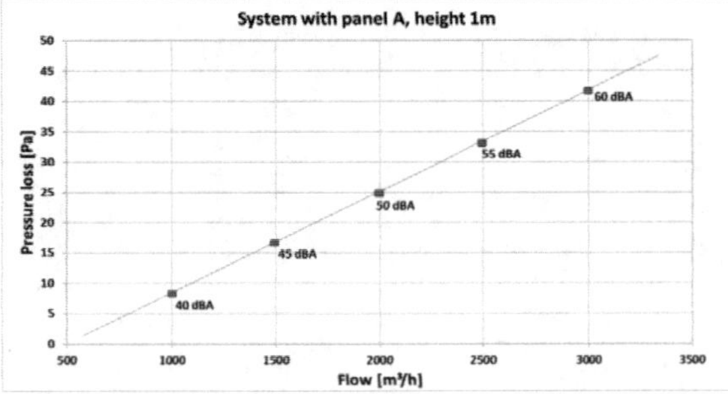

Figure 6. Ventilation system with panel type A, noise 1m

Presenting the data in this manner it's easier to read and allows the user to determine the other two parameters by only knowing one of them. A first comparation regarding the noise level measurements was made between the values obtained at 1.8m and 1m height. All systems have charts with the same allure, that's why only some comparations are presented here (Fig. 7÷10). It was observed that there is only a slight difference between the values measured at these two heights, with a difference of 2dB in average. The peaks that can be observed in the left part of the charts, at low flow rates, are due to environmental disturbances that happened outside. At this flow rates the ventilation system does not generate a lot of noiseand any small noisesfrom outside the climatic chamber,generated in the same time with the measurements,are adisturbing factor.

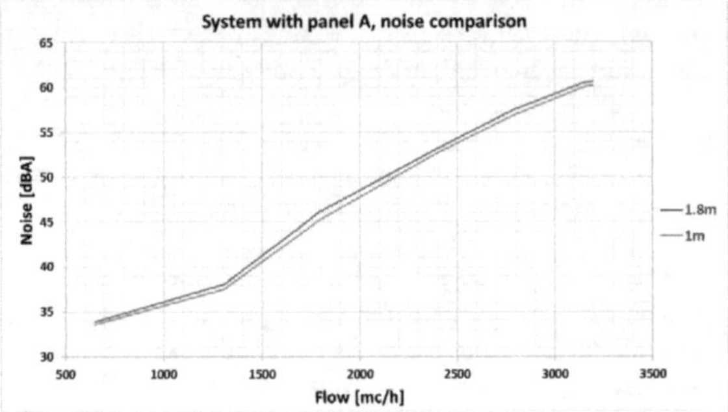

Figure 7. Noise comparationwith panel A, 1.8m and 1m

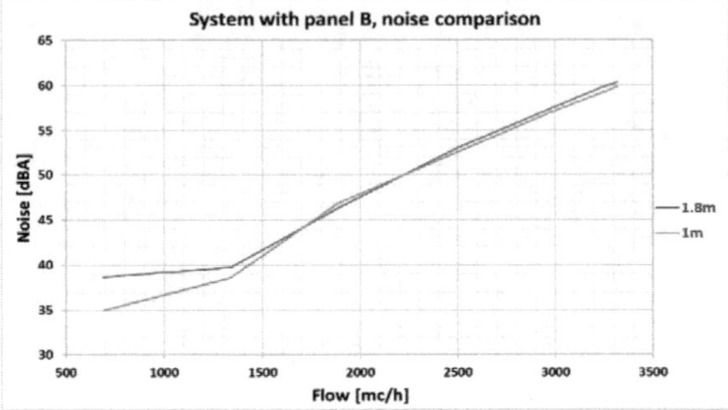

Figure 8. Noise comparation with panel B, 1.8m and 1m

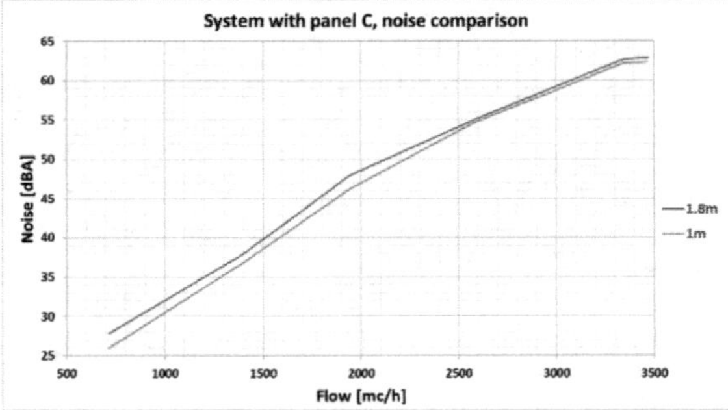

Figure 9. Noise comparation with panel C, 1.8m and 1m

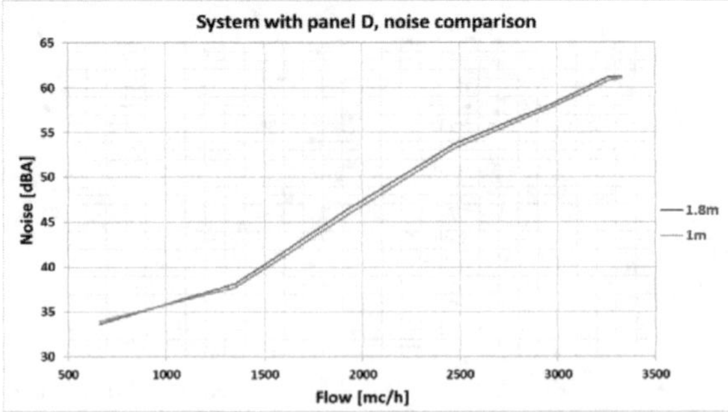

Figure 10. Noise comparation with panel D, 1.8m and 1m

Another interesting comparation was made with the noise level values fromall systems, including the values without panels. This comparation can be observed in Fig. 11 and Fig. 12. A first comment here will be regarding the role of the panelsin these acoustic measurements. It can be observed that the panels attenuate the noise and the valuesare decreasing greatly with them mounted on the UAF diffuser. Here we can see again the influence of the perturbing factors. Despite these perturbations, the measured values are roughly the same for all systems, with slight differences. Also, the differences between the valuesare almost equal across the entire range of flows. Thus, one can assume that either system will generate almost the same noise and it has noreason to choose one instead of another.

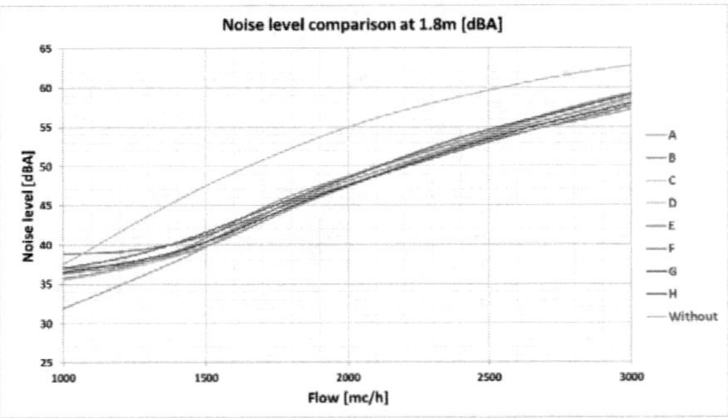

Figure 11. Noise comparation for all systems, 1.8m height

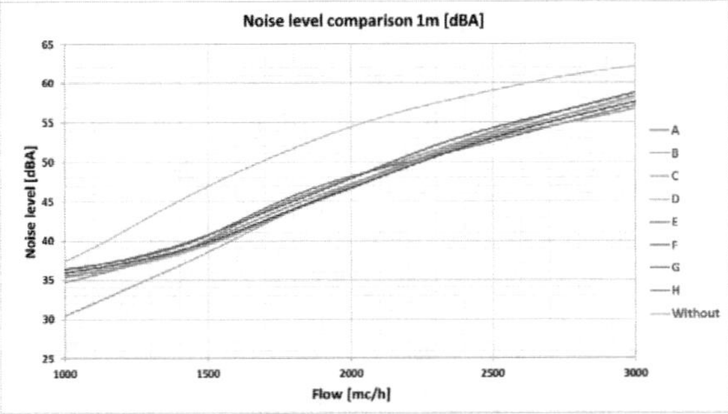

Figure 12. Noise comparation for all systems, 1m height

The pressure loss values obtained for all systems were compared and can be seen in Fig. 13. It can beobserved here that there is a threshold around the value of 2000 mc/h. Before this value, there are very small differences between the systems. After this value, the system with panel type E(„Δ")and D(„□") tends to generate more pressure loss then the others. The difference between the system with panel type E, D and the next one, panel type A („+ intercalated"), is around9-10Pa.

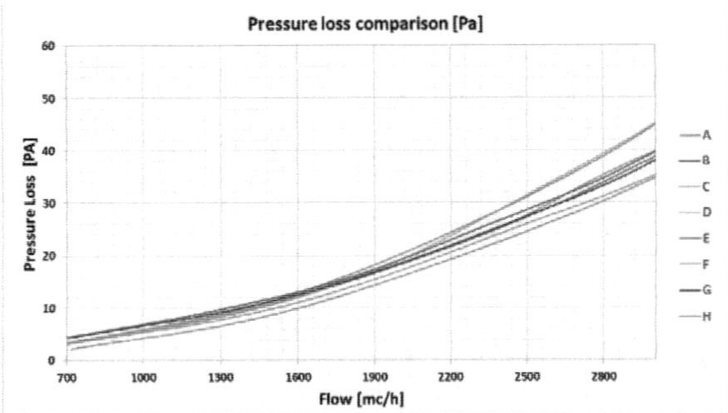

Figure 13. Pressure loss comparation

The other 6 systems tend to keepthe same differences between values even after this threshold. For this parameter, one can assume that you should avoid a system with type E panel, while a better solution will be a system with type C („O") or F („x intercalated").

A comparation with all the technical parameters was made in Table II and III. For determining the pressure loss and the noise level which corresponds to the 3 flow rates that were chosen, interpolations were made based on the values obtained from measurements. The velocity was calculated for eachflow by knowing the free flow surface of a geometry and the total number of geometries in a panel, resulting free flow area.

TABLE I. Technical parameters, same flow, panels A÷D

V (m/s)	Q (m^3/h)		Panel A („+" i)	Panel B („x" L)	Panel C („O")	Panel D („□")
0.43	700	h (Pa)	3.1	3.4	2.3	3.0
		Lp (dBA 1.8m)	33.9	38.7	28.0	33.7
		Lp (dBA 1m)	33.6	35.0	26.2	34.0
0.93	1500	h (Pa)	10.4	11.2	9.7	11.7
		Lp (dBA 1.8m)	40.5	41.7	40.9	40.6
		Lp (dBA 1m)	39.8	41.1	39.5	40.1
1.85	3000	h (Pa)	35.8	38.8	36.7	44.8
		Lp (dBA 1.8m)	57.7	57.6	59.9	58.3
		Lp (dBA 1m)	57.2	57.1	59.5	58.0
Aff (m^2) =	0.45	Sff (mm^2) =	19.625			

TABLE II. Technical parameters, same flow, panels E÷H

V (m/s)	Q (m^3/h)		Panel E („Δ")	Panel F („x" i)	Panel G („+ x")	Panel H („+" L)
0.43	700	h (Pa)	3.1	3.2	4.2	4.2
		Lp (dBA 1.8m)	35.1	36.4	35.1	35.5
		Lp (dBA 1m)	31.9	33.7	34.2	34.7
0.93	1500	h (Pa)	11.6	10.5	11.9	11.3
		Lp (dBA 1.8m)	40.6	40.7	41.0	41.0
		Lp (dBA 1m)	40.1	40.3	40.4	40.2
1.85	3000	h (Pa)	42.5	35.8	37.2	35.7
		Lp (dBA 1.8m)	57.9	57.4	57.6	57.7
		Lp (dBA 1m)	57.5	57.0	57.3	57.3
Aff (m^2) =	0.45	Sff (mm^2) =	19.625			

4. Conclusions

This study allowed us to obtain the technical parameters of this panelsand the ventilation system and represents a first step in the attempt to research how these geometries influences the air distribution in the enclosure.

Some conclusions can be drawn on the values obtained from the measurement campaign presented in this paper.

Regarding the noise measurement values, one can say that no significantdifferences are between the values obtained at different heights,and the noise produced by the system will be perceived almost the same for both patient (≈1m) and surgeon (≈1.8m).Another conclusion that can be drawn is that almost no differences were observed for noise levels generated by the different systems.Anobservationwhich cannot be overlooked is the fact that any panelattenuates the noise produce by the fan.

Regarding the pressure loss measurements, it was confirmed that some geometries, which were expected to generate highpressureloss („Δ", „□"), had higher values. Also, it was observedthat the special geometries didn't generate high pressure loss at high flow rates, some even getting good results („x"i. or „+" L.).

More research must be done in this fieldfor abetter understanding.Future research can study the influences of perforated panels on thermal comfort felt by the occupants, aerosol distribution in the room, energy consumption of the system.

This work was supported by a grant of the Romanian space agency ROSA, QUEST - Advanced air diffusion system of the crew quarters for the ISS and deep space habitation systems, STAR-CDI-C3-2016-577

References

1. Koufi L, Younsi Z, Cherif Y, Naji H. Numerical investigation and analysis of indoor air quality in a room based on impinging jet ventilation. Energy Procedia;139:710-7 (2017)
2. Kandzia C, Kosonen R, Melikov AK, Nielsen PV. Mixing Ventilation. Guide on mixing air distribution design. Müller D, editor: Federation of European Heating and Air-Conditioning Associations, REHVA; (2013)
3. Awbi HB. Ventilation for Good Indoor Air Quality and Energy Efficiency. Energy Procedia;112:277-86 (2017)
4. Fischer S, Thieves M, Hirsch T, Fischer K-D, Hubert H, Bepler S, et al. Reduction of Airborne Bacterial Burden in the OR by Installation of Unidirectional Displacement Airflow (UDF) Systems. Medical Science Monitor : International Medical Journal of Experimental and Clinical Research;21:2367-74 (2015)
5. Yang C, Yang X, Zhao B. The ventilation needed to control thermal plume and particle dispersion from manikins in a unidirectional ventilated protective isolation room. Building Simulation;8(5):551-65 (2015)
6. Heather C. Willis Whitfield, inventor of modern-day laminar-flow clean room, passes away. Sandia Lab News, Vol 64, 16:8 (2012.11)
7. Nastase I, Croitoru C, Dan M, Ursu I, Meslema A. Experimental Study for the integration of an Innovative Air Distribution System in Operating Rooms. Sustainable Solutions for Energy and Environment, EENVIRO 2016; 26-28 October 2016; Bucharest, Romania (2016)
8. Nastase I. Strategii performante pentru creşterea calităţii ambientale în sălile de operaţie (EQUATOR). UTCB (2016)
9. Meslem A, Nastase I, Allard F. Passive mixing control for innovative air diffusion terminal devices for buildings. Building and Environment;45:2679-88 (2010)

Preliminary Study on a Reduced Scaled Model Regarding the Air Diffusion inside a Crew Quarter on Board of the ISS

Mihnea Sandu[1*], *Ilinca Nastase*[1], *Florin Bode*[1,2], *CristianaVerona Croitoru*[1], *Laurentiu Tacutu*[1]

[1]CAMBI Research Center, Technical University of Civil Engineering Bucharest,021414Bucharest, Romania
[2]Technical University of Cluj Napoca, Department of Mechanical Engineering 400020 Cluj - Napoca, Romania

Abstract. The paper focus on the air quality inside the Crew Quarters on board of the International Space Station. Several issues to improve were recorded by NASA and ESA and most important of them are the following: noise level reduction, CO_2 accumulation reduction and dust accumulation reduction. The study in this paper is centred on a reduced scaled model used to provide simulations related to the air diffusion inside the CQ. It is obvious that a new ventilation system is required to achieve the three issues mentioned above, and the solutions obtained by means of numerical simulation need to be validated by experimental approach. First of all we have built a reduced scaled physical model to simulate the flow pattern inside the CQ and the equipment inside the CQ has been reproduced using a geometrical scale ratio. The flow pattern was considered isothermal and incompressible. The similarity criteria used was the Reynolds number to characterize the flow pattern and the length scale was set at value 1/4. Water has been used inside the model to simulate air. Velocity magnitude vectors have been obtained using PIV measurement techniques.

1.Introduction

Currently, on the International Space Station (ISS) several personal Crew Quarters (CQ) are installed for the crew members. Functionally a CQ provides an acoustically quiet and visually isolated area in which crew members can sleep, relax, and retreat to a private area (Figure 1).

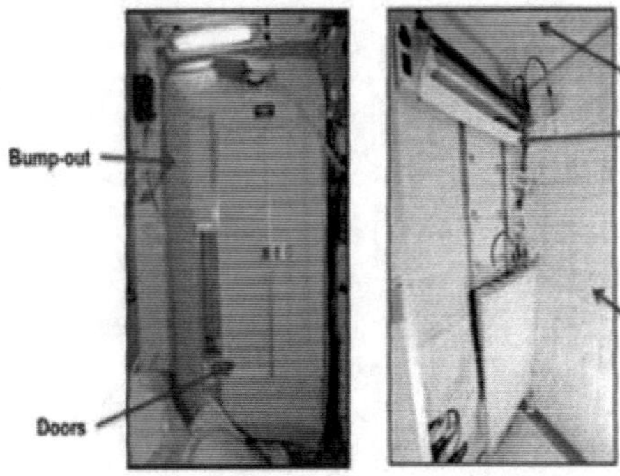

Fig. 1. CQ on-orbit exterior view (left) and interior view (right)

Design and development efforts for long mission crew member's accommodations started with the US Skylab CQ [1]. Skylab featured visual private space for each crew member but lacked of acoustic and light isolation and had very limited ventilation control [2, 3]. The Russian CQ Kayutas were introduced on Salyut 6 and the same configuration was used on MIR and on the ISS Service Module. Kayutas CQ provided an increased visually private volume and a 20 cm diameter window. However the window increased crewmember space radiation exposure. The Kayutas CQ draw air from the module cabin but were generally found to be warm and not to provide enough acoustic attenuation of the cabin noise [3]. The crew uses the CQ primarily for sleeping, but also for performing tasks such as put on or remove clothing, using laptops, private communication, and some minimal personal hygiene.

Each USOS CQ was designed with an individual ventilation system, acoustic mitigation materials, laptop connections, and internet connection to allow crew members personal communication with family and friends [4]. Since their deployment in 2008, the CQ performance has been closely monitored to validate that the design continues to meet requirements. At the beginning, the USOS CQs have received generally favorable crew comments with respect to acoustics [5].
For the control of the airflow, the CQ provides a fan speed switch with three positions and adjustable guide vanes to allow the crew to customize the airflow levels and direction. The crew reported that, in general, the variable fan speed and the guide vanes provided good adjustability although most crew members set the fan speed and the guide vanes to direct air in one direction and leave them in that position. Nominal fan speed varied from crewmember to crew member, but most crew members reported that the fans were kept on medium or high speed formost of the time. Some crew members commented that fans on high speed produced

*Corresponding author: mihnea.sandu@utcb.ro

too much noise inside the CQ, indicating the need for reducing the acoustic signature of fans at middle and high speed levels of the fans[5]. Most crew members felt that the fan low speed did not provide enough airflow and, due to dust accumulationover time in the CQ ventilation system, the fans at low speed produced a low airflow situation that resulted in caution alarms in CQ.

Finally, the following three issues were reported by NASA and ESA as the most important:

- **Noise levels**: Since their development [2, 3, 6, 7] engineers deposed a lot of efforts to reduce the acoustic signature of the chosen ventilation solution. In the present concept they had to allocate around13% and 6%, respectively of the total mass and volume of a CQ to acoustic reductions. The total impact of the current ventilation solution, in terms of acoustic reduction is respectively 23kg and 0.3 m^3 for one CQ unit.
- **CO_2 accumulation issues:** Some crew members have been waking up with headaches [8, 9]. It is believed the CQ has good air flow but there is high CO_2 in the station when the Carbon Dioxide Removal Assembly (CDRA) is not functioning to its capacity. Currently NASA is looking for a local solution for scrubbing CO_2 from the local air of the CQ to help crew rest better if the CDRA is working poorly [8].
- **Dustaccumulation issues**: The ISS environment in general and the CQ environment in particular are getting more polluted in terms of particulate matter than expected originally by the engineers [8]. Most of this dust comes from the people themselves. This dust gets into the ducting of the CQ's ventilation system and causes problems with the air flow sensors that are monitoring the flow variation to prevent asphyxiation hazards.

This paper is part of a larger study which implies as main objective the proposal of an optimized concept of the existing CQ on board of the ISS with improved air distribution systems in terms of fans and acoustics. The improved CQ ventilation system will allow the possibilityof a direct adaptation of the existing CQ on the USOS Node 2 module of the ISS to eliminate the current problems of the crew members of the ESA and NASA teams. This will be translated through the improvement of the acoustical performance of the existing CQ on board of the ISS allowing future developments for new concepts of crew accommodations for future long duration and through the increasing of the space dedicated to the crew member in the current CQ design with a reduction of the total mass. The new configuration will also cover the pressure drop related to the introduction of a filter assembly for fine aerosols.

In this respect numerical simulations for the air diffusion system and CO_2 diffusion inside the CQ must be performed and these simulations need experimental validation. This paper deals precisely with the first attempts to create a reduced scaled model for the CQ made by our team. We have focus to model at a smaller scale the air flow pattern and the diffusion of CO_2 inside the CQ. We have considered that the velocities and the pressure drops are small enough to consider the fluid incompressible and the temperature is constant so that the main driving force for the CO_2 diffusion is the concentration gradient, natural convection being not considered. Another important point is that all experiments will be done in gravitational space which is not present on ISS. Below there is a short description of several similarity criterions and the opportunity to use them for our study.

2.Similarity Criteria

Complete similitude between two physical phenomena means that they must be of the same nature and both geometric, dynamic, kinematic similitude and boundary conditions should be accomplished[10]. The geometric similitude requires the proportionality of the corresponding lengths and the equality of angles. The result of the geometric similitude is the length scale. The kinematic similitude requires that if the flow is made of particles, the corresponding particles take the same place inside the flow pattern at same moments in time. As a result, if the kinematic similitude is accomplished between the two phenomena there are a constant length scale and a constant velocity scale. The dynamic similitude implies also the proportionality of the corresponding forces between the phenomena in addition to the geometric and kinematic similitude. The result is obviously a constant force scale.

For this study the dynamic similitude has been found as appropriate. However, the complete similitude between two phenomena of the same nature is impossible to achieve so one must consider only the main forces that govern the flow. It results a restraint dynamic similitude criterion as follow:

- The Froude similitude: the leading forces are the inertia and the gravity forces.

$$Fr = \frac{u}{\sqrt{gL}} \tag{1}$$

Froude similitude implies the condition Fr_N=Fr_M. The relation between the scales in case of the Froude similitude is:

$$\frac{S_u^2}{S_g S_l} = 1 \tag{2}$$

- The Euler similitude: the leading forces are the inertia and the pressure forces.

$$Eu = \frac{p}{\rho u^2} \tag{3}$$

Euler similitude implies the condition Eu_N=Eu_M. The relation between the scales in case of the Euler similitude is:

$$\frac{S_p}{S_\rho S_u^2} = 1 \tag{4}$$

- The Reynolds similitude: the leading forces are the inertia and the viscosity forces.

https://doi.org/10.1051/e3sconf/20183201015

$$Re = \frac{ul}{\vartheta} \quad (5)$$

Reynolds similitude implies the condition $Re_N=Re_M$. The relation between the scales in case of the Reynolds similitude is:

$$\frac{S_u S_l}{S_\vartheta} = 1 \quad (6)$$

- The Peclet similitude: the leading forces are the inertia and the difusivity forces.

$$Pe = \frac{ul}{D} \quad (7)$$

Peclet similitude implies the condition $Pe_N=Pe_M$. The relation between the scales in case of the Peclet similitude is:

$$\frac{S_u S_l}{S_D} = 1 \quad (8)$$

It must be mentioned here that the Peclet number is also found in the literature under the form $Pe = \frac{ul}{\alpha}$ where α is called thermal diffusivity and the physical signification in this case is represented by the intensity of the mass convective transfer.

- The Schmidt similitude: the physical meaning of this criteria is represented by the ratio between the intensity of the transfer of the substance by pressure or by concentration differences.

$$Sc = \frac{\nu}{D} \quad (9)$$

Schmidt similitude implies the condition $Sc_N=Sc_M$. The relation between the scales in case of the Schmidt similitude is:

$$\frac{S_\nu}{S_D} = 1 \quad (10)$$

It is obvious that ***Pe=Re Sc.***

3.Experimental Setup Conception

The next step is to establish the similarity criteria used. Considering the nature of the phenomena and the fact that not all the criteria of similarity can be satisfied simultaneously, we considered that the determinant for the studied phenomenon is the Reynolds criterion defining the flow structure with major implications in the heat transfer as well as the Peclet criterion with major implications in the diffusion mass transfer. First, the interest was tosimulate the CO_2 diffusion in air. Here we have identified two issues: the first issue is to find a way to simulate the microgravity, more precisely the fact that there is a difference between the CO_2 density and the air density (CO_2 is heavier) which is very important in gravitational field but not in microgravity space. This issue can be solved by using two fluids with the same density to simulate the diffusion of CO_2in the air in microgravity conditions. In this case the effects of gravity are canceled by the buoyancy forces. If the temperature is kept constant inside the fluid body, natural convection is also avoided. Thus, the only sources of motion between two liquids are forced convection, molecular diffusion and, in some cases, gradients of surface tension. Research on fluid behavior in space is oriented towards the behavior of liquids in the absence of gravity. On Earth, gravitational effects often cancel other influences, but in space, in an essentially no-gravitational environment, the importance of other factors increases. Such a factor is the gradient of surface tension that exists between two different liquids. This gradient can be caused by differences in concentration or temperature and, in both cases, causes mass transfer between two fluids (the Marangoni effect).

Since the experimental scale model will introduce a liquid jet into an existing body of water, a question arises as to the influence of other factors on the liquid behavior. The bibliographic studies on the movement induced by the surface tension gradient in the near-density systems lead to the conclusion that these effects are not relevant to our case.

Therefore, we have tried a low-scale model in which CO_2 airborne diffusion in air was modeled by means of Rhodamine diffusion in water. The choice of rhodamine diffusion in water was imposed by the possibility to use laser induced fluorescence techniques to measure level of concentration. In this respect, using the Peclet criterion and considering that S_D in this case is equal to about 3,38 x 10^{-5} for a scale of S_L = 1/4 lengths, the scale of the $S_v$should be about 1,35 x 10^{-4}. It means that the velocity values on the reduced model should be about 10^4times smaller than the real values which was not possible to measure with common PIV techniques.

We drop off for the moment the idea of modelling the CO_2 diffusion and we have tried to find a way to simulate the flow pattern inside the CQ using Reynolds criterion. Water was kept as working fluid because of technical reasons. In this case considering that the viscosity scale is S_ν=1/15 and using S_L = ¼ for the length it results a 4/15 scale for the velocity.

The plans and the main dimensions of the reduced scaled model are showed in figure 2

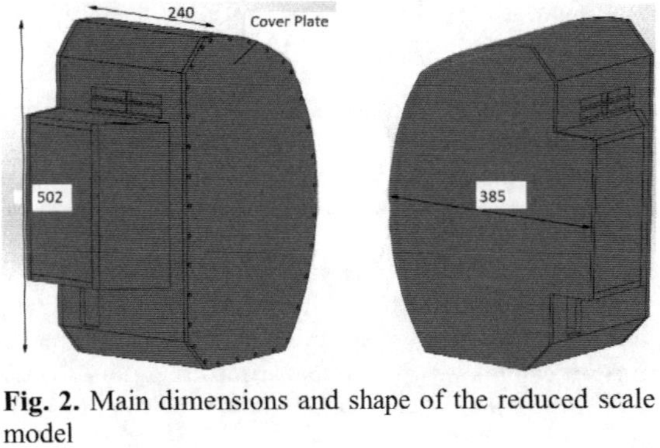

Fig. 2. Main dimensions and shape of the reduced scale model

 https://doi.org/10.1051/e3sconf/20183201015

The model was made of polycarbonate plates with a suitable refraction index to allow velocity measurement using PIV techniquesas shown in figure 3a,b,c:

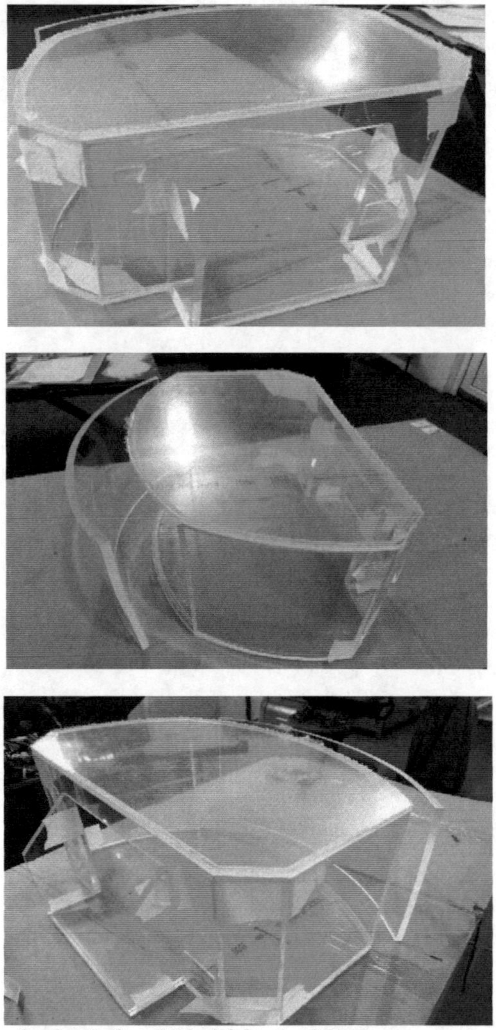

Fig. 3.a,b,c Reduced Model for the CQ.

The model was integrated in a closed water loop using a small pump for ensuring the water flow rate. The main purpose was to prove the feasibility of the model, precisely to demonstrate how to determine the velocity field inside the reduced scaled model.

In this study, we have employed a Dantec Particle Image Velocimetry (PIV)system for the velocity field measurements. This system is composed of one high sensitivity Flow Sense 4M camera of $4x10^6$pixels resolution and of a Dual Power 200 mJ laser having the wave length of 532nm. The acquisition frequency of the PIV system was 7.5 Hz. The water flow was seeded with hollow glass spheres with a mean diameter of 50µm. The images calibration gave a spatial resolution of 100 µm per pixel which is corresponding to a 200 x 200 mm^2 field of view.

The laser has been placed above the model and the camera from a lateral side of the model as in the figure 4 below.

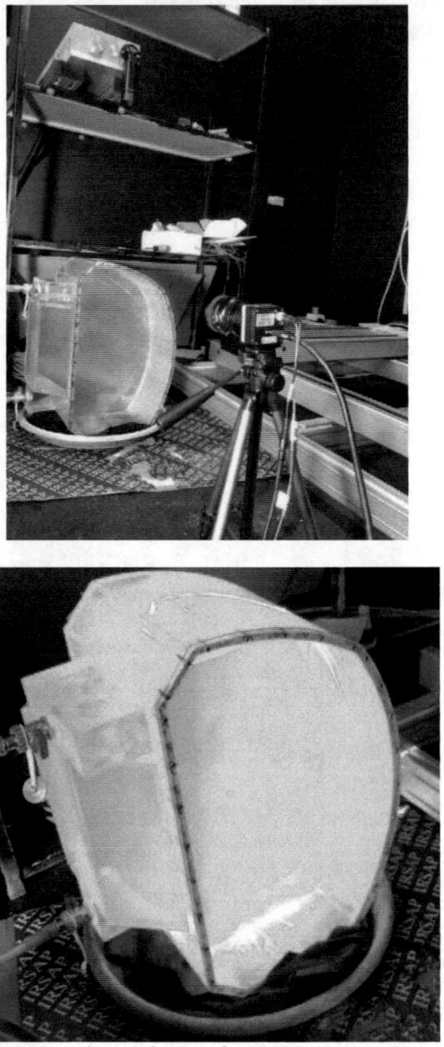

Fig. 4.a,b Experimental setup for PIV measurements

4.Results

The velocity field has been measured in a plane parallel to the lateral face of the model as indicated in figure 5. Several other planes and interesting configurations should be investigated but given the fact that the main purpose was to prove the feasibility of the model all these experiments will be accomplished following the future numerical simulations that will be part of this study.

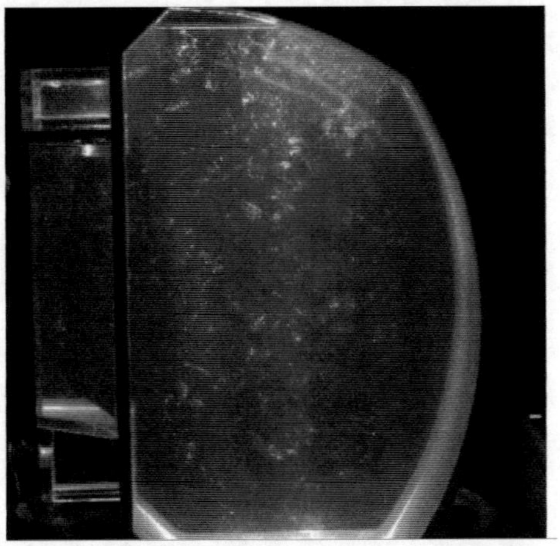

Fig. 5. Flow visualisation in plane parallel to the lateral face of the model

The velocity vectors measured are shown on the figure 6 below. The measured velocity magnitudes are less the 0.05 m/s.

Fig. 6. Velocity vectors

5.Conclusions and perspectives

Even if the study is a preliminary one, some conclusions and some interesting perspectives can be outlined of it. It is obvious from the beginning that all the experiments have been done in gravitational field which is not the case in space. This is the main reason to use two substances with nearly the same density for modelling the CO_2 diffusion in air but due to our limited measurement possibilities we had no results. However, we have continued using water inside the model to simulate the flow pattern with respect to Reynolds criterion. This preliminary study guide us towards the following conclusions:

- The CO_2 diffusion in air should be modelled on a reduced scaled model using the Peclet criterion. Two gases with nearly the same density are needed in order to avoid the gravity effects. Using water as the main fluid is not appropriate because the diffusivity coefficients values for different substances (tracers) in water are 10^5 times smaller than the CO_2 diffusion coefficient in air and the corresponding scales for length and velocity are not technically feasible.
- The results have shown that it is possible to use water on the model to simulate the flow pattern with respect to Reynolds criterion. The use of the water is indicated for this case because it is easier to measure using PIVtechnique and the insemination with particles is better than for gases.
- The similitude regarding Peclet and Reynolds criterion cannot be achieved at the same time, the physical phenomena must be splitted.
- A reduced scaled model using gases as working fluids need to be tested also with respect to the Reynolds criterion. The advantage here is that the microgravity could be simulated better for this case.

This work was supported by a grant of the Romanian space agency ROSA, QUEST - Advanced air diffusion system of the crew quarters for the ISS and deep space habitation systems, STAR-CDI-C3-2016-577

References

1. P.O.Wieland, Living Together in Space: The Design and Operation of the Life Support Systems on the International Space Station NASA/TM—1998–206956/**1**. (1998).
2. J.L.Broyan, M.A. Borrego, and J.F. Bahr. International Space Station USOS Crew Quarters Development. in 08ICES-0222. (2008).
3. C.Adams, Four Legs in the Morning: Issues in Crew-Quarter Design for Long-Duration Space Facilities. in 28th International Conference on Environmental Systems, Danvers, MA. (1998).
4. NASA Reference guide to the International SpaceStationhttps://www.nasa.gov/pdf/508318main_ISS_ref_guide_nov2010.pdf. (2010).
5. T.P.Schlesinger, B.R. Rodriguez, and M.A. Borrego, International Space Station Crew Quarters On-Orbit Performance and Sustaining Activities. AIAA Journal, (2013).
6. J.L.Broyan, S.M. Cady, and D.A. Welsh, International Space station Crew Quarters Ventilation and Acoustic Design Implementation. AIAA Journal, (2010).
7. S.Fairburn, and S. Walker, 'Sleeping With the Stars'–The Design of a Personal Crew Quarter for

https://doi.org/10.1051/e3sconf/20183201015

the International Space Station. 2001, SAE Technical Paper.
8. 2015-2016 HUNCH Design & Prototyping program https://www.nasahunch.com/design-prototype/. (2015).
9. J.James, The Headache of Carbon Dioxide Exposures in SAE Technical Paper 2007-01-3218, 2007, doi:10.4271/2007-01-3218. (2007).
10. M.Degeratu, Analiza dimensionala, similitudine si modelare. (2015), Ed. Academiei Oamenilor de Stiinta Romania.

Potential of HVAC and solar technologies for hospital retrofit to reduce heating energy consumption

Octavian G. Pop[1,*], *Ancuta C.* Abrudan[2], *Dan S.* Adace[3], *Adrian G.* Pocola[1], and *Mugur C.* Balan[1*]

[1]Technical University of Cluj-Napoca, Dept. of Mechanical Engineering, Bd. Muncii 103-105, 400641, Cluj-Napoca, Romania
[2]Technical University of Cluj-Napoca, Dept. of Building Services, Bd. 21 Decembrie 1989 128-130, 400604, Cluj-Napoca, Romania
[3]Arhi Details, Str. A. Iancu 37/A, 401163, Turda, Romania

Abstract. The study presents a combination of several energy efficient technologies together with their potential to reduce the energy consumption and to increase the comfort through the retrofit of a hospital building. The existing situation is characterized by an old and inefficient heating system, by the complete missing of any ventilation and by no cooling. The retrofit proposal includes thermal insulation and a distributed HVAC system consisting of several units that includes air to air heat exchangers and air to air heat pumps. A condensing boiler was also considered for heating. A solar thermal system for preparing domestic hot water and a solar photovoltaic system to assist the HVAC units are also proposed. Heat transfer principles are used for modelling the thermal response of the building to the environmental parameters and thermodynamic principles are used for modelling the behaviour of HVAC, solar thermal system and photovoltaic system. All the components of the heating loads were determined for one year period. The study reveals the capacity of the proposed systems to provide ventilation and thermal comfort with a global reduction of energy consumption of 71.6 %.

1 Introduction

The study is enrolled in the global effort dedicated to the reduction of energy consumption in buildings, which according to [1] accounts 40% of final energy, in Europe.

Due to their low thermal efficiency the existing buildings are responsible of most of this energy consumption [2-4] and 95% of the existing buildings do not achieve energy efficiency standards and their retrofit provides a lot of potential for reducing the energy consumption [4]. In the particular case of Romania 80% of buildings need to be renovated from thermal considerations [5].

Important energy savings can be realised through the simple retrofit of the envelope and by mounting heating, ventilation and air conditioning system (**HVAC**) systems [6].

In the actual European context regulated by the directive on the energy performance of buildings [1], the near zero energy buildings (**nZEB**) concept should be considered for buildings retrofit. There most important strategies for the retrofit of existing buildings to **nZEB** are to reduce the energy loads and to implement technologies based on renewable energy sources [6-8].

In Romania the energy requirements set for new buildings of **nZEB** level are of (93–217) kWh/m^2/year for residential buildings and of (50–192) kWh/m^2/year for non-residential buildings [8].

Hospitals represent 7% of the non-residential buildings in Europe [8]. Different strategies for the thermal retrofit of a hospital building envelope are analysed in [9]. Different technologies were considered in hospitals retrofit: substation climatic 3-way valve, radiators thermostatic valves and air handling unit (**AHU**) with programmable regulation [10], or demand control ventilation (**DCV**) of the existing heating, ventilation and air conditioning (**HVAC**) units [11] or control systems, cogeneration, heating regulation, new lighting systems and solar panels [12]. Renewable energies considered in the hospitals retrofit are solar panels [12] or a combination of fuel cells, solar thermal system and solar photovoltaic system [13]. A trigeneration system is considered in [14].

The goal of the study is to investigate the potential of different energy efficient technology to be integrated in a retrofit plan of an existing hospital building in Romania to reduce energy consumption. The considered retrofit technologies are a new condensing heating boiler, new radiators, a distributed **HVAC** system based on several units that integrates air to air heat recovery (**AAHR**) and air to air heat pumps (**AAHP**), a solar thermal system and a photovoltaic system. This study continues the concerns regarding to the energy efficiency in buildings at the Technical University of Cluj-Napoca [15-20]).

* Corresponding author: mugur.balan@termo.utcluj.ro

 https://doi.org/10.1051/e3sconf/20183201016

2 Material and method

The study refers to a building of the Municipal Hospital of Sighetu Marmației in the northern part of Romania at the coordinates: 47.927957 N, 23.876285 E. The position of the city on the map of Europe is presented in figure 1 and the position of the Municipal Hospital on the map of the city is presented in figure 2.

Fig. 1. Location of the city on the map of Europe (Google Maps).

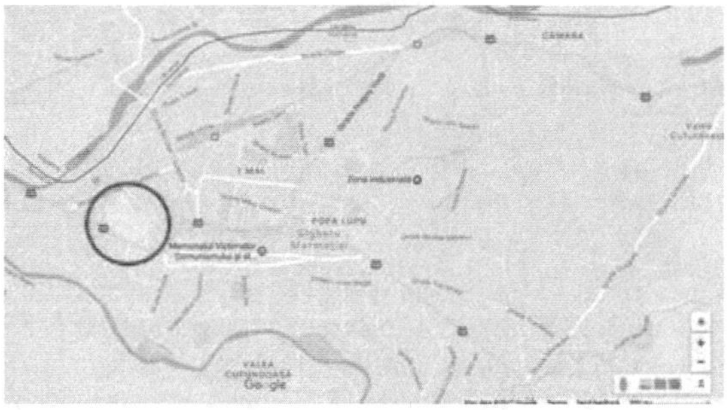

Fig. 2. Location of the city on the map of Europe (Google Maps).

The building considered for refurbishing is the Maternity located as indicated in figure 3.

Fig. 3. Location of the hospital building considered in the study (Google Maps).

The frontal and the rear view of the building are presented in figure 4 and figure 5.

Fig. 4. Frontal view of the building.

Fig. 5. Rear view of the building.

The built area of the building is of 1236 m^2, the unfolded built area of the building is of 4576 m^2 and the heated area of the building is of 4079 m^2.

The walls are realized of autoclaved cellular concrete bricks with the width of 30 cm.

The existing energetically situation is characterized by the following aspects: the building is not thermally insulated; the windows frames are of polyvinyl chloride (**PVC**) type and the glazing are double; windows frames presents a lot of leakage problems; the thermal agent for heating and for domestic hot water (**DHW**) is prepared in a gas boiler in advanced state of deterioration; the building is connected to the thermal plant through an old thermal network; there is no mechanical ventilation and no air conditioning in the building.

The functional scheme of the existent heating system is presented in figure 6.

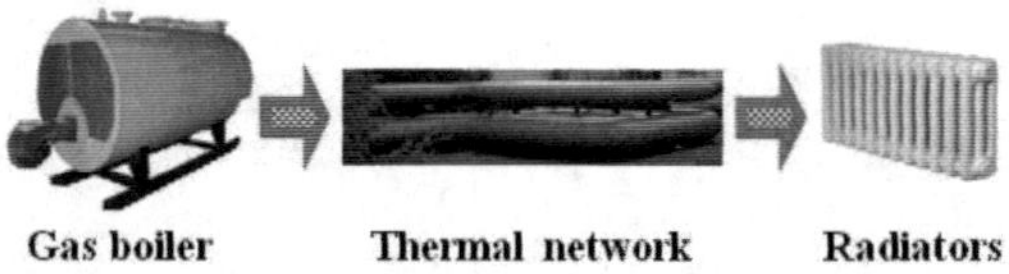

Fig. 6. Functional scheme of the existent heating system

The efficiency of the existing old boiler is estimated at only 80% due to several technical problems and the efficiency of the existing thermal network from the boiler to the building is estimated at the same value of 80% due to the existing leaks and to the degraded insulation. According to these efficiencies, the global efficiency of the existing heating system is of only 64%.

The proposed retrofit consists in the following updates: insulation of the building with an outer layer of rock wool of 15 cm width (0.036 W/m^2K); windows replacement with thicker PVC frames and triple glazing (0.8 W/m^2K); replacement of existing old radiators with sanitized radiators in agreement with hospital regulations [21]; dedicated to the building thermal plant of container type, mounted in the immediate vicinity of the building and based on efficient condensing type gas boiler; distributed **HVAC** system each unit including an **AAHR** (efficiency of 50%) and an **AAHP** (nominal COP=6.3 at interior temperature of 24 °C and exterior temperature of 0 °C); a solar thermal system based on flat thermal collectors (**ST**); and a photovoltaic system (**PV**).

ST is used to assist the **DHW** preparation system and **PV** is used to assist the **HVAC** system. Each of **ST** and **PV** are placed on 50 % of the available flat attic of the building, resulting 119 thermal collectors and 157 PV collectors.

The **HVAC** units are equipped with electric preheaters to provide adequate air temperature if the outside temperature decreases below -10 °C representing the lower operating limit of the equipment and with additional heating coils if the outlet air temperature is not sufficient.

The functional scheme of the proposed ventilation, heating and cooling system is presented in figure 7.

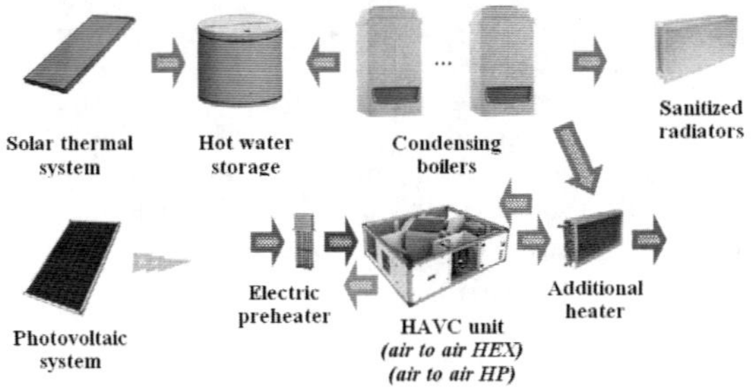

Fig. 7. Functional scheme of the proposed ventilation, heating and cooling system

The global purpose of the proposed improvements is to provide both high comfort and also reduced energy consumption.

To evaluate the thermal behaviour of the building, it was considered simplified as single volume with constant interior temperature of 24 °C.

The thermal loss through walls was determined based on heat transfer through inertial elements equations, according with Romanian regulations.

The thermal load through inertial elements such as walls ($\dot{Q}_{in}$ [W]) was determined as:

$$\dot{Q}_{in} = S \cdot [k \cdot (t_{ac} - t_i) + \eta \cdot \alpha_i \cdot (t_c - t_{ac})] \qquad (1)$$

where:

S [m^2] is the lateral area of the inertial element of the building envelope;

k [W/m^{2}K] is the global heat transfer element through the inertial element of the building envelope;

t_{ac} [°C] is the average conventional exterior temperature, influenced by the solar radiation;

t_i [°C] is the interior temperature;

η [-] is the coefficient of temperature oscillation amortization effect;

α_i [W/m^{2}K] is the interior coefficient of convection;

t_c [°C] is the conventional exterior temperature, influenced by the solar radiation.

The thermal load through non-inertial elements such as windows ($\dot{Q}_n$ [W]) was determined as:

$$\dot{Q}_n = \dot{Q}_{td} + \dot{Q}_{sr} \qquad (2)$$

where:

$\dot{Q}_{td}$ [W] is the thermal load due to the temperature difference between outside and inside;

$\dot{Q}_{sr}$ [W] is the thermal load due to the solar radiation.

Heat transfer through the triple windows is considered to calculate the corresponding thermal load.

The thermal load corresponding to the fresh air ($\dot{Q}_{fa}$ [W]) was determined as:

$$\dot{Q}_{fa} = \dot{m} \cdot c \cdot \Delta t \qquad (3)$$

where:

$\dot{m}$ [kg/s] is the fresh air mass flow rate;

c [J/kgK] is the specific heat of the air;

Δt [°C] is the temperature difference between the outside and the inside air, considering the air blowing temperature of 24 °C during winter and 23 °C during summer.

In the heating mode with constant interior temperature of 24 °C, the following variations of the **HVAC** units parameters were determined by modelling the thermal behaviour of the equipment, as function of the exterior temperature (t_e [°C]) in the normal operating range of (-10…15) °C:

Thermal power of the **AAHR** ($\dot{Q}_{AA}$ [kW]):

$$\dot{Q}_{AA} = 19.05 - 0.7939 \cdot t_e \qquad (4)$$

Thermal power of the **AAHP** ($\dot{Q}_{HP}$ [kW]):

$$\dot{Q}_{HP} = 19.477 + 0.252 \cdot t_e \qquad (5)$$

Electrical power absorbed by the **AAHP** compressor (P_{HP} [kW]):

$$P_{HP} = 3.0932 + 0.0384 \cdot t_e \qquad (6)$$

Coefficient of performance of the **AAHP** (COP_{HP} [kW]):

$$COP_{HP} = 6.2979 + 0.0035 \cdot t_e - 7e^{-5} \cdot t_e^2 \qquad (7)$$

Global coefficient of performance of the **HVAC** unit (COP_g [kW]):

$$COP_g = 12.487 - 0.3324 \cdot t_e + 0.0035 \cdot t_e^2 \qquad (8)$$

In the cooling mode with constant interior temperature of 25 °C, the following variations of the HVAC units parameters were determined by modelling the thermal behaviour of the equipment, as function of the exterior temperature (t_e [°C]) in the normal operating range of 28…38) °C:

Thermal power of the **AAHR** ($\dot{Q}_{AA}$ [kW]):

$$\dot{Q}_{AA} = 0.7938 \cdot t_e - 19.847 \qquad (9)$$

Thermal power of the **AAHP** ($\dot{Q}_{HP}$ [kW]):

$$\dot{Q}_{HP} = 18.974 + 0.2969 \cdot t_e \qquad (10)$$

Electrical power absorbed by the **AAHP** compressor (P_{HP} [kW]):

$$P_{HP} = 2.9385 + 0.0488 \cdot t_e \qquad (11)$$

Coefficient of performance of the **AAHP** (COP_{HP} [kW]):

$$COP_{HP} = 5.2898 + 0.0047 \cdot t_e - 0.0001 \cdot t_e^2 \quad (12)$$

Global coefficient of performance of the **HVAC** unit (COP_g [kW]):

$$COP_g = 1.50720 + 0.1574 \cdot t_e \quad (13)$$

The calculation of solar geometry, relative position between the sun and the plane of the collectors (**ST** or **PV**), solar thermal energy production and solar electric production, were realised based on previously successful used algorithms [22-24], [17], [25].

The meteorological parameters representing input data for the calculations were taken from the available typical meteorological year (**TMY**) for the climatic zone of Sighetu Marmaţiei. The solar radiation and the ambient temperature are presented in figure 8.

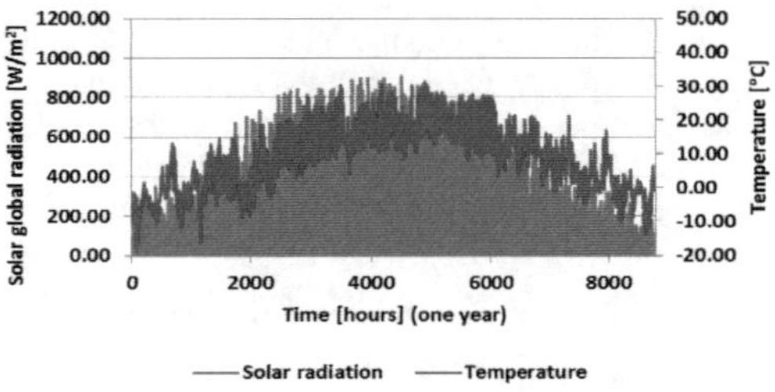

Fig. 8. Global radiation on horizontal plane and ambient temperature for the climatic zone of Sighetu Marmaţiei

The components of the thermal load are the following: for **DHW** (H), through walls (W); through glazing (G) and due to the fresh air (F).

In the existing situation, the energy is distributed in the building only as hot water (HW) and through the existing radiators (ER) while in the proposed situation it has the following components: hot water (HW) through proposed sanitary radiators (SR); through the electric preheaters of the **HVAC** units (EP); through the air to air heat exchangers (AA); through the air to air heat pumps (HP) and through the supplementary heating coils of the **HVAC** units (SH).

In the existing situation the primary energy is represented only by the consumption of natural gas in the existing boiler (NGB) while in the proposed situation it has the following components: consumption of natural gas for hot water (NGH); consumption of natural gas in condensing boiler (NGB); consumption of natural gas in the supplementary coils of the **HVAC** units (NGS); consumption of electricity in the preheaters (EEP); consumption of electricity in the compressors of the heat pumps (EHP) and consumption of electricity in the fans (CEF).

2 Results and discussions

The components of the thermal load are determined hourly for both existent and proposed situation as presented for the proposed situation in figure 9.

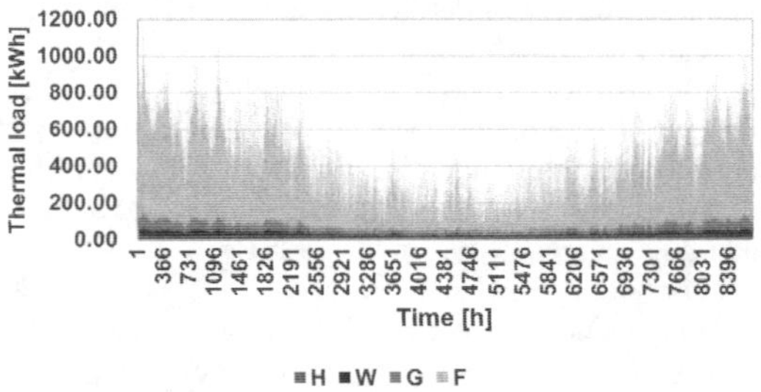

Fig. 9. The hourly components of the thermal load for the proposed situation

It can be observed that the thermal load due to the fresh air is the most important one in the context of reduced loads due to the proposed retrofit measures.

The monthly components of the thermal load are presented in figure 10.

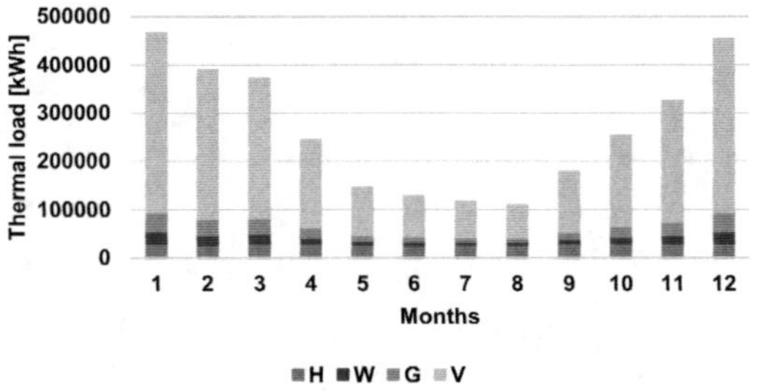

Fig. 10. The monthly components of the thermal load for the proposed situation

It can be observed the difference between the thermal loads in summer and in winter. The total thermal load varies between 109000 kWh in August and 467500 kWh in January.

The shares of the yearly thermal loads are presented in figure 11.

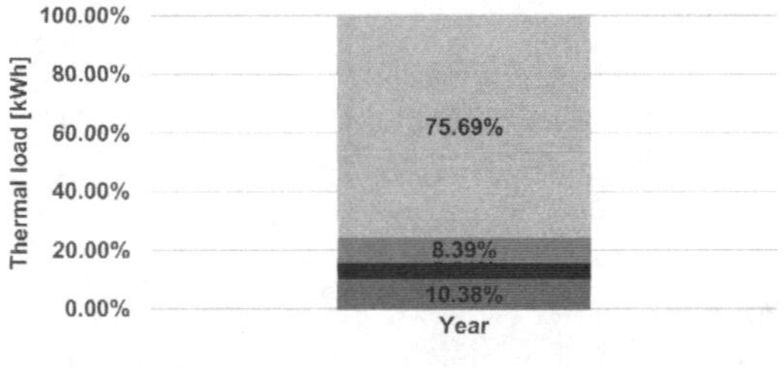

Fig. 11. The yearly shares of the thermal load for the proposed situation

E3S Web of Conferences **32**, 01016 (2018)
EENVIRO 2017
https://doi.org/10.1051/e3sconf/20183201016

The **DHW** share of 10.38 % is higher than the ones corresponding to the walls and to the windows, but the higher share of 75.69 % corresponds to the fresh air.

The monthly components of the distributed thermal loads are presented in figure 12.

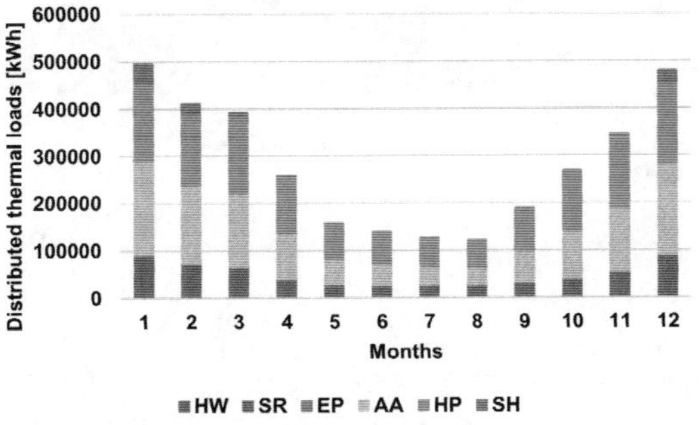

Fig. 12. The monthly components of the distributed thermal loads for the proposed situation

The electric preheaters are needed only in the months of December, January and February, while the supplementary heating coils are used in October – December and January – April.

The shares of the yearly distributed thermal loads are presented in figure 13.

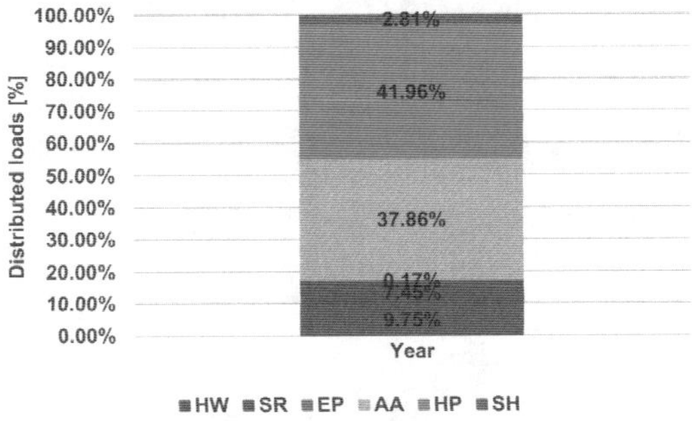

Fig. 13. The yearly shares of the distributed thermal loads for the proposed situation

The most important shares of the distributed thermal loads correspond to the air to air heat recovery (AA: 37.86 %) and to the air to air heat pumps (HP: 41.96 %).

The annual operating time for each type of heat distribution is presenting in figure 14.

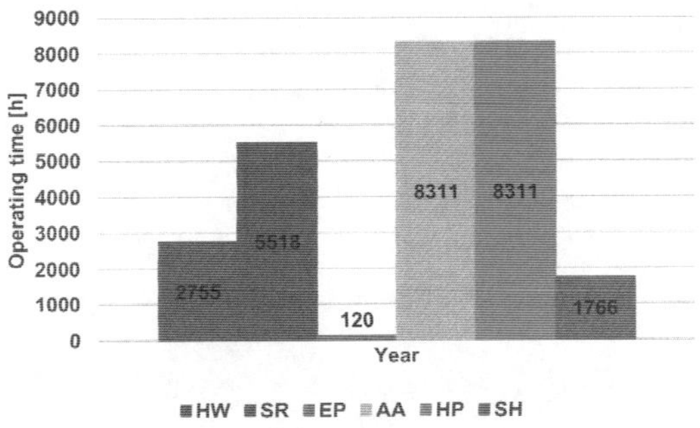

Fig. 14. The annual operating time for each type of heat distribution for the proposed situation

The air to air heat recovery (AA) operates the same number of hours with the air to air heat pumps (HP), because AA can always provide only 50 % of the needed thermal power and HP is needed for the rest of 50 %. Supplementary over the 8311 hours of yearly operation in heating mode, the **HVAC** systems that includes both AA and HP equipment, operates also 319 hours in cooling mode and 130 hours in "only ventilation" mode (no heating or cooling is needed because the outside temperature differs with less than ±3 °C than the inside temperature). The shortest operation period corresponds to the electric preheater (EP 120 hours).

The consumed energy needed to provide the thermal load through the heat distribution equipment is of both thermal type (represented by consumption of natural gas) and of electric type (represented by electricity to power the fans, the heat pump compressors and the electric preheaters).

The monthly components of the consumed energy are presented in figure 15.

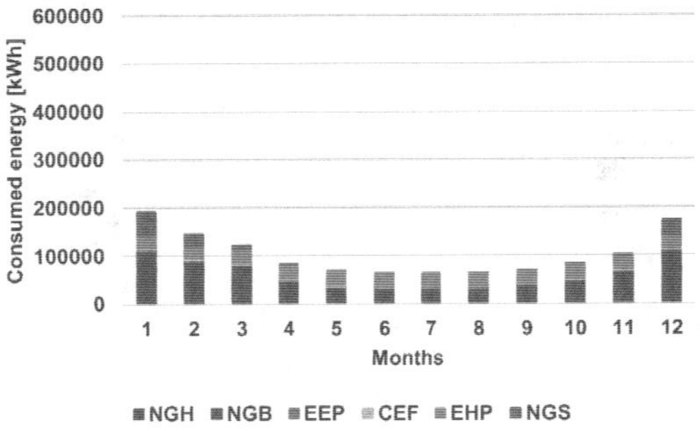

Fig. 15. The monthly components of the consumed energy for the proposed situation

The absolute values of the consumed energy are significant lower than the values of the thermal load and of the distributed thermal load because an important share of thermal load is provided by heat recovery in the air to air heat exchangers and the distributed thermal load of the heat pump is provided with low energy consumption by the air to air heat pump that operates with very high coefficient of performance in the range of (6.25 – 6.34) for all the operating domain of ambient temperatures (-10 – +15) °C.

The shares of the yearly energy consumptions are presented in figure 16.

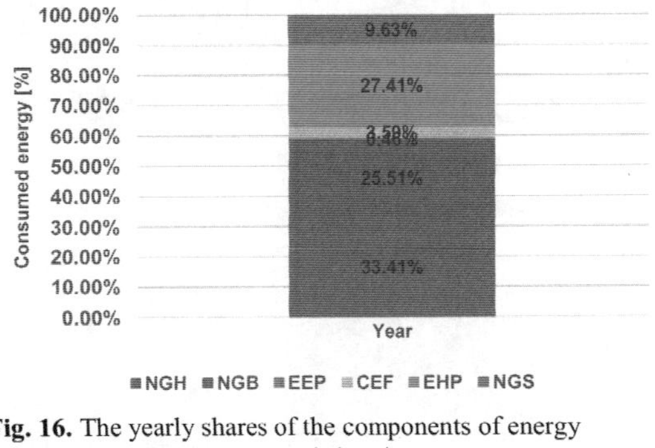

Fig. 16. The yearly shares of the components of energy consumptions for the proposed situation

https://doi.org/10.1051/e3sconf/20183201016

The natural gas consumption represents 68.55 % (NGH + NGB + NGS) and the consumption of electricity represents 31.45 % (EEP + CEF + EHP).

The global solar radiation in the plane of the tilted solar collectors, both thermal and photovoltaic, is presented in figure 17.

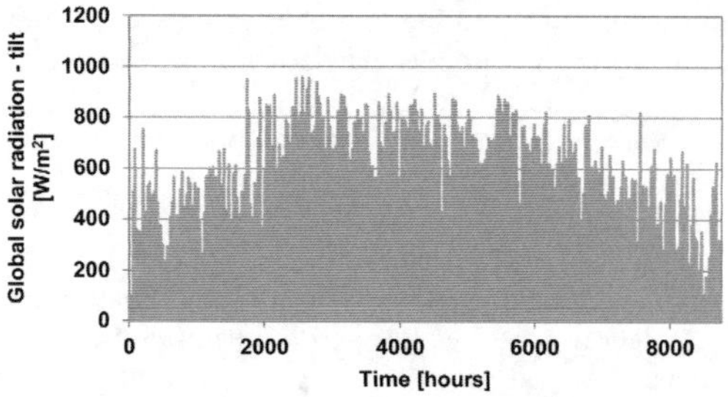

Fig. 17. The global solar radiation in the tilted plane

Due to the tilted plan of the collectors the available solar radiation in this plan is higher than the global radiation in the horizontal plan, available in the **TMY**.

The thermal efficiency of the thermal collectors is presented in figure 18.

The higher value of the considered thermal collectors is of almost 60%.

The solar thermal production of the ST system is presentd in figure 19.

The solar thermal production is contributing to the **DHW** preparation as it is presented in figure 20.

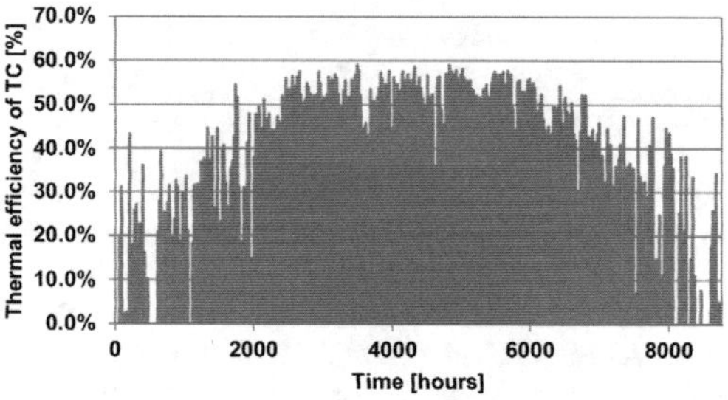

Fig. 18. The thermal efficiency of the thermal collectors

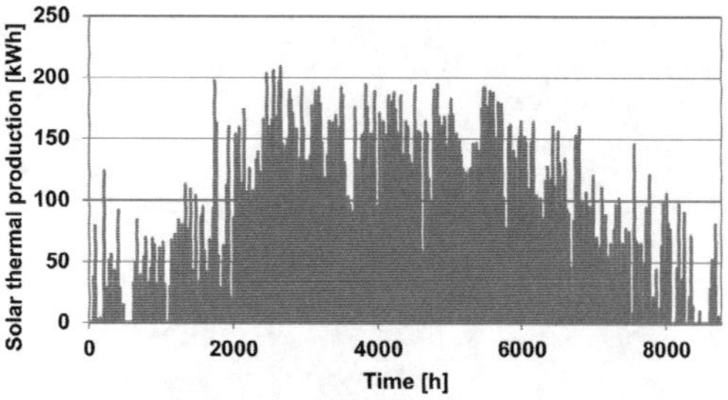

Fig. 19. The solar thermal production of the thermal collectors

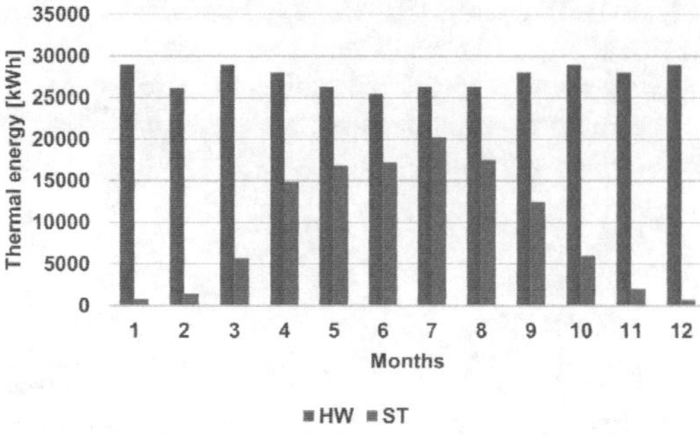

Fig. 20. DHW thermal load and solar thermal production

The total yearly amount of solar thermal production represents about 35 % of the total **DHW** thermal load.

The global efficiency of PV to provide electricity of alternative current (**AC**) type is presented in figure 21.

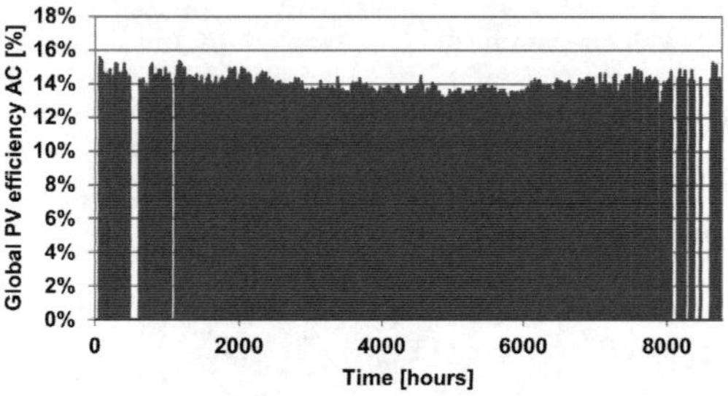

Fig. 21. The global efficiency of the **PV**

The solar electric production is presented in figure 22.

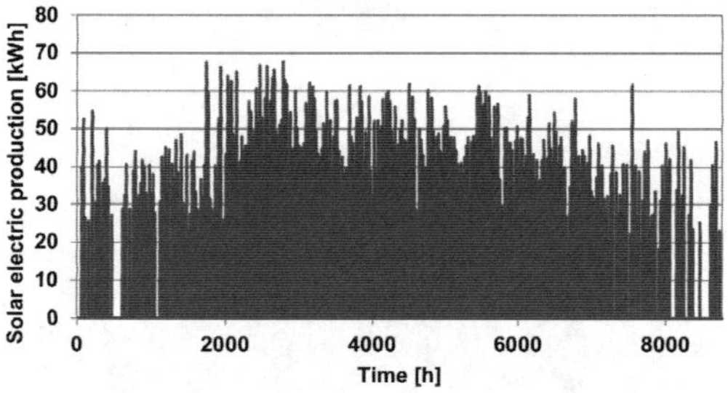

Fig. 22. The solar electric production

The maximum hourly provided electric energy is of 67.8 kWh and is distributed for the building.

The comparative electric consumption of the fans of the distributed HVAC system and the provided electricity is presented in figure 23.

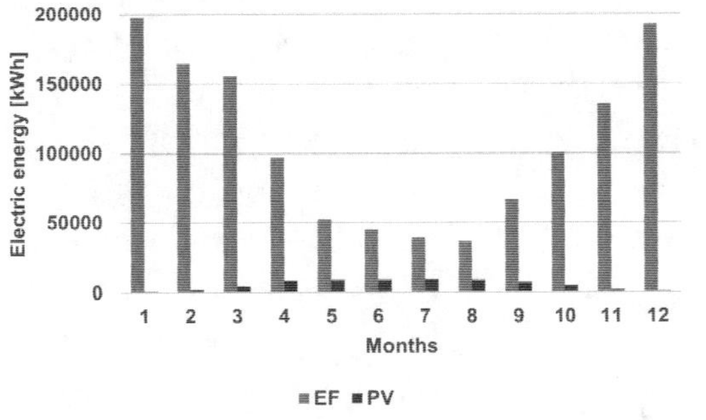

Fig. 23. Fans consumption and solar electric production

The total yearly amount of solar electrical production represents about 5.2 % of the total electrical consumption of the fans.

The thermal and electrical solar energy represents 14.9 % of the global energy consumption of the building.

The comparative global energy consumption in the existent and proposed situations is presented in figure 24.

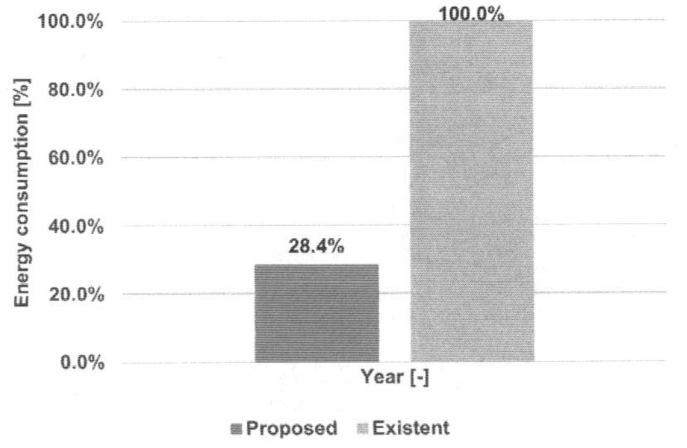

Fig. 24. Existent and proposed global energy consumptions

The global reduction of energy consumption through the proposed retrofit improvements can be obtained in the context of the increased energy demand with 31.1%, determined by the distributed **HVAC** system capable of ventilation with sufficient air flow rate and also capable of cooling the fresh air.

3 Conclusions

The study proposed a set of retrofit measures to reduce the energy consumption of an existent hospital building.

The proposed retrofit includes classic improvements of building envelope like thermal insulation and windows replacement, but also very efficient improvements from the energy consumption point of view like a distributed **HVAC** system and two solar systems, one thermal and one photovoltaic.

Despite the increasing of the global energy demand of 31.1% comparing to the existing situation, determined by the need to heat and cool the ventilated fresh air, the global energy consumption of the building can be reduced by 71.6 %.

The specific energy consumption of the building, reported to the heated area, was reduced from 901.3 kWh/m^2/year in the existing situation, at 301.0 kWh/m^2/year in the proposed situation not taking into account the solar technologies and at 256.3 kWh/m^2/year in the proposed situation including the solar technologies. The last value is close to the upper limit of nZEB non-residential building in Romania (192 kWh/m^2/year). Other solutions to reach the status of nZEB can be the use of a ground air heat exchanger of to use the facades of the building to produce solar heat or solar electricity.

The solar technologies alone, provides 14.9 % of the global energy consumption of the building.

Under these circumstances the study proved the potential of distributed **HVAC** system including air to air heat recovery and air to air heat pump, together with **ST** and **PV**, to increase the energy efficiency of an existent hospital building.

References

1. Directive 2010/31/EU, Official Journal of the European Union, L153 (2010)
2. S. Barlow, D. Fiala, Energy and Buildings **39** 837-846 (2007)
3. S. Roberts, Energy Policy **36** 4482-4486 (2008)
4. Z. Ma, P. Cooper, D. Daly, L. Ledo, Energy and Buildings **55** 889-902 (2012)
5. A.A. Muresan, S. Attia, Renevable and Sustainable Energy Reviews **74** 349-363 (2017)
6. F. Ascione, N. Bianco, R.F. De Masi, G.M. Mauro, G.P. Vanoli, Energy and Buildings **144** 303-319 (2017)
7. T.D. Mora, F. Cappelletti, F. Peron, P. Romagnoni, F. Bauman, Energy Procedia **78** 1359-1364 (2015)
8. D. D'Agostino, P. Zangheri, L. Castellazzi, Energies **10** 117 (2017)
9. J. Staljanssens, S. Mange, N.V. Den Bossche, J. Moen, Energy Procedia **78** 961-966 (2015)
10. A. Buonomano, F. Calise, G. Ferruzzi, A. Palombo, Energy **78** 555-572 (2014)
11. A. F. Radwan, A.A. Hanafy, M. Elhelw, A.E-H.A. El-Sayed, Alexandria Engineering Journal **55** 3061-3071 (2016)
12. A. Carbonari, R. Fioretti, M. Lemma, P. Prinicipi, Energy Procedia **78** 1033-1038 (2015)
13. G. Bizzarri, G.L. Morini, Applied Thermal Engineering **26** 161-169 (2006)
14. G. Pagliarini, C. Corradi, S. Rainieri, Applied Thermal Engineering **44** 150-158 (2012)
15. T.P. Todoran, M.C. Balan, Energy and Buildings **133** 799-809 (2016)

16. B.A. Porumb, M.C. Balan, R.A. Porumb, Energy Procedia **85** 433-441 (2016)
17. R.A. Botu, B.A. Porumb, M.C. Balan, Energy Procedia **85** 442-451 (2016)
18. B.A. Porumb, P.V. Unguresan, L.V. Fechete Tutunaru, A. Serban, M.C. Balan, Energy Procedia **85** 461-471 (2016)
19. B.A. Porumb, P.V. Unguresan, L.V. Fechete Tutunaru, A. Serban, M.C. Balan, Energy Procedia **85** 452-460 (2016)
20. L. Jäntschi, M.C. Balan, E. Podar, S.D. Bolboaca, IEEE Region 8 Eurocon 2007, 2009-2014 (2007)
21. A. Bahtia, Continuing Education and Development (N/A)
22. J.A. Duffie, W.A. Beckman, Wiley & Sons (1980)
23. V. Quaschning, Earthscan (2007)
24. R.A. Porumb, B.A. Porumb, M.C. Balan, Energy Procedia **112** 108-117 (2017)
25. P.V. Unguresan, D. Petreus, A.G. Pocola, M.C. Balan, Energy Procedia **85** 584-593 (2017)

Preliminary investigation of thermal behaviour of PCM based latent heat thermal energy storage

Octavian G. Pop[1], *Lucian* Fechete Tutunaru[2], *Florin* Bode[1] and *Mugur C.* Balan[1,*]

[1]Technical University of Cluj-Napoca, Dept. of Mechanical Engineering, Bd. Muncii 103-105, 400641, Cluj-Napoca, Romania
[2]Technical University of Cluj-Napoca, Dept. of Automotive Engineering and Transports, Bd. Muncii 103-105, 400641, Cluj-Napoca, Romania

Abstract. Solid-liquid phase change is used to accumulate and release cold in latent heat thermal energy storage (**LHTES**) in order to reduce energy consumption of air cooling system in buildings. The storing capacity of the **LHTES** depends greatly on the exterior air temperatures during the summer nights. One approach in intensifying heat transfer is by increasing the air's velocity. A **LHTES** was designed to be integrated in the air cooling system of a building located in Bucharest, during the month of July. This study presents a numerical investigation concerning the impact of air inlet temperatures and air velocity on the formation of solid **PCM**, on the cold storing capacity and energy consumption of the **LHTES**. The peak amount of accumulated cold is reached at different air velocities depending on air inlet temperature. For inlet temperatures of 14°C and 15°C, an increase of air velocity above 50% will not lead to higher amounts of cold being stored. For Bucharest during the hottest night of the year, a 100 % increase in air velocity will result in 5.02% more cold being stored, at an increase in electrical energy consumption of 25.30%, when compared to the reference values.

1 Introduction

Worldwide effort is being dedicated to the reduction of energy consumption in the buildings sector, as about 40% of the global energy consumption is used in buildings [1]. The energy efficiency of air cooling systems can be enhanced by storing the cold available during summer nights in latent heat thermal energy storage systems (**LHTES**). These technologies use phase change materials (**PCM**), and their capability of enhancing the energy efficiency of fresh air conditioning systems have been the subject of many reviews [2-8].

Cooling applications using **PCM** are still in a phase where demonstration and research is required [8]. Considering that, the solidification process takes place due to the low ambient temperatures during the night the selection of the appropriate **PCM** plays a very important role. In [6] thermo-physical properties that the **PCM** must poses are presented. In [9] a selection algorithm is presented that takes into account the climatic parameters.

Organic **PCM** such as paraffin waxes are available on the market [10, 11].

Paraffin waxes are known to have a range of temperatures in which phase change occurs, and their thermo-physical properties and thermal behaviour of commercially available paraffin during solidification and melting are studied in [12-16]. Modern colorimetric techniques are necessary in order to accurately determine the variation of the heat flux with temperature during phase change. Once this variation is known, the specific heat, or apparent heat capacity variation with temperature, of the **PCM** can be determined, through thermal balance equations [12, 13].

For accumulation of maximum amount of cold, both the air inlet temperature and the air velocity present a high influence [17]. It was found that for the same flow rate and solidification time, with inlet air temperatures of (20-24) °C the solid mass fraction of (0.4-1) was obtained.

In [18] the melting process of a salt hydrate PCM was carried out at two different air inlet temperatures (melting temperature +2°C and melting temperature +9°C) and the solid fraction in pouches was determined during melting between (0.62-1).

For this study an **LHTES** consisting of a packed bed of spheres with simple cubic structure, was considered. Each sphere is filled with **PCM**. The advantage that spherical shells have is the high ratio between the outer surface area of the sphere and the interior volume [19].

The goal of the study is to investigate the influence of the air velocity and air inlet temperature on the thermal behaviour of the **LHTES** and on the electrical energy consumption. For this purpose a mathematical model was developed to describe the thermal behaviour of the **LHTES**. The model takes into account the variation of the apparent heat capacity with temperature for the selected **PCM**.

* Corresponding author: mugur.balan@termo.utcluj.ro

2 Material and method

The **LHTES** was designed to reduce the energy consumption of an office building situated in Bucharest, Romania at 44.511° N latitude, 26.078° E longitude and 90 m altitude during the month of July. The climatic data was taken from the Typical Meteorological Year (**TMY**) [20].

The considered office building has two floors, with surface area per floor of 1500 m^2, the height of the total building is 7m and the number of occupants is 180. The specific air flow rate for this type of activity per occupant (25 m^3/h/pers.) and per unit surface area (1.26 m^3/h/pers.) were taken from [21].

2.1. Selection and thermal properties of the PCM

The **PCM** considered for this study was adopted by implementing the selection algorithm, presented in [9], considering the particular climatic conditions of Bucharest, according to the **TMY**. The resulting phase change temperature is in the vicinity of 20°C. Thus the commercially available paraffin **RT20** was chosen [10].

Similar with other paraffin waxes, **RT20** undergoes phase change in a temperature range. Differential scanning calorimetry (**DSC**), is a method that allows the determination of the thermo-physical properties of **PCM**, in small samples. The **DSC** results provide the variation of the heat flux with temperature of the **PCM** during melting and solidification [22], the start (t_s°C) and the end (t_e°C) temperatures of the phase change process, the onset (t_{on}) and the endset (t_{end}) temperatures of the measurement, the peak temperature (t_p°C) where the value of the heat flux assumes a maximal value, and the latent heat for melting and solidification (L [kJ/kg]), which represents the surface area beneath the heat flux curve.

The **DSC** measurements for **RT20** are presented in [12], at three different cooling and heating rates. It is also specified that the cooling and heating rate of the measurement should match the rates of the application, concluding that for a rate of 0.1 K/min the numerical predictions are the most accurate for air conditioning applications [12].

The thermal properties determined in [12] for **RT20**, at a cooling and heating rate of 0.1K/min are presented in table 1.

Table 1 thermal properties of **RT20** at heating and cooling rate of 0.1K/min

Process	t_{on} [°C]	t_s [°C]	t_p [°C]	t_e [°C]	t_{end} [°C]	L [kJ/kg]
Melting	6.8	17.9	21.8	23	23.4	138.3
Solidification	21.8	21.7	21.5	21.1	2.9	143.4

The density of **RT20** is 880 kg/m3 and 770 kg/m3 in solid phase and liquid phase respectively [10].

A thermal hysteresis effect between melting and solidification is observed in the considered **PCM**. The thermal behaviour of the **PCM** during melting and solidification is not identical. There is also a slight difference between the measured latent heat, for solidification and melting. The encapsulation of the **PCM** also contributes to this effect [23]. This effect generates difficulties when estimating the mass fraction of solid and liquid **PCM** as phase change is can be interrupted in free cooling applications. To eliminate the hysteresis effect, for this study, it was assumed that the **PCM** behaves identically during both solidification and melting, considering a theoretical variation of the apparent heat capacity with temperature as a mean value between the apparent heat capacity for solidification and melting. Fig. 1 presents the experimentally determined variation of the apparent heat capacity with temperature for **RT20** and the theoretical mean value considered.

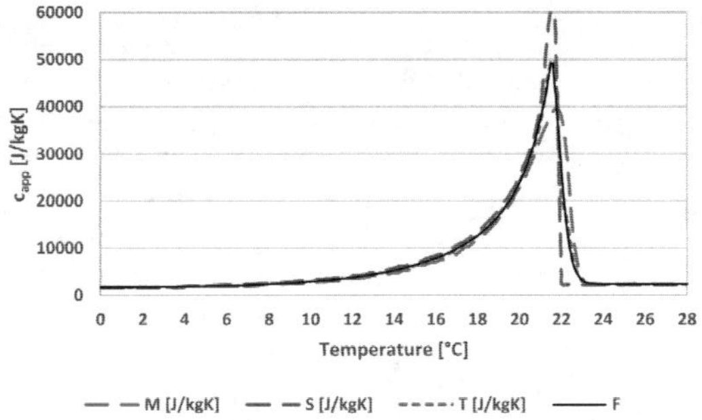

Fig. 1 The apparent heat capacity of **RT20** for melting (M), solidification (S), theoretical mean value (T) and the fitted curve (F)

To take into account the variation of the apparent heat capacity with temperature in the governing thermal balance equations, a curve was fitted that accurately matches the variation. The curve is mathematically described by a discontinuous function as follows:

- An exponential growth of the second order:

$$c_{app,m}(t_{PCM}) = y_0 + A_1 \cdot exp(t_{PCM}/b_1) + A_2 \cdot exp(t_{PCM}/b_2), for\ t_{PCM} < t_p \quad (1)$$

- A Boltzmann variation:

$$c_{app,m}(t_{PCM}) = B_2 + (B_1 - B_2) \cdot (1 + exp((t_{PCM} - x_0)/d)),\ \ for\ t_{PCM} \geq t_p \quad (2)$$

where $c_{app,m}(t_{PCM})$ is the mean apparent heat capacity as a function of temperature, t_{PCM} is the temperature of the **PCM**. The values of the coefficients A_1, A_2, B_1, B_2, b_1, b_2, d, x_0, y_0, where determined with Origin Lab, and are presented in table 2 and 3.

Table 2 Values of the coefficients -Exp. growth of the second order: $t_{PCM}<t_p$

y_0	A_1	A_2	b_1	b_2
1612.07	4.29E-6	93.27	0.96	3.81

Table 3 Values of the coefficients -Boltzmann variation: $t_{PCM}>t_p$

B_1	B_2	x_0	d
64397.50	2361.30	21.90	0.26

The peak temperature for both solidification and melting was considered to be 21.5°C, and the theoretical latent heat was determined with the following equation:

$$L_{PCM,m} = \int_{t_{on}}^{t_{end}} c_{app,m}(t_{PCM})dt \quad (3)$$

where $L_{PCM,m}$ is the mean latent heat [kJ/kg] for both solidification and melting, assuming a value of 140,50 kJ/kg, determined considering t_{on} and t_{end} at 0°C and 28°C respectively.

2.2 The sizing of the LHTES

A rectangular parallelepiped shape is considered for the **LHTES**, the spheres filled with **RT20** are considered uniformly distributed inside de **LHTES**. The **LHTES** and the characteristic dimensions are presented in fig. 2.

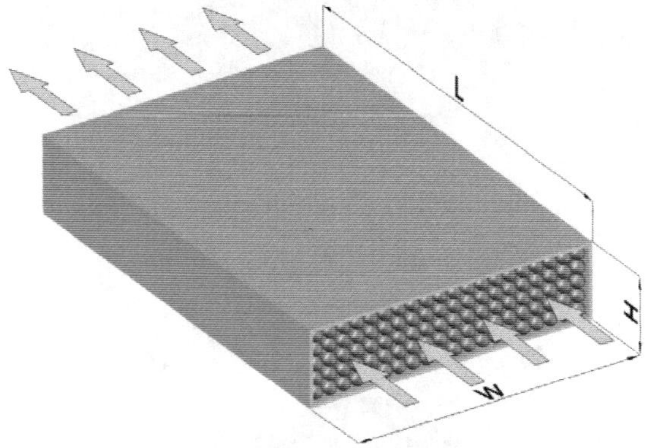

Fig. 2 The **LHTES** and characteristic dimensions

In figure 2, H [m] represents the height, W[m] represents the width and L[m] represents the length of the **LHTES**. The encapsulation material is polyethylene, as in [12].

The **LHTES** was design based on the cooling potential of the exterior air during the nights of July, so to ensure that the mass fraction of the solid **PCM** after the charging period is as high as possible.

The total mass of the **PCM** ($m_{t,PCM}$ [kg]) for the considered **LHTES** was determined as:

$$m_{t,PCM} = \frac{\dot{V}_a \cdot \rho_a \cdot c_a \cdot (t_{s,c} - t_{ext}) \cdot \Delta\tau_h}{L} \quad (4)$$

where $\dot{V}_a$ [m³/s] is the volumetric air flow rate of the air $\Delta\tau_h$ = 1 hour, ρ_a [kg/m³] is the density of the air, c_a is the specific heat capacity of the air, t_{ext} [°C] is the exterior air temperature during the night, $t_{s,c}$ [°C] is the temperature at which the solidification starts. For each night of the considered period the total mass of **PCM** was calculated, and an average value was adopted.

2.3 The mathematical model of the LHTES

A mathematical model was developed in order to predict the thermal behavior of the **LHTES**. The governing thermal balance equations are as follows:

$$\dot{m}_a \cdot c_a \cdot \Delta t_a \cdot d\tau = \left[m_{PCM} \cdot c_{app,m}(t_{PCM}) + \rho_s \cdot V_s \cdot c_s\right] \cdot dt_{PCM} \quad (5)$$

$$h \cdot \Delta t_{a,PCM} \cdot S_{PCM} \cdot d\tau = \dot{m}_a \cdot c_a \cdot \Delta t_a \cdot d\tau \quad (6)$$

where $\dot{m}_a$ [kg/s] is the mass air flow rate of the air, t_a [°C] is the exterior air temperature, x [m] is the position of the row of spheres, τ [s] is the time, m_{PCM} is the mass of the **PCM**, h [W/m²K] is the coefficient of forced convection, ρ_s [kg/m³] is the density of the encapsulation material, V_s [m³] is the volume of the encapsulation material, c_s [kJ/kgK] is the specific heat capacity of the encapsulation material, $\Delta t_{a,PCM}$ [°C or K] is the mean logarithmic temperature difference between the air temperature and the **PCM** temperature, and S_{PCM} [m²] is the heat transfer surface area representing the surface of the **PCM** spheres.

The coefficient of convection was determined as:

$$h = \frac{Nu \cdot \lambda}{d} \quad (7)$$

where Nu [-] is the Nusselt number for forced convection over spheres, calculated with the correlation presented in [24, 25], as follows:

$$Nu = 0.33 \cdot Re^{0.6} \quad (8)$$

where Re [-] is the Reynolds number in the free flowing section.

This correlation is valid for a large range of Re = (20…150000). In the model the variation of the thermo-physical properties of the air, with the temperature, was taken into account.

The following assumptions that are at the basis of the numerical solution are at the basis of the numerical solution:

- It is considered that the mass of all **PCM** spheres on one row is concentrated in one equivalent sphere;
- Spheres are evenly distributed, therefore porosity is constant;
- The temperature of **PCM** spheres from the same row is considered constant;
- Heat transfer through natural convection between the **LHTES** and surrounding environment is neglected;
- The air temperature on the flow section is considered constant

The same assumptions where used in [19].

The mass fraction of the solid **PCM** (γ[-]) was calculated as:

$$\gamma = \frac{\int_{t_{on}}^{t_{PCM}} c_{app,m}(t_{PCM})dt}{\int_{t_{on}}^{t_{end}} c_{app,m}(t_{PCM})dt} \quad (9)$$

2.4 Pressure drop of the LHTES

The local pressure drop of the **LHTES** (Δp [Pa]) was determined:

$$\Delta p = \xi \cdot \frac{\rho_a \cdot w^2}{2} \quad (10)$$

where w [m/s] is the air velocity, ξ is the local pressure drop coefficient of the **LHTES** calculates as [26]:

$$\xi = \frac{1.53}{\varepsilon^{4.2}} \cdot \left(\frac{30}{Re} + \frac{3}{Re^{0.7}} + 0.3 \right) \cdot \frac{L}{d_s} \quad (11)$$

where ε is the porosity of the bed of spheres:

$$\varepsilon = 1 - \frac{\pi}{6 \cdot (1 - cos\theta \cdot \sqrt{1 + 2 \cdot cos\theta})} \quad (12)$$

where θ = 90° is the angle of relative placement of the spheres [26] and d_s = 50 mm is the considered exterior diameter of the spheres. The wall thickness of the sphere is 1mm. The considered material from which the spheres were manufactured is polyethylene with a specific heat capacity of c_s=1.55 [kJ/kgK] and a density of ρ_s=925 [kg/m^3] [27].

According to the **TMY**, the ambient air temperatures during the nights of the month of July in Bucharest are ranging from 13°C to 21°C. In order to solidify as much mass as possible the convective heat transfer must be enhanced by increasing the air velocity. This however results in additional energy required for the circulation of air through the **LHTES**. Using the mathematical model, a numerical analysis was carried out, considering a series of constant air inlet (t_i [°C]) temperature in the interval of (13-21) °C, with an increment of 1°C. The air velocity was also increased up to 300%, with a step of 25%, to study both the effect on the heat transfer intensity as well as the impact on the consumed energy.

The electric energy consumed by the fan (E_f [Wh]) was determines as:

$$E_f = \dot{V}_a \cdot \Delta p \cdot \tau / \eta_v \quad (13)$$

where τ [hours] is the operating time and η_v is the efficiency of the fan considered 0.8 for this study.

Each numerical simulation was carried out for a time period of 5h, and the initial **PCM** temperature was considered to be 28°C.

The **LHTES** sizing algorithm was implemented in C++ and the results were exported in Excel.

3 Results and discussions

3.1 The size of the LHTES

The required air flow rate for the considered office building is 8280 m^3/h. The reference air velocity corresponding to this air flow rate is 0.77 m/s. The mass of the **PCM** integrated in the **LHTES** was determined using eq. (4), resulting in a total quantity if 2.065t of **PCM**. The Entire mass of **PCM** was distributed in the **LHTES** rectangular parallelepiped shaped resulting in four heat exchangers with the following dimension: H=0.50 m, W=1.50 m and L=2.20 m.

During cyclic operating conditions, for air cooling application, melting of the entire mass of available **PCM** in the **LHTES** occurs, especially after peak operating conditions. Solidification of the entire mass of **PCM** is however problematic as after peak operating conditions the temperatures of the following night are high. As the mass, heat transfer surface area and size of the **LHTES** are constant parameters, the only method of intensifying the heat transfer between air and **PCM** is by increasing the velocity of the heat transfer fluid, in this case air.

3.2 Influence of air inlet temperature and air velocity

Fig. 3 and 4 present the correlation between the accumulated cold, the mass fractions of solid **PCM** and air velocity, for air inlet temperatures of 13°C, 17°C and 21°C.

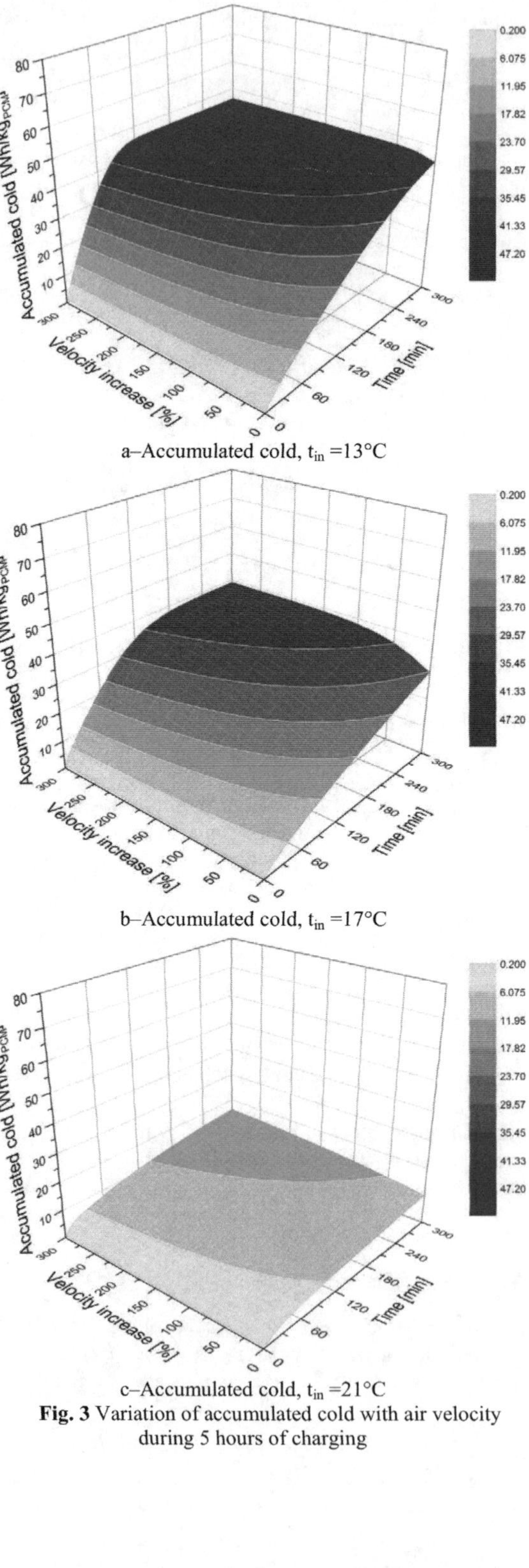

a–Accumulated cold, t_{in} =13°C

b–Accumulated cold, t_{in} =17°C

c–Accumulated cold, t_{in} =21°C

Fig. 3 Variation of accumulated cold with air velocity during 5 hours of charging

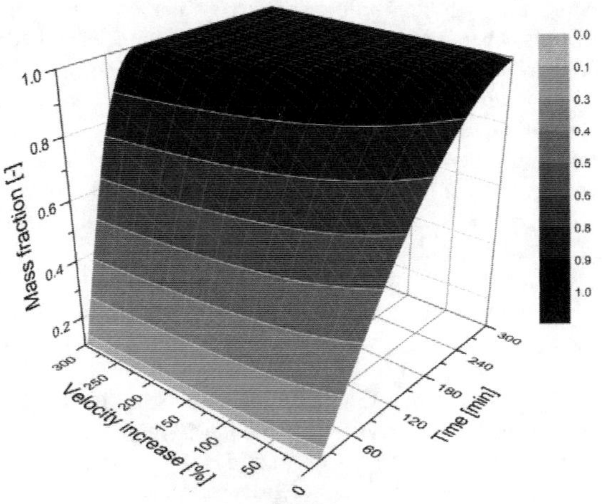

a–Mass fraction of solid **PCM**, t_{in} =13°C

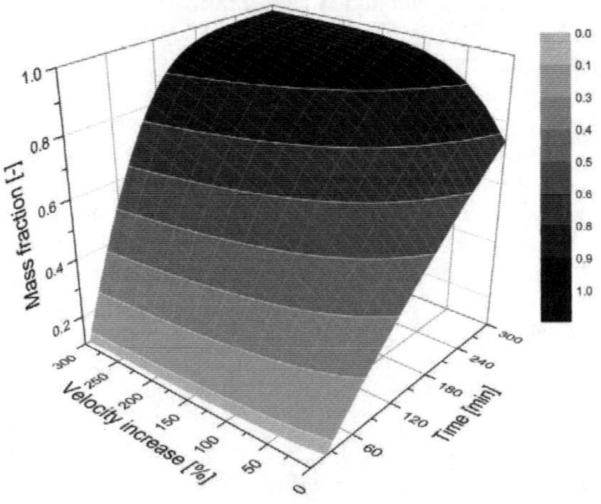

b–Mass fraction of solid **PCM**, t_{in} =17°C

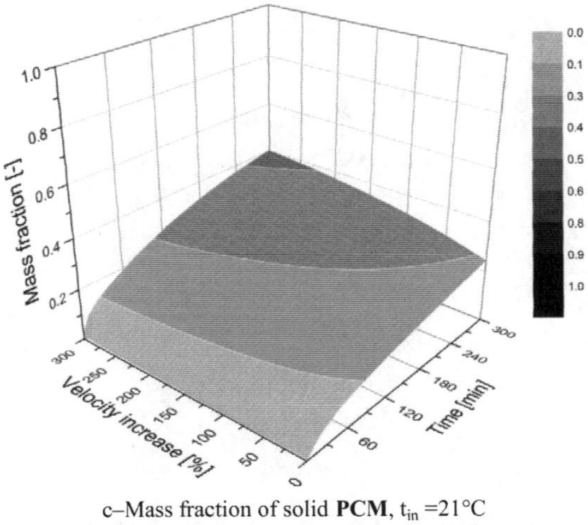

c–Mass fraction of solid **PCM**, t_{in} =21°C

Fig. 4 Variation of mass fraction of the solid **PCM** with air velocity during 5 hours of charging

The influence of air inlet temperature on the electrical energy consumed by the fan is negligible. Considering the air inlet temperature of 13°C as reference, at an air inlet temperature of 21°C the energy consumed by the fan drops with a maximum of only 2.62%. Fig. 5 presents the variation of the accumulated energy consumption of the fan at different velocities during 5h of operation, for an air inlet temperature of 13°C.

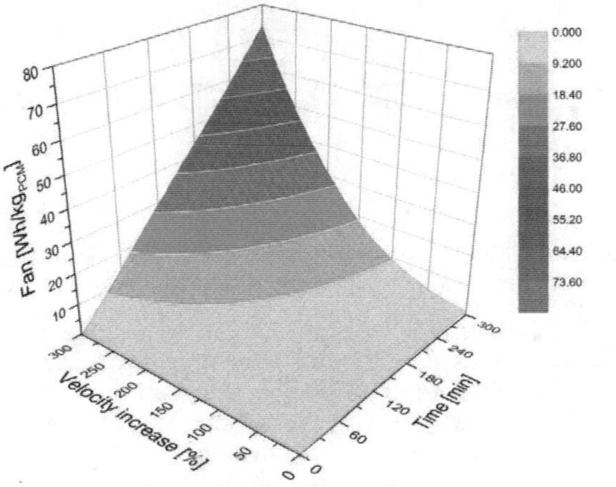

Fig. 5 The variation of electrical Energy absorbed by the Fan with air velocity during 5 hours of charging

Fig. 6 presents the total accumulated cold, at different air velocities and different air inlet temperatures, after five hours of charging. The total accumulated cold [Wh/kg_{PCM}] and the electrical energy consumed by the fan [Wh/kg_{PCM}] was calculated considering the maximum value of cold that can be stored at the studied air inlet temperature interval, velocities and charging time, as reference.

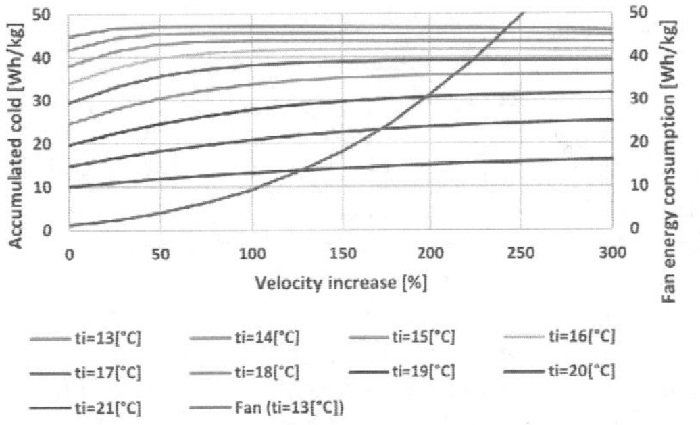

Fig. 6 Variation of accumulated cold and electrical energy consumed by the fan with air velocity, at different air inlet temperatures

The maximum value of cold that can be stored is 47.09 [Wh/kg_{PCM}] at an air inlet temperature of 13°C and air velocity of 3.067 m/s. The minimum value of cold that can be stored is 10.06 [Wh/kg_{PCM}] at an air inlet temperature of 21°C and at the reference air velocity of 0.77 m/s.

Fig. 7 shows the formation of solid mass of **PCM** at different air velocities and different air inlet temperature, after five hours of charging.

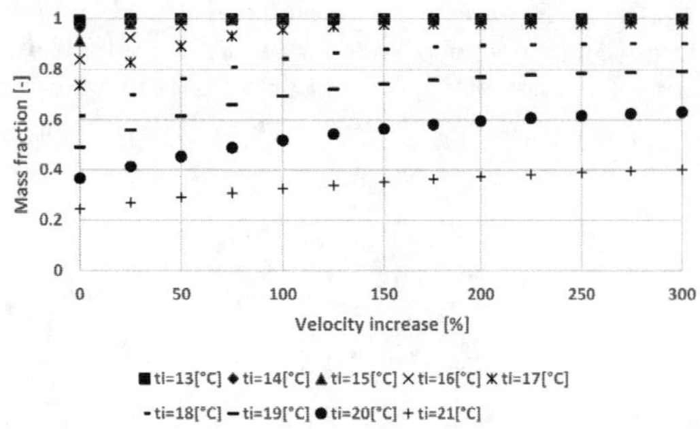

Fig. 7 Evolution of mass fraction of solid **PCM** with air velocity, at different air inlet temperature

A stagnation on the variation curves of accumulated cold and of solid mass fraction of **PCM** is observed in the following cases: for the air temperature of 13 °C, above a velocity increase of 25 %; for the air temperature of (14-15) °C, above a velocity increase of 50 %; for the air temperature of 16 °C, above a velocity increase of 100 %; for the air temperature of 17 °C, above a velocity increase of 150 %.

For air inlet temperature of 18°C the stagnation of the variation of accumulated cold and solid mass fraction of **PCM** is observed at the intersection between the fan's electrical energy consumption curve and the accumulated cold curve. Above this point the fan consumes more energy than the energy stored a cold in the **LHTES**. An air velocity increase of maximum 50% considering the variation of the fan's electrical energy consumption curve is a good compromise between accumulated cold, solidified mass of **PCM** and electrical energy consumption.

For air inlet temperatures between (19-21) °C, for the considered operating conditions, a stagnation zone in this interval is inexistent. An air velocity increase of maximum 50 % considering the variation of the fan's electrical energy consumption curve is a good compromise between accumulated cold, solidified mass of **PCM** and electrical energy consumption.

Table 4 summarizes the variations of the accumulated cold, solid mass formed and energy consumption of the fan, determined by the variations of the inlet air temperature and velocity.

Table 4 Response of the **LHTES** at the variation of the inlet air temperature and velocity

t_i [°C]	↑w [%]	↑Q_a [%]	↑γ [%]	↑E_{f1} [%]	↑E_{f2} [%]
13	25	4.20	0.69	92.83	2.50
14	50	8.81	3.63	230.23	6.65
15	50	13.18	9.33	230.19	7.26
16	100	22.36	18.98	673.46	23.78
17	150	32.58	32.83	1399.27	56.79

↑$E_{f1}$is the surplus electric energy consumption of the fan reported to the same value at the reference air velocity
↑$E_{f2}$is the surplus electric energy consumption of the fan reported to the accumulated cold at the reference air velocity
↑Q_a is the increase in accumulated cold

For air inlet temperature of 16°C, the stagnation zone starts at air velocity increase of 100%, where the percentage of energy consumed by the fan and cold stored, when compared to the reference accumulated cold are very close (Q_a=22.36% and E_{f2}=23.78%).

For an air inlet temperature of 17°C, to reach the stagnation zone of the cold accumulation curve at a minimum increase in air velocity of 150%, the electrical energy consumed by the fan increases with 56.79% when compared to the reference accumulated cold, while the cold stored increases with only 32.58% when compared to the reference accumulated cold.

In these cases accumulating the maximum amount of cold at the given parameters requires higher amounts of electrical energy absorbed by the fan.

For the air inlet temperature of 16°C, an increase in air velocity of 75% will lead to 20.80% more cold being accumulated, 18.46% more solid mass being formed while the while the increase in electrical energy consumed by the fan is only 14.86%, when compared to the reference accumulated cold. In this case **PCM** is not fully solidified, the maximum value of the solid mass fraction being 0.995.

The same compromise can be considered for air inlet temperature of 17°C, by reducing the increase in air velocity from 150% to 100%, the resulting increase in accumulated cold is 29.79% with an increase in electrical energy consumption of 27.33%, compared to the reference value, while the maximum value of the solid mass fraction is 0.957. In these cases the stagnation zone of the cold accumulation curve is not reached, however the increase in electrical energy consumption of the **LHTES** is lower than the increase in stored cold.

Fig. 8 and 9 show the **PCM** temperature distribution and mass fraction of solid **PCM** distribution, in the **LHTES** at 17°C and 21°C and the reference air velocity and an air velocity increased by 50%, after five hours of charging.

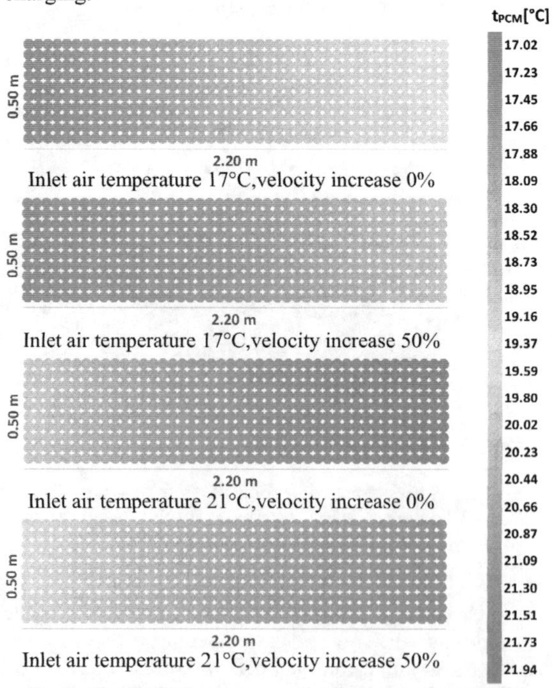

Fig. 8 The **PCM** Temperature distribution in the **LHTES**

 https://doi.org/10.1051/e3sconf/20183201017

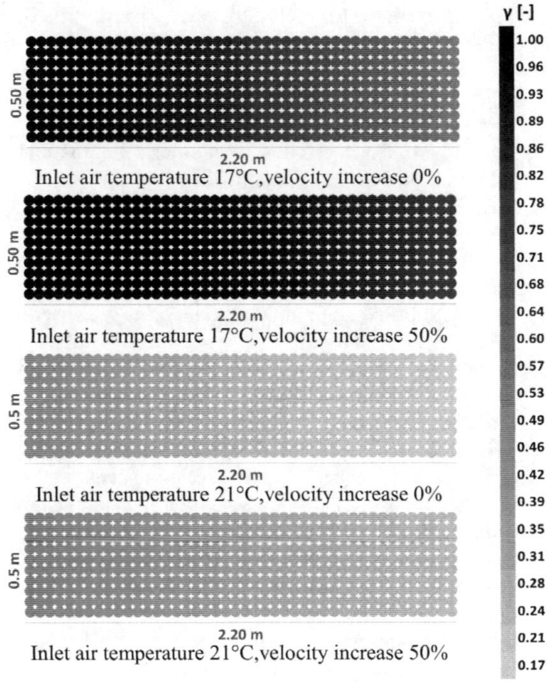

Fig. 9 The mass fraction of solid **PCM** distribution in the **LHTES**

A simulation was carried out (as a case study) for the hottest day of the year in Bucharest, according to the **TMY**, (6th of July), representing the most unfavorable situation for the cold accumulation in the **PCM**. For the air inlet temperatures of (16-19) °C the air velocity and consequently the air flow rate was increased up to 100 % with an increment of 10 %.

Fig. 10 shows the variation of the exterior temperature over a period of 24h according to the **TMY** and end temperature of the solidification process.

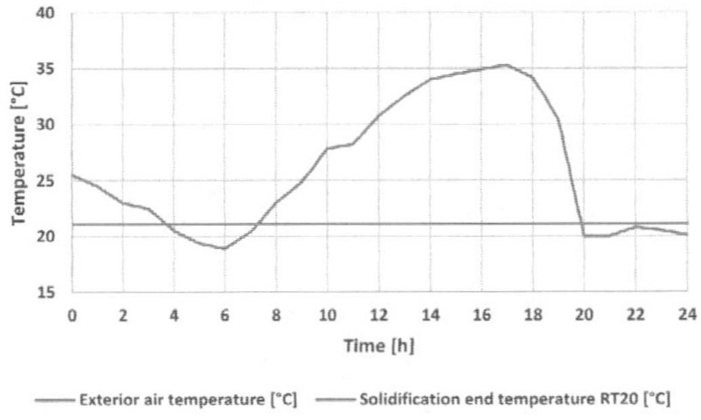

Fig. 10 Variation of the exterior air temperature during the hottest day of July according to the **TMY** and the **PCM** temperature at which solidification ends

As can be seen from fig. 10 the potential for the **PCM** to solidify is limited by the low ambient air temperatures during the charging period, due to the high ambient temperatures. Therefore for air inlet temperatures between (16-19) °C different simulations were carried out where the air velocity was increased up to 100%, with an increment of 10%.

Table 5 presents the influence of air velocity increase during cold accumulation for the hottest day in Bucharest (6th of July) according to the **TMY**.

Table 5 Impact of velocity increase on the accumulated cold (Q_a), and on the corresponding energy absorbed by the fan during cold accumulation (E_f)

↑ w [%]	w [m/s]	↑Qa [%]	↑E_f [%]
0.0%	0.77	0.00%	0.00%
10.0%	0.84	0.54%	1.23%
20.0%	0.92	1.06%	2.69%
30.0%	1.00	1.58%	4.41%
40.0%	1.07	2.09%	6.42%
50.0%	1.15	2.59%	8.72%
60.0%	1.23	3.09%	11.35%
70.0%	1.30	3.57%	14.32%
80.0%	1.38	4.05%	17.66%
90.0%	1.46	4.56%	21.17%
100.0%	1.53	5.02%	25.30%

It can be observed that by increasing the air velocity up to 100 % when compared to the reference air velocity, the accumulated cod can be increased up to 5.02 % with a supplementary energy consumption of the fan of 25.3 %. These values correspond to an increase of accumulated cold of 59.86 Wh/kg from 57.00 Wh/kg at the reference velocity and an increase of fan energy consumption of 8.89 Wh/kg from 7.10 Wh/kg at the reference velocity.

4 Conclusions

This article presents a numerical analysis of thermal behavior of **LHTES** using **RT20** a commercially available paraffin wax as **PCM**.

The thermal behavior of the **LHTES** was analyzed at different air inlet temperatures and air velocities. The air velocities were increased up to 300% with an increment of 25%. The studied temperature domain was (13-21) °C.

It was observed that at certain values of the air velocity the variation of the cold stored in the **LHTES** stagnate. The stagnation zone starts at an increase in air velocity of: 25% for air inlet temperature of 13°C, 50% for air inlet temperature of 14°C and 15°C, 100% for air inlet temperature of 16°C and 150% for air inlet temperature of 17°C. For the temperature interval of (18-21) °C the stagnation zone is not clearly defined.

The impact on the formation of solid mass of **PCM** and the fan's electrical energy consumption was also investigated.

The numerical investigation identified values of the air velocity where the amount of cold stored in the **LHTES** reaches peak values for certain temperatures.

The study evaluated the favorable impact of the inlet air temperature on the accumulated cold that increase with 346 % from 10.05 Wh/kg at the inlet air temperature of 21 °C to 44.82 Wh/kg at the inlet air temperature of 13 °C, and reference air velocity.

Following the study it was highlighted that a reasonable increase of the air velocity between (25-100) %, depending on air inlet temperatures, determines a benefic impact on the accumulated cold. For the air inlet temperature of 15 °C the amount of accumulated cold solidified mass of **PCM** increases with 13.18% and 9.33% respectively, while the increase in energy consumption of the fan represents 7.26% of the accumulated cold at the reference velocity.

The numerical simulation carried out for the hottest day of July in Bucharest, revealed that the accumulated cold can be increased by 5.02 % with an increase in electrical energy consumption of the fan by 25.30 %. Due to the low air velocities (0.77-1.53) m/s the electrical energy consumed by the fan is relatively low in comparison to the accumulated cold. To store 59.86 Wh/kg of cold in the **LHTES** 8.89 Wh/kg of electrical energy is required, representing 14.91 % energy from the total accumulated cold.

With higher quantities of cold stored in the **LHTES** the operating hours of classic air cooling system can be reduced.

These results must be correlated with the operating hours and energy consumption of chiller which will be the subject of future research activities.

The results presented in this paper were obtained with the support of the Technical University of Cluj-Napoca through the research Contract no. 2013/12.07.2017, Internal Competition CICDI-2017.

References

1. Directive 2010/31/EU of the European Parliament and of the Council of 19 May 2010 on the energy performance of buildings 2010.
2. V.A.A. Raj, R. Velraj, Renewable and Sustainable Energy Reviews, **14**, 2010
3. F.A. Regin, S.C. Solanki, J.S. Saini, Renewable and Sustainable Energy Reviews, **12**, 2008
4. R. Parameshwaran, S. Kalaiselvam, S. Harikrishnan, A. Elayaperumal, Renewable and Sustainable Energy Reviews, **16**, 2012
5. V. Basecq, G. Michaux, C. Inard, P. Blondeau, Advances in Building Energy Research, **7**, 2013
6. E. Osterman, V.V. Tyagi, V. Butala, N.A. Rahim, U.Stritih, Energy and Buildings, **49**, 2012
7. H. Akeiber, P. Nejat, M.Z.A. Majid, M.A. Wahid, F. Jomehzadeh, I.Z. Famileh, J.K. Calautit, B.R. Hughes, S.A. Zaki, Renewable and Sustainable Energy Reviews, **60**, 2016
8. M. Alizadeh, S.M. Sadrameli, Renewable and Sustainable Energy Reviews, **58**, 2016
9. O. Pop, M.C. Balan, Termotehnica, **1** 2016
10. *www.rubitherm.de*
11. *www.pcmproducts.net*
12. C. Arkar, S. Medved, Thermochimica Acta, **438**, 2005
13. M. Iten, S. Liu, A. Shukla, P.D. Silva, Applied Thermal Engineering, **117**, 2017
14. X. Jin, H. Hu, X. Shi, X. Zhang, Energy Conversion and Management, **106**, 2015
15. C. Barreneche, A. Solé, L. Miró, I. Martorell, A.I. Fernández, L.F. Cabeza, Thermochimica Acta, **553** 2013
16. L. Huang, C. Doetsch, C. Pollerberg, International Journal of Refrigeration, **33**, 2010
17. A. Waqas, S. Kumar, Energy and Buildings, **43**, 2011
18. A. Lazaro, P. Dolado, J.M. Marín, B. Zalba, Energy Conversion and Management, **50**, 2009
19. O. Pop, L.F. Tutunaru, M. Balan, Energy Procedia, **112**, 2017
20. MRDPA (Ministry of Regional Development and Public Administration). "Typical meteorological data for Bucharest." (2013) Bucharest, Romania (in Romanian)
21. Normativ pentru proiectarea, executarea si exploatarea instalatiilor de ventilare si climatizare, Indicativ I5. 2010 (in Romanian)
22. A. Hasan, S.J. McCormack, M.J. Huang, B. Norton, Energy Conversion and Management, **81**, 2014
23. K. Kumarasamy, J. An, J. Yang, E.-H. Yang, Energy, **132**, 2017
24. E. Cao, Heat transfer in process engineering, ed. T.M.-H. Companies, 2010
25. N. Leonăchescu, Termotehnică, ed. E.d.ș.p. București, 1981 (in Romanian)
26. I.E. Idelcik, Editura tehnică 1984 (in Romanian)
27. *http://www.firebid.umd.edu/material-database.php*

CFD simulation of a cabin thermal environment with and without human body – thermal comfort evaluation

Paul Danca[1, 2], *Florin* Bode[3], *Ilinca* Nastase[2,*] and *Amina* Meslem[1,*]

[1]IUT Rennes, Laboratoire Génie Civil Génie Mécanique 35704 Rennes Cedex 7 France
[2]CAMBI Research Center, Technical University of Civil Engineering Bucharest, 021414 Bucharest, Romania
[3]Technical University of Cluj Napoca, Department of Mechanical Engineering 400020 Cluj - Napoca, Romania

Abstract. Nowadays, thermal comfort became one of the criteria in choosing a vehicle. In last decades time spent by people in vehicles had risen substantially. During each trip, thermal comfort must to be ensured for a good psychological and physical state of the passengers. Also, a comfortable environment leads to a higher power concentration of the driver thereby to a safe trip for vehicle occupants and for all traffic participants. The present study numerically investigated the effect of human body sited in the driver's place, over the air velocity distribution and over the thermal comfort in a passenger compartment. CFD simulations were made with different angles of the left inlet grill, in both cases, with and without driver presence. In majority of the actual vehicles environment studies, are made without consideration of human body geometry, in this case, the results precision can be affected. The results show that the presence of human body, lead to global changing of the whole flow pattern inside the vehicular cabin. Also, the locations of the maximum velocities are changing with the angle of the guiding vanes. The thermal comfort PMV/PPD indexes were calculated for each case. The presence of human body leads to a more comfortable environment.

1 Introduction

Thermal comfort is defined by ASHRAE [1] as the condition of mind that expresses physical and psychological satisfaction with the thermal environment. The first approach was made at the beginning of the 19th century, for the buildings environment by Haldane [2], then, during the time several evaluation models were developed. In 1970 Fanger [3] developed in controlled laboratory conditions and with measurement tools available at that time, PMV and PPD indexes which further was embraced by thermal comfort standards. Fanger's indexes were developed in quasi steady-state conditions and were designed for evaluation of the buildings thermal environment.

Researchers focused their attention in direction of thermal comfort of vehicular occupants only in last decades. Thermal comfort of vehicular occupants is gaining more and more importance due to the rising attention towards comfortable mobility, in addition to the growing time that people spend in vehicles (private or public transport)[4]. A comfortable vehicular climate in many cases not only help to reduce the driver stress but also guarantee good visibility by avoiding the fog phenomenon, and contributing to a safer driving. In addition, many independent studies provide evidence of improper thermal environment and its negative influence on the human body [5, 6].

The present standards for evaluation of vehicle thermal environment, EN ISO 14505 [7-9], proposed buildings evaluation models, which do not meet the demands of for cabin environment evaluation. Vehicle environment is very different form buildings environment, because on the one hand is affected by more factors (i.e. solar radiation, high air velocity, high differences between interior surfaces)[10]. These factors are neglected by proposed standardized evaluation models. On the other hand vehicle environment is fast transient and non-uniform compared to buildings environment that may be considered quasi steady-state. Other factors who may affect thermal state of vehicular passengers is little space affected to each passenger and complex geometry of this environment. Due to the confined environment (small volume) of the vehicle cabin space, almost any detail of interior geometry is very important. Human body affect vehicles thermal environment though heat dissipation and as an obstacle for airflow. His presence can change temperature and velocity distribution map.

The understanding of flow patterns is still a challenge for nowadays researchers due to the interior cabin complex geometry and also to the ventilation

* Corresponding author: ilinca.nastase@gmail.com

system complexity (flow rate, location and geometry of vents) [8]. Additionally, the thermal non-uniformity of the surfaces directly affects the air flow distributions through convective effects. Flow patterns and their effect on thermal sensation are not completely considered by manufacturers or by users, given that flow trajectory might substantially differ from the direction imposed by the guiding vanes of the air vents. This is related on one hand to these previously mentioned convective effects but might be also an in intrinsic characteristic of the air vent itself. Indeed, Lezovic et al [11] studied the flow patterns generated by different positions of the horizontal and vertical guiding vanes of a regular air vent. In their study, the authors used hot wire measurements and smoke visualizations. The results showed that the overall performance of the air diffuser is apparently affected by the individual vanes geometry and their position inside the vent and by the geometry of the supply duct. The influence of different shapes of vents with the same discharge aria over the cabin thermal environment was studied numerically by Ahirrao et al [12]. The results showed that for the same initial conditions a round shape vent might be more efficient than regular square or elliptical vents, leading to a lower mean air temperature and to a better distribution of air velocity and temperature, under certain circumstances. The effects on thermal sensation of different positions of the air vents inside the passenger car were also numerically investigated by Kilic [13]. Three different positions, for the same cooling power, were analyzed under transient conditions. The best results in terms of heat dissipation from the driver's body were achieved using the air vents on the dashboard. The effect of an innovative dynamic air diffusers system over the surfaces temperatures of thermal manikins was studied by Limaye [14]. The authors showed that this active control method might ensure more uniform heat dissipation from the driver's body. Zhang [15] compared in his numerically study the air velocities and temperatures at head, knee and foot level for two cases: in the first case there was no passenger in the car, while in the second case was considered four passengers in cabin. In the second case, higher velocities was found in the rear part of the compartment at the head and foots level and low temperatures, while in the front part at all the levels air velocity was higher for the first case as well as the temperature values.

All these previously cited papers show that the global distribution of the air flows inside the vehicular space plays an important role on thermal sensation, directly via the possibility of heat dissipation of the human body and its ambient, and indirectly through the homogeneity of air temperature and velocity distributions.

This article is a part of a larger study, intended to deepen the knowledge on thermal comfort inside vehicles and its numerical methods of prediction. In this paper the effect of the driver presence and different angles of airflow from an air vent, over the thermal sensation was investigated. In previous papers [16, 17], we presented the setup of the model of a cabin car and we showed that we can reproduce the in-cabin environment without passengers, finding values of indoor thermal variables from the CFD model that were similar to those of experimental measurements. We present in this article, results obtained for the air vent located on the dashboard at the driver's position (left side air vent).

2 Assessment protocol

In order to evaluate numerically the cabin thermal environment, the interior of a vehicle was modelled, using dimensions of a real car Renault Megane. After validation of the reference numerical model in the previous study presented in [16], we decided to investigate the influence of the driver presence, over the thermal environment of a vehicle cabin. Therefore, in the vehicle cabin was added a thermal manikin. As a second goal, the air flow direction imposed at the discharge over the global air velocity fields and the air flow patterns in the cabin car is investigated. We present in this article, results obtained for the air vent located on the dashboard at the driver's position (left side air vent) (see Fig. 1).

In this paper we will call Case I – the numerical simulations without a virtual manikin; and Case II – the numerical simulations with a virtual manikin in the cabin. For each of these cases left grill airflow angle was modified on horizontal plane from 0° to 30°, with one step of 5° towards the interior cabin. In conclusion 14 numerical simulations were made for each case, each having a different angle.

To evaluate the effects of these different boundary conditions, 16 comparison points was considered (Fig. 2), these points being positioned in at the head, chest, knee and foot levels in the place of the passengers.

Fig. 1. Renault Megane dashboard, the considered air vent for the parametric variation of the angle of guiding vanes

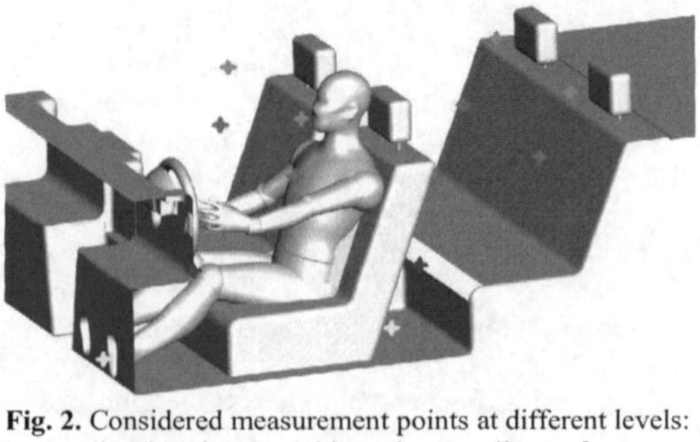

Fig. 2. Considered measurement points at different levels: green – head; red – chest; blue – knee; yellow – foot;

2.1. Numerical models

2.1.1. Modelling geometry

The Renault Megane cabin was designed in Catia after drawings from Renault and then imported in Design Modeler from Ansys 18. Several meshes with tetrahedral elements where tested to check the independence of the solution. For the first Case the chosen mesh has 6.14 million elements and is presented in Fig. 3.

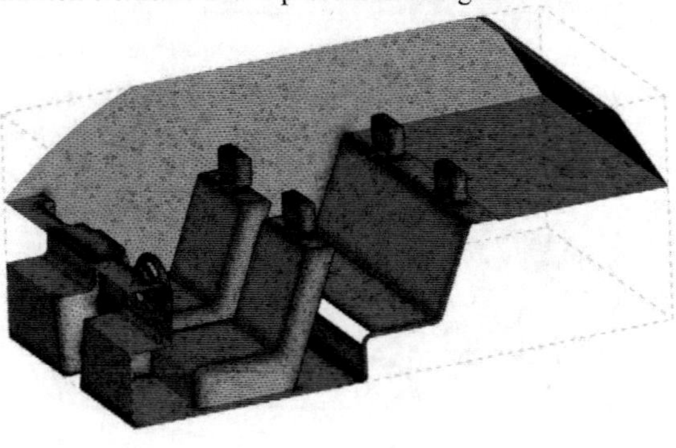

Fig. 3. Mesh detail corresponding to the 6.14 million elements grid for the Case I

For the second Case, we used the vehicle geometry from the Case I, which it had added the geometry of the manikin. After several trials a human body was modified from a freeware version found on internet. The body was modified in Catia and then imported to Design Modeler.

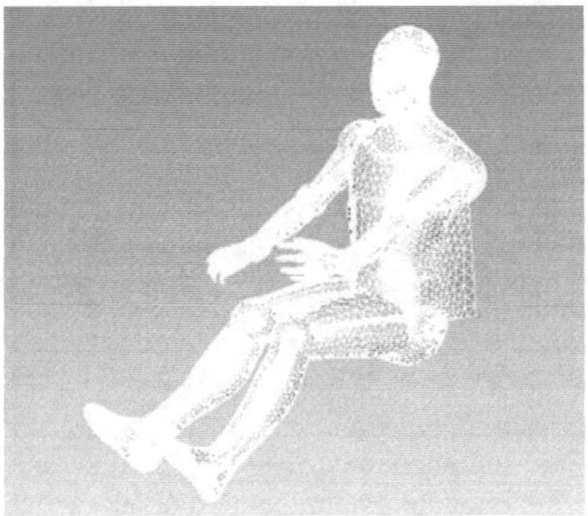

Fig. 4. Virtual thermal manikin

The manikin was first tested in a virtual box (Fig. 5) in order to check if it is suitable for meshing, therefore, the evolution of the thermal plume was simulated. Only after several tests it was introduced in the car.

After the insertion of the manikin in the vehicle cabin, it was necessary to re-mesh this case. Several meshes with tetrahedral elements where tested to check the independence of the solution. For the second case the proper mesh have 6.54 million elements and is presented in Fig. 6.

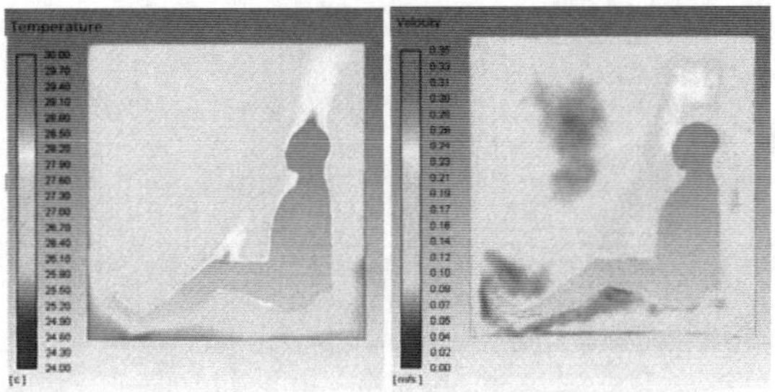

a).

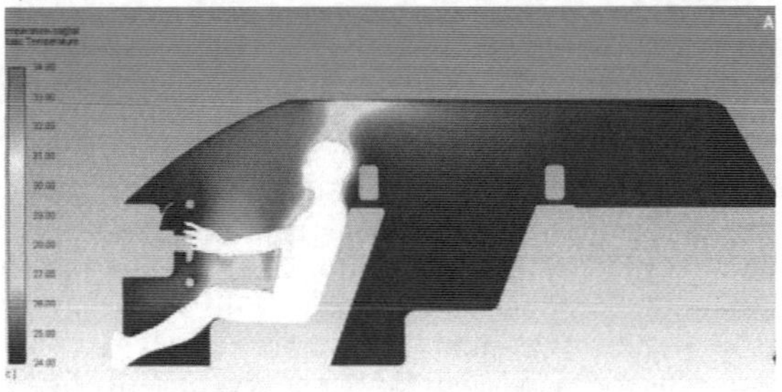

b).

Fig. 5. Natural convection of the body plume: a). – in a test box; b). in vehicle cabin

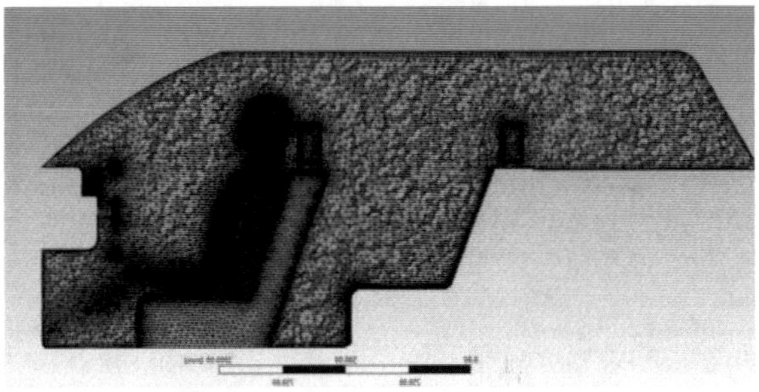

Fig. 6. Mesh detail corresponding to the 6.54 million elements grid for the Case II

For this case with manikin, the very next step will be conceiving a UDF for the direct calculation of t_{eq} in Fluent. It will serve further to evaluate the local thermal comfort of the human body at the local level. This manikin have a standard height of 1.70 m and its surface area is 1.81m^2 being suitable for a standing posture and had a total surface area (1.20 m^2) for a sitting posture. The rest of the total surface was contact with the solid surfaces of the automobile cabin.

2.1.2. Boundary conditions

As the boundary conditions, we imposed a mass flow rate of 0.042 kg/s on the central air vents and a mass flow rate of 0.039 kg/s on the side air vents. These values are corresponding to the measured values for the 2nd position on the manual control of the ventilation fan speed. We considered this study as isothermal; therefore the value of 23°C was imposed for the inlet air temperature, internal and external ambient temperature and on the cabin and manikin surfaces temperature.

The Reynolds number at the exit of the central air vents based on stream wise mean velocity and on the equivalent diameter (De_1= 0.093m) was Re_1= 17800, and for the side air vents Re2=14000 (De2= 0.073m). The inlet turbulence intensity was imposed for all vents as 4.11%, being calculated using the empirical relation proposed by Jaramillo [13]. For the left side air vent, we modified the angle of the vertical guiding vanes in the horizontal plane from 0° to 30°, with a 5° step angle.

3 Results and discussions

Fig. 7 and Fig. 8 are presenting the distributions of the velocity magnitude and of the in-plane vectors for the median plane of the driver for both cases. One could observe that the global pattern of the flow inside the cabin is changing dramatically with the variation for only one air vent, and only in the horizontal plane.
Secondly the introduction of the human body, obstruct the air passing on the rear part of cabin, producing vortices in the neck and head zone of the driver . ,Starting with an angle of 20°, the velocity magnitude values in the drivers' chest region are higher than 1m/s. In the region of the face and chest for the rear left passenger the velocity magnitude displays values over 0.5m/s.

The same observations could be done for the median plane of the passenger Fig. 9 and Fig. 10. The flow pattern changes dramatically with the angle changing, while the presence of the manikin body diminishes air velocity in the rear part of the cabin.
Comparing the velocities found in rear part of the vehicle, high velocities was found in the rear right part of the car comparing with the rear left part. The causes are the inlet angle modification and the presence of the driver.

As a general observation, we note that the maximum values of the velocity magnitude are rather high compared to the limits of the PMV-PPD models from the standards. Moreover, the locations of these maximum values are changing with the angle of the guiding vanes and the presence of the manikin geometry. However, given the high values recorded for the air velocities, the non-suitability of the PMV-PPD model is obvious, giving as a qualitative method of comparison in this case. These values are rather high at several locations when compared to the limits of the PMV-PPD model given in EN ISO 7730[18]. In the same time, the same study should be performed taking into account the thermal components, like the radiative transfer since the creation of recirculation regions with high velocities, resulting from the impinging of side flows on the windows could be benefic through the perspective of an enhanced heat transfer.

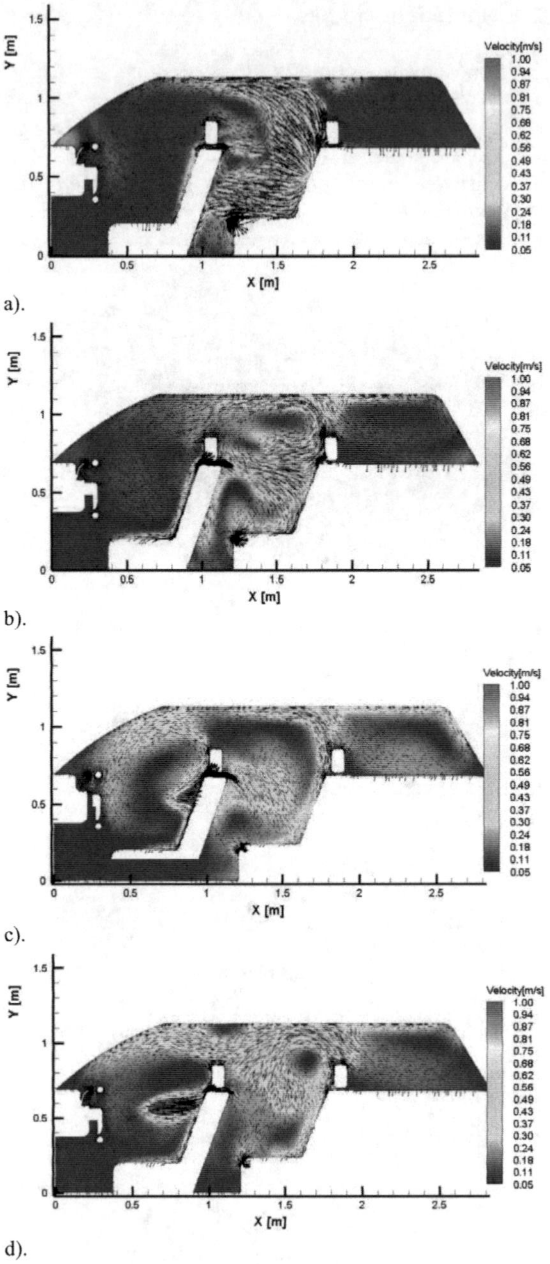

Fig. 7. Distribution of the velocity magnitude and of the in plane vectors for the median plane of the driver: a). 0°; b). 10°; c) 20°; d) 30°, for Case I

https://doi.org/10.1051/e3sconf/20183201018

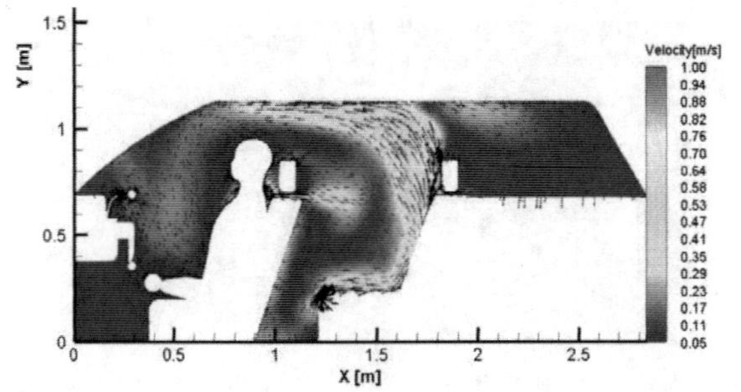

a).

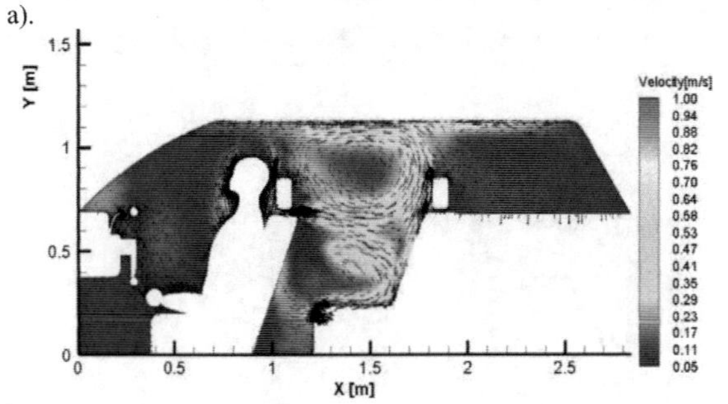

b).

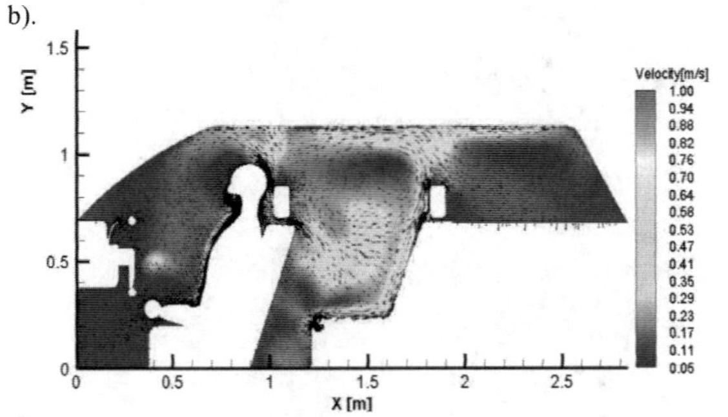

c).

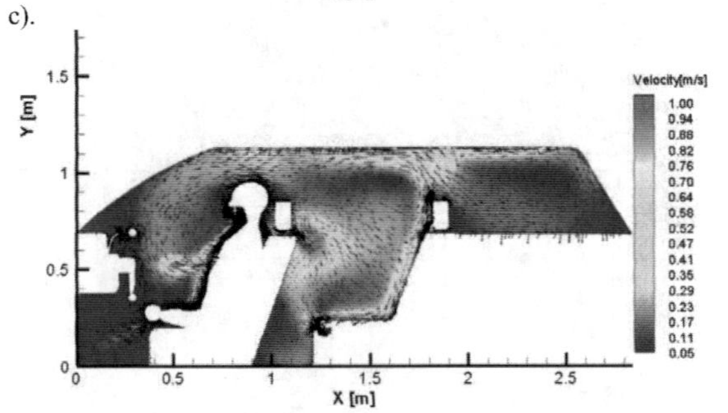

d).

Fig. 8. Distribution of the velocity magnitude and of the in plane vectors for the median plane of the driver: a). 0°; b). 10°; c) 20°; d) 30°, for Case II

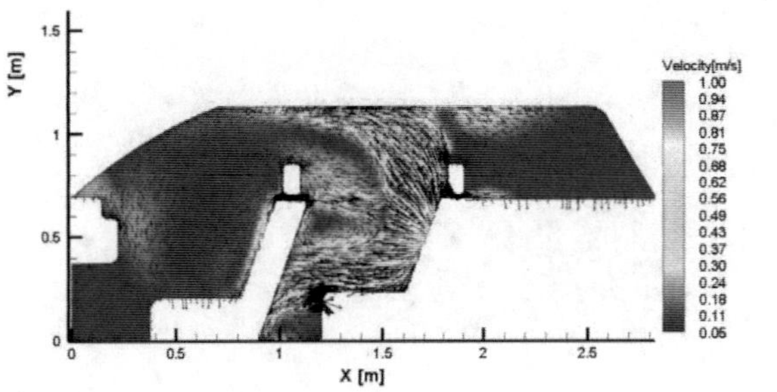

a).

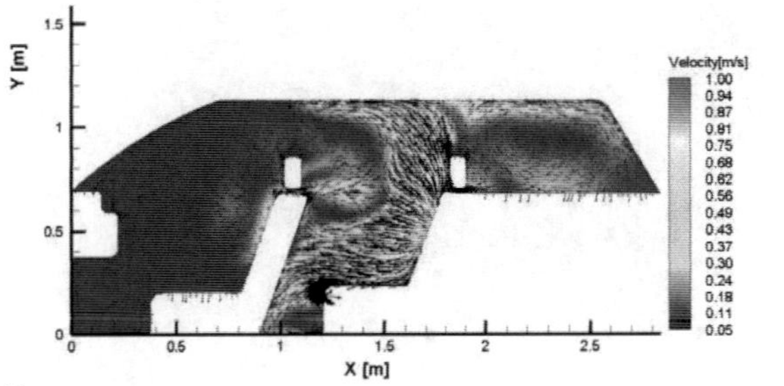

b).

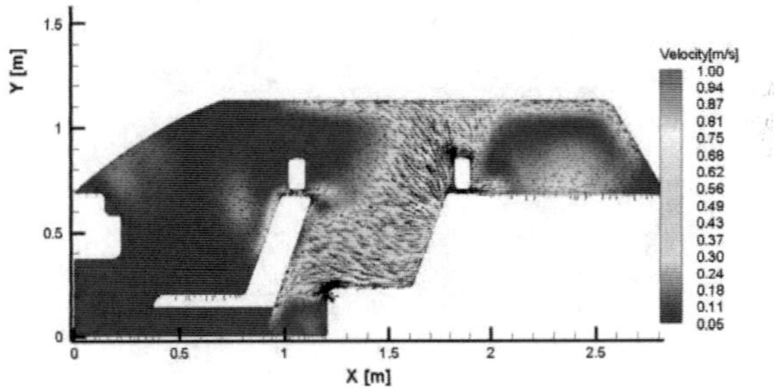

c).

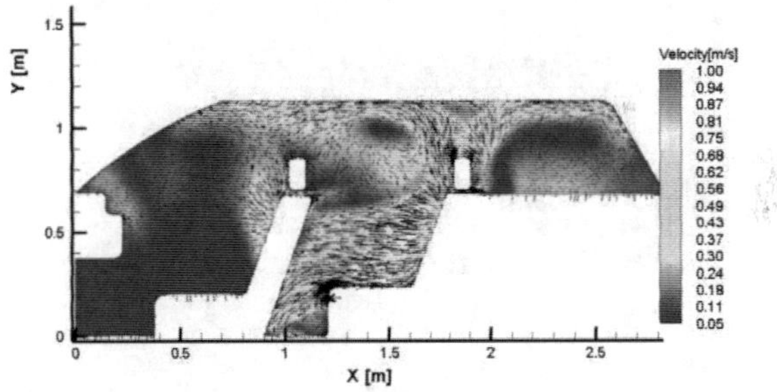

Fig. 9. Distribution of the velocity magnitude and of the in plane vectors for the median plane of the passenger: a). 0°; b). 10°; c) 20°; d) 30°, for Case I

https://doi.org/10.1051/e3sconf/20183201018

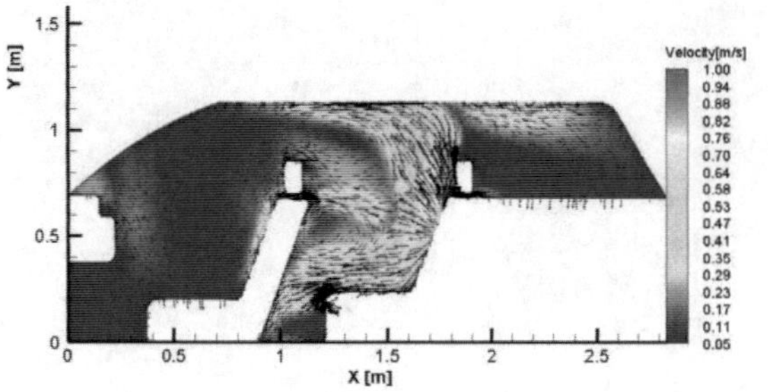

a).

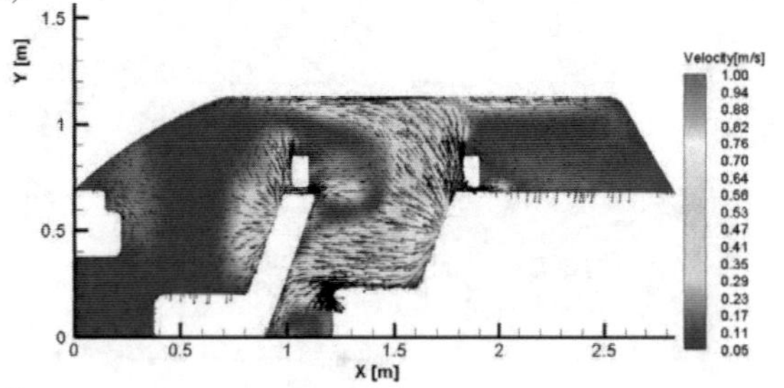

b).

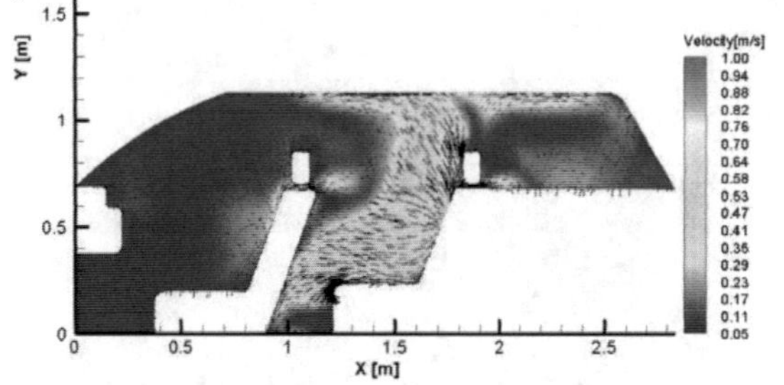

c).

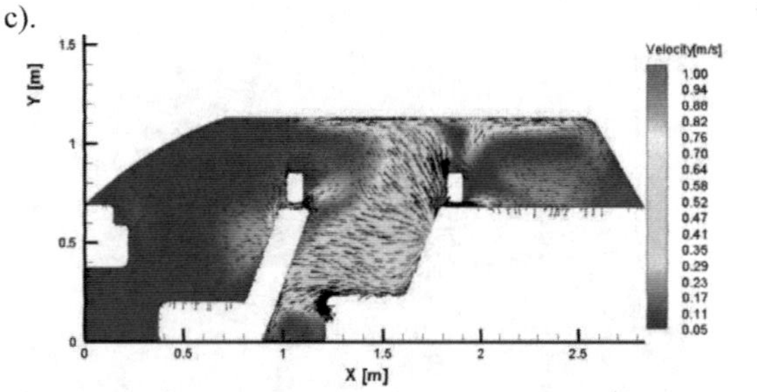

Fig. 10. Distribution of the velocity magnitude and of the in plane vectors for the median plane of the passenger: a). 0°; b). 10°; c) 20°; d) 30°, for Case II

The horizontal planes from the chest level help us to understand better the effects of the changes.

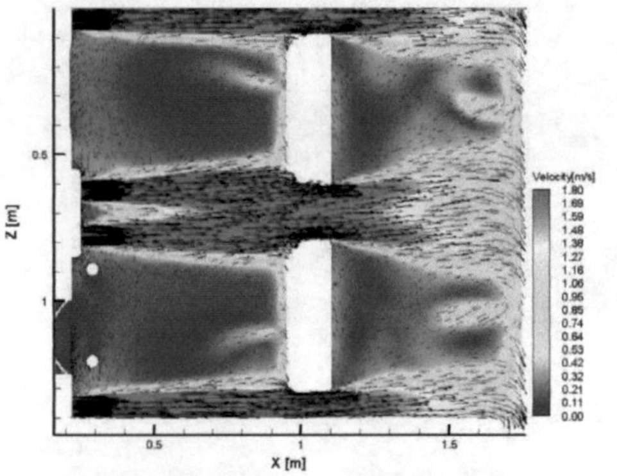

a).

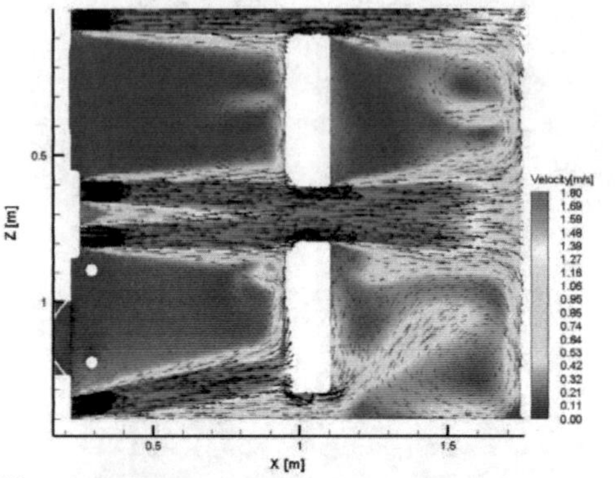

b).

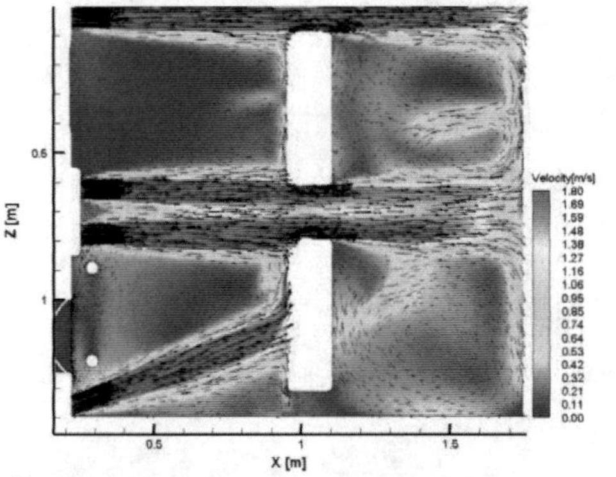

c).

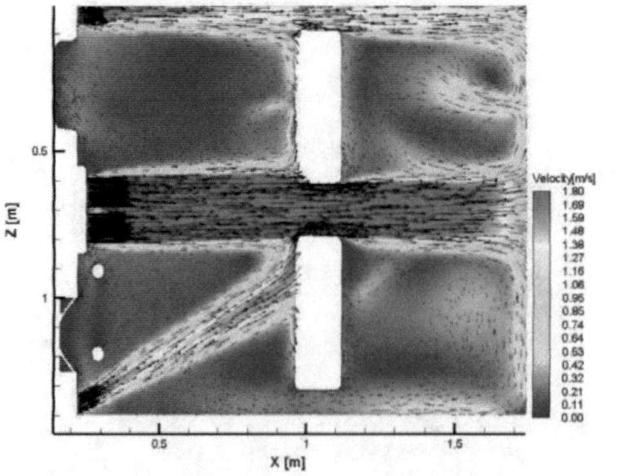

d).

Fig. 11. Distribution of the velocity magnitude and of the in plane vectors for the median plane of the passenger: a). 0°; b). 10°; c) 20°; d) 30°, for Case I

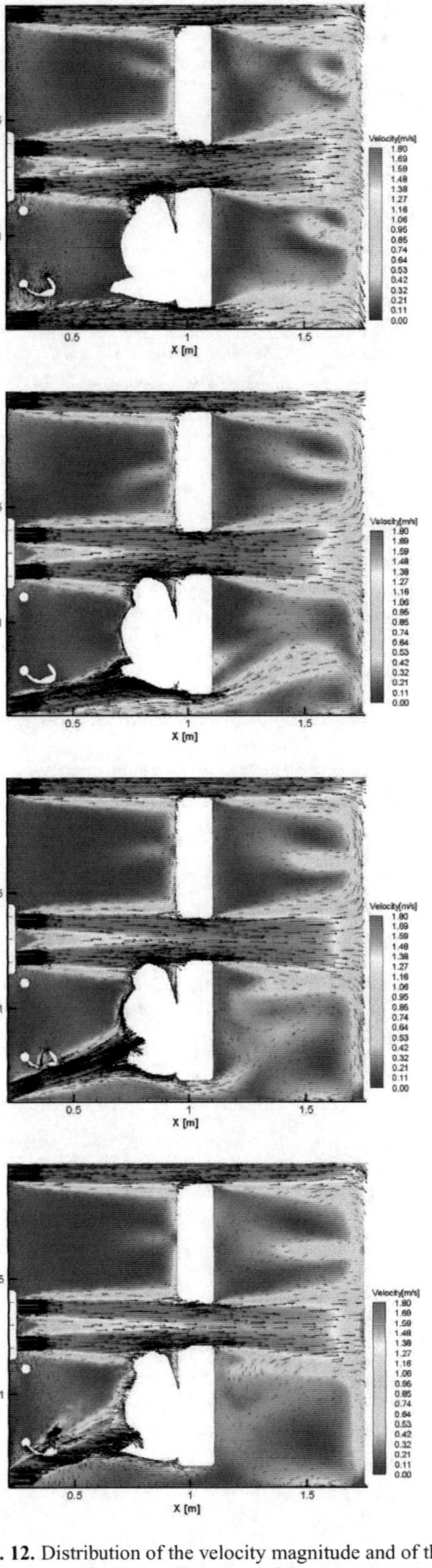

a). b). c). d).

Fig. 12. Distribution of the velocity magnitude and of the in plane vectors for the median plane of the passenger: a). 0°; b). 10°; c) 20°; d) 30°, for Case II

Table 1. PMV index comparison

Point/	0°		10°		20°		30°	
parameter	Case I	Case II	Case I	Case II	Case I	Case II	Case I	Case II
D. Head	-0.54	-0.54	**-0.54**	**-1.00**	-0.54	-0.54	-0.54	-0.58
P. Head	-0.54	-0.54	-0.54	-0.58	-0.65	-0.54	**-0.71**	**-0.54**
R.R. Head	-1.44	-1.45	-1.12	-1.15	-0.94	-0.93	**-1.08**	**-0.85**
R.L. Head	-1.50	-1.56	-1.53	-1.54	-1.59	-1.50	**-1.23**	**-1.50**
D. Chest	-0.54	-0.54	**-0.58**	**-1.02**	**-0.80**	**-1.95**	**-2.08**	**-1.92**
P. Chest	-0.54	-0.54	-0.74	-0.72	-0.72	-0.71	**-0.60**	**-0.74**
R.R. Chest	**-1.53**	**-0.90**	**-1.19**	**-0.92**	-0.94	-0.99	**-1.07**	**-0.92**
R.L. Chest	-1.49	-1.39	-1.56	-1.51	-1.47	-1.45	**-1.61**	**-1.34**
D. Knee	-0.54	-0.54	-0.54	-0.54	-0.54	-0.54	-0.54	-0.55
P. Knee	-0.63	-0.54	-0.54	-0.54	-0.54	-0.54	-0.54	-0.54
R.R. Knee	**-0.98**	**-0.68**	**-0.77**	**-0.95**	**-1.02**	**-0.87**	**-0.83**	**-1.00**
R.L. Knee	-1.44	-1.41	-1.46	-1.46	-1.45	-1.39	**-1.48**	**-1.26**
D. Foot	-0.54	-0.54	-0.54	-0.54	-0.54	-0.54	-0.54	-0.54
P. Foot	-0.54	-0.54	-0.54	-0.54	-0.54	-0.54	-0.54	-0.54
R.R. Foot	**-0.58**	**-0.82**	-0.72	-0.68	-0.72	-0.80	**-0.54**	**-0.77**
R.L. Foot	-0.80	-0.76	-0.80	-0.72	**-0.54**	**-0.74**	-0.54	-0.54

Table 2. PPD [%] index comparison

Point/	0°		10°		20°		30°	
parameter	Case I	Case II	Case I	Case II	Case I	Case II	Case I	Case II
D. Head	11.23	11.23	**11.24**	**26.16**	11.23	11.23	11.23	11.99
P. Head	11.23	11.23	11.23	11.99	13.81	11.23	**15.55**	**11.23**
R.R. Head	48.05	48.51	31.45	32.83	23.75	23.25	**29.94**	**20.36**
R.L. Head	51.16	54.27	52.81	53.01	55.98	51.03	**36.77**	**50.99**
D. Chest	11.23	11.23	**12.18**	**26.94**	**18.59**	**74.45**	**80.41**	**73.40**
P. Chest	11.23	11.23	16.49	15.84	15.93	15.65	**12.57**	**16.68**
R.R. Chest	**52.81**	**22.08**	**34.82**	**22.92**	23.75	25.69	**29.21**	**22.92**
R.L. Chest	50.73	45.04	54.38	51.37	49.42	48.60	**57.06**	**42.73**
D. Knee	11.23	11.23	11.23	11.23	11.23	11.23	11.23	11.32
P. Knee	13.33	11.23	11.23	11.23	11.23	11.23	11.23	11.23
R.R. Knee	**25.37**	**14.90**	**17.68**	**24.16**	**26.94**	**20.93**	**19.48**	**26.16**
R.L. Knee	48.05	46.19	48.97	48.97	48.51	45.39	**49.86**	**38.22**
D. Foot	11.23	11.23	11.23	11.23	11.23	11.23	11.23	11.23
P. Foot	11.23	11.23	11.23	11.23	11.23	11.23	11.23	11.23
R.R. Foot	**11.99**	**19.39**	15.84	14.90	15.84	18.59	**11.23**	**17.68**
R.L. Foot	18.59	17.32	18.59	15.84	**11.23**	**16.77**	11.23	11.23

Knowing the air velocity magnitude values from the numerical model and considering the metabolic rate value of 1 met and clothing insulation of 0.7 clo, the Predicted Mean Vote (PMV) and the Predicted Percentage of Dissatisfied (PPD) indexes were calculated. Tables Table 1 and Table 2 present values for the PMV and PPD indexes obtained at the right of each passenger place in the cabin for respectively the head, the chest, the knees and the feet. It appears from these tables that the most favorable position of the guiding vanes of the studied air vent is for an angle of 0°, knowing that most of PMV and PPD values fall in the acceptable range close to thermal neutrality. As the angle of the guiding vanes of the left air vent increases, values of PMV decreases showing an increased sensation of cooling.

This results in higher percentages of dissatisfied.

For the front passenger head, we could not distinguish a clear trend for the variation of PMV/PPD values, most of them indicating much higher values for the velocity magnitude of air. As for the head region located at the position of the two rear passengers, all values of PMV fall between slightly cool and cool range. From the Table 1 it appears that the PMV values have the tendency to a more comfortable state when the manikin in present than without manikin. Table 2 shows that the percentage of dissatisfied is smaller for the second case at 0° and 30°, with de high difference of 30,73% for R.R. Chest for 0°; and 14,33% for R.L. Chest. These differences are due to the manikin hand which intersects the flow from the left air vent. Among the simulations including the human geometry, more comfortable predicted values were achieved with the 10° angle with the respect of the sensations at tke levels of the head and of the knees.

4 Conclusions

One of the conclusions is that even a slight modification of the flow direction, only in the horizontal plane and only for one air vent is leading to the global changing of the whole flow pattern inside the vehicular cabin.

The second general conclusion is related to the presence of human body. As was sown the manikin geometry have a big effect over the air velocity distribution in vehicle compartment. Also any detail of cabin interior that can influence interior thermal environment must to be considered even in parametrical studies.

For air flow rates corresponding to the second position of the manual control of the fan speed, air velocities are rather high compared to the limits of the PMV-PPD models from the standards. Moreover, the locations of the maximum velocities are changing with the presence of the manikin geometry and with the angle of the guiding vanes. In these cases, in our opinion, it appears that global measurement methods for assessing thermal comfort, i.e. the use of a comfort meter placed in the specific point (as in the center) of the cabin is not anymore suitable.

This work was supported by a grant of the Romanian space agency ROSA, QUEST - Advanced air diffusion system of the crew quarters for the ISS and deep space habitation systems, STAR-CDI-C3-2016-577

References

1. ASHRAE, *''Thermal environmental conditions for human occupancy,'' ANSI/ASHRAE Standard 55-2013, American Society of Heating, Refrigerating and Air-Conditioning Engineers, Atlanta, GA*. 2013.
2. Haldane, J.S., *The Influence of High Air Temperatures*. The Journal of Hygiene, 1905. **5**: p. 494-513.
3. Fanger, P.O., ed. *Thermal Comfort-Analysis and Applications in Environmental Engineering*. ed. C.D.T. Press. 1970.
4. Cristiana CROITORU, et al., *Thermal comfort models for indoor spaces and vehicles - current capabilities and future perspectives*. Renewable & Sustainable Energy Reviews, accepted for publication, 2015.
5. Chen, A. and V.W.C. Chang, *Human health and thermal comfort of office workers in Singapore*. Building and Environment, 2012. **58**(Supplement C): p. 172-178.
6. Ormandy, D. and V. Ezratty, *Health and thermal comfort: From WHO guidance to housing strategies*. Energy Policy, 2012. **49**(0301-4215): p. 116-121.
7. ISO, *Ergonomics of the thermal environment - Evaluation of thermal environments in vehicles Part 3: Evaluation of thermal comfort using human subjects*, in *ISO 14505-3:2006*. 2006.
8. ISO, *Ergonomics of the thermal environment - Evaluation of thermal environments in vehicles - Part 2: Determination of equivalent temperature* in *ISO 14505-2:2006*. 2006.
9. ISO, *Ergonomics of the thermal environment - Evaluation of thermal environments in vehicles Part 1: Principles and methods for assessment of thermal stress*, in *ISO 14505-1:2007*. 2007.
10. jonsson, J., *Including Solar Load in CFD Analysis of Temperature Distribution in a Car Passenger Compartment*. 2007, Lulea University of Technology: Sweden.
11. Ležovič, T., et al., *HVAC automotive vents evaluation and their performance*. HVAC&R Research, 2013. **19**(8): p. 1073-1082.
12. Ahirrao, A.K., et al., *Effect of Vent Shape on Thermal Comfort of Passengers in a Car*. SASTECH 2011. **10**(2): p. 21-28.
13. Kiliç, M. and G. Sevilgen, *The effects of using different type of inlet vents on the thermal characteristics of the automobile cabin and the human body during cooling period*. The International Journal of Advanced Manufacturing Technology, 2012. **60**(5): p. 799-809.
14. Limaye, V.M., et al., *Design of Dynamic Airvents and Airflow Analysis in a Passenger Car Cabin*. SASTECH, 2012. **11**(1): p. 41-48.
15. Zhang, H., et al., *Studies of air-flow and temperature fields inside a passenger compartment for improving thermal comfort and saving energy. Part II: Simulation results and discussion*. Applied Thermal Engineering, 2009. **29**(10): p. 2028-2036.
16. Danca, P., et al., *On the Possibility of CFD Modeling of the Indoor Environment in a Vehicle*. Energy Procedia, 2017. **112**(Supplement C): p. 656-663.
17. Bode, F., et al. *The influence of the Inlet angle of vehicle air diffuser on the thermal comfort of passengers*. in *2017 International Conference on ENERGY and ENVIRONMENT (CIEM)*. 2017.
18. ISO, E., *ISO 7730 - Ergonomics of the thermal environment —Analytical determination and interpretation of thermal comfort using calculation of the PMV and PPD indices and local thermal comfort criteria*. 2005.

Evaluation of indoor temperature for various building envelopes damaged

Viktor Petrenko[1,*], *Kostiantyn* Dikarev[1], *Denys* Volchok[1] and *Oleksandra* Kuzmenko[1]

[1]Prydniprovska State Academy of Civil Engineering and Architecture, 49000, 24-A Chernishevskogo str., Dnipro, Ukraine

Abstract. The purpose of this paper is to represent inner air temperature calculation algorithm provided that interior space is formed by several building envelopes which transmit thermal losses. The chosen external wall includes an area with damaged structural layers. We examine one-dimensional thermal transfer between hot (inner) and cold (outside) space in steady-state conditions. The derived equation allows calculating the estimated inner temperature of the air taking into consideration the various combination of building envelopes on condition that one of them is damaged.

1 Introduction

The problem is to analyze the process of microclimate formation considering variable properties of building envelopes with damaged zone.

Analysis of previous studies shows the engineering solutions based on one-dimensional thermal transfer in steady-state conditions [1, 2]. However, there are no investigations devoted to the estimation of indoor air temperature under the assumption that building envelope contains the damaged area.

Some investigations [3, 4, 5, 6, 7, 8] are focused on the problem of microclimate parameters evaluation in old buildings with visibly damaged zones on building envelopes. Others refer to the problem of buildings envelope modernization but generally, they oriented to the choice of convenient technical solution for building reconstruction and analysis of thermal insulation impact on microclimate [9, 10, 11].

The problem of indoor temperature determination for various building envelopes on condition that one of them (external wall) contains damaged zone with fractured structural layers leads us to the following tasks:

- to specify the initial data of the problem concerned;
- to derive a mathematical expression for inner air temperature determination provided that one of considered building envelopes (the external wall) is damaged by a certain area of erosion;
- to calculate and analyze the results of estimation for a particular case.
- to generalize an analytic expression;
- to carry out a numerical modeling using ELCUT software.

2 Initial data

Previous investigations [12, 13] allowed us to define initial data and boundary conditions for inner air temperature determination in case of the only heat transfer through an external wall with damaged zones.

Analytic determination of indoor temperature for various building envelopes on condition that one of them (external wall) contains damaged zone with fractured structural layers requires the following assumptions:

- we examine plane wall, window, floor slab or roof slab containing several construction layers with different thicknesses $\delta_{1,2...n}$, mm, this solid is infinite in y and z direction ($\frac{\partial t}{\partial y}=\frac{\partial t}{\partial z}=0$ - the temperature is depending on one variable only in case of one-dimensional heat conduction)
- material properties are constant ($\lambda_{1,2...n}=const$, $\rho_{1,2...n}=const$, $c_{1,2...n}=const$);
- the outside surface of the wall is exposed to ambient air with temperature t_{ext}, ^{o}C for the cold season while inside surface has heating air next to it with defined interior temperature t_{int}, ^{o}C according to the building standard;
- there are no time dependence of the temperatures for one-dimensional steady state heat transfer ($\frac{\partial t}{\partial \tau}=0$);
- the coefficient of heat-transfer from warm indoor air to the wall denoted by α_{int}, $\frac{W}{m^2\cdot K}$, from wall to cool outside ambience - α_{ext}, $\frac{W}{m^2\cdot K}$, (those values have no time dependence as well).

3 Derivation of a formula

* Corresponding author: petrenko@mail.pgasa.dp.ua

If a study air space is limited by a combination of building envelopes (Fig. 1), we calculate an estimated inner temperature of the air taking into consideration that one of enclosing structures (external wall) contains the damaged area. Enclosing structures (with appropriate properties) of standard buildings of 70-80th (XX century) were chosen for study.

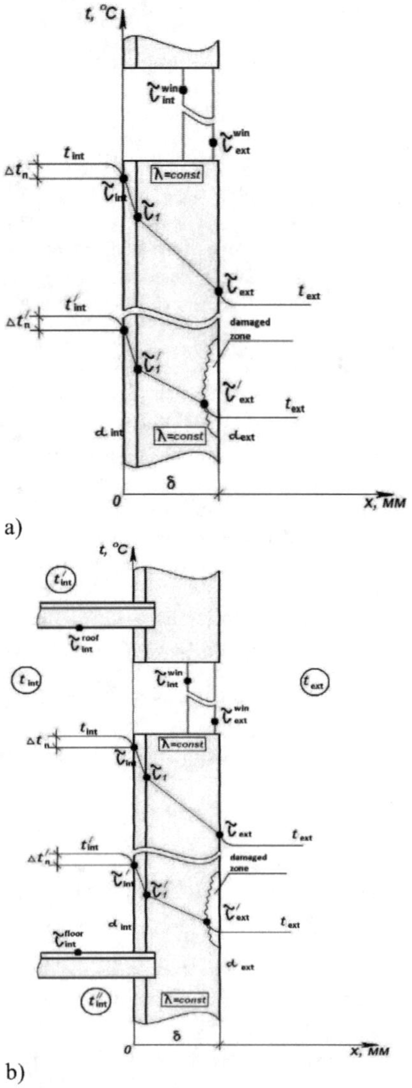

Fig. 1. Variants of heat transfer through:
a) external wall with damaged zone and installed window; b) floor slab, roof slab, an external wall with damaged zone and installed window

According to the assumptions given in [13], we assume that the plane multi-layer wall contains damaged area caused by erosion. The thickness of this area $\delta_{d.z.}$, (mm) is ranging from 0 to $\sum \delta_{wall}$ (Fig. 1).

It is plausible to assume that the presence of damaged area on the external side of the plane wall results in the temperature change of the both inner and exterior surfaces ($\tau^{/}_{\text{int}}$, ^{o}C, $\tau^{/}_{ext}$, ^{o}C), the same phenomenon happens to the temperature between layers $\tau^{/}_{1}$, ^{o}C.

Let us consider two variants of indoors with several types of building envelopes. For the first variant, we consider a room with external wall wich contains damaged zone and installed window (Fig. 1a). As for the second variant, we study a room enclosed by the floor slab, the roof slab and the external wall with damaged zone and installed window (Fig. 1b).

If we take in consideration mentioned above for the first case the heat flows through different building envelopes can be given as the following mathematical expression:

- heat flux through the plane multi-layer wall with the surface F, m^2, can be written as:

$$Q = k_1 \cdot F \cdot (t_{\text{int}} - t_{ext}), W \quad (1)$$

k_1 - coefficient of heat transmission through the wall which does not contain damaged zones, $\frac{W}{m^2 \cdot K}$;

t_{int} - defined interior temperature, ^{o}C;

t_{ext} - defined outside temperature for the cold season, ^{o}C;

- the heat transfer through multi-layer damaged wall F, m^2, can, therefore, be given as:

$$Q^{/} = k_1 \cdot (F - F_{dam}) \cdot (t^{/}_{\text{int}} - t_{ext}) + k_{dam} \cdot F_{dam} \cdot (t^{/}_{\text{int}} - t_{ext}), W \quad (2)$$

k_{dam} - the coefficient of heat transmission for the damaged wall, $\frac{W}{m^2 \cdot K}$;

F_{dam} - area of the affected surface on the wall, m^2;

$t^{/}_{\text{int}}$ - estimated temperature of the inner air, ^{o}C.

After the transformation we can obtain the following formula:

$$Q^{/} = (k_1 \cdot F + F_{dam} \cdot (k_2 - k_1)) \cdot (t^{/}_{\text{int}} - t_{ext}), W; \quad (3)$$

- according to [14], heat flux through the window with a surface F_w, m^2, which is installed in the external wall without damages, can be defined as:

$$Q_w = F_w \cdot (k_w - k_1) \cdot (t_{\text{int}} - t_{ext}), W. \quad (4)$$

k_w - the coefficient of heat transmission for the window, $\frac{W}{m^2 \cdot K}$;

- the heat flux through the window with a surface F_w, m^2, which is installed in an external wall with damaged zone can be given as:

$$Q^{/}_{w} = F_w \cdot (k_w - k^{av}_{wall}) \cdot (t^{/}_{w} - t_{ext}), W. \quad (5)$$

k^{av}_{wall} - the averaged coefficient of heat transmission for the wall with damaged zone, $\frac{W}{m^2 \cdot K}$, which can be written as:

$$k_{wall}^{av} = \frac{k_1 \cdot (F - F_{dam}) + k_{dam} \cdot F_{dam}}{F} = \frac{k_1 \cdot (1 - x_{dam}) \cdot F + k_{dam} \cdot x_{dam} \cdot F}{F} = k_1 + x_{dam} \cdot (k_{dam} - k_1) \quad (6)$$

x_{dam} - the part of damaged area in the whole wall and which (according to [13])can be expressed as:

$$x_{dam} = \frac{F_{dam}}{F}, \quad (7)$$

The value of x_{dam} is ranging from 0 to 1.

Let us suppose that the heating equipement produces the equal quantity of thermal energy whether the wall (with window) contains damaged area or not. The condition $Q = Q^{/}$ implies that:

$$(k_1 \cdot F + F_{dam} \cdot (k_2 - k_1)) \cdot (t_{int}^{/} - t_{ext}) + F_w \cdot (k_w - k_{wall}^{av}) \cdot (t_w^{/} - t_{ext}) = \quad (8)$$
$$= k_1 \cdot F \cdot (t_{int} - t_{ext}) + F_w \cdot (k_w - k_1) \cdot (t_{int} - t_{ext})$$

The estimated temperature of the interior air can be derivated from (8)

$$t_{int}^{/} = \frac{k_1 \cdot F \cdot (t_{int} - t_{ext}) + F_w \cdot (k_w - k_1) \cdot (t_{int} - t_{ext})}{k_1 \cdot F + F_{dam} \cdot (k_{dam} - k_1) + F_w \cdot (k_w - k_{dam}^{av})} + t_{ext}, \ ^oC. \quad (9)$$

Taking into consideration (7) an expression (9) can be written as:

$$t_{int}^{/} = \frac{(k_1 \cdot F + F_w \cdot (k_w - k_1)) \cdot (t_{int} - t_{ext})}{k_1 \cdot F + x_{dam} \cdot F \cdot (k_{dam} - k_1) + F_w \cdot (k_w - k_{dam}^{av})} + t_{ext}, \ ^oC \quad (10)$$

Concerning the second variant, heat flux through the building envelopes including installed window in a plane external wall with or without damaged zone can be given in the same way as for (1), (2), (3), (4). For the other building envelopes heat flux can be calculated using the following expression:

- heat flux through the floor slab with the surface F_{floor}, m^2, can be calculated as:

$$Q_{floor} = k_{floor} \cdot F_{floor} \cdot (t_{int} - t_{ext}) \cdot n_{floor}, \ W, \quad (11)$$

k_{floor} - the coefficient of heat transmission for the floor slab, $\frac{W}{m^2 \cdot K}$;

n_{floor} - a coefficient which depends on the positioning of the building envelope (floor slab) against the external air;

- heat flux through the roof slab with the surface F_{roof}, m^2 can be written as:

$$Q_{roof} = k_{roof} \cdot F_{roof} \cdot (t_{int} - t_{ext}) \cdot n_{roof}, \ W, \quad (12)$$

k_{roof} - the coefficient of heat transmission for the roof slab, $\frac{Bm}{м^2 \cdot K}$;

n_{roof} - a coefficient which depends on the positioning of the building envelope (roof slab) against the external air;

- heat flux through the floor slab with the surface area F_{floor}, m^2 of the room enclosed by an external wall which contains damaged zone, can be represented as:

$$Q_{floor}^{/} = k_{floor} \cdot F_{floor} \cdot (t_{int}^{/} - t_{ext}) \cdot n_{floor}, \ W; \quad (13)$$

- heat flux through the roof slab with the surface area F_{roof}, m^2, of the room enclosed by an external wall which contains damaged zone can be represented as:

$$Q_{roof} = k_{roof} \cdot F_{roof} \cdot (t_{int} - t_{ext}) \cdot n_{roof}, \ W. \quad (14)$$

Similarly to the 1st variant let us assume that $Q = Q^{/}$, then the estimated temperature can be defined as

$$t_{int}^{/} = \frac{\begin{pmatrix} k_1 \cdot F \cdot n_{wall} + F_w \cdot (k_w - k_1) \cdot n_w + \\ + k_{floor} \cdot F_{floor} \cdot n_{floor} + k_{roof} \cdot F_{roof} \cdot n_{roof} \end{pmatrix} \cdot (t_{int} - t_{ext})}{[k_1 \cdot F + x_{dam} \cdot F \cdot (k_{dam} - k_1)] \cdot n_{wall} + F_w \cdot (k_w - k_{dam}^{av}) \cdot n_w + k_{floor} \cdot F_{floor} \cdot n_{floor} + k_{roof} \cdot F_{roof} \cdot n_{roof}} + t_{ext}, \ ^oC \quad (15)$$

n_{wall}, n_w - coefficient which depends on the positioning of the building envelope (wall, window) against the external air.

Expressions (10) and (15) are quite lengthy. Its obviously if we consider more building envelopes we will obtain larger expression. Therefore the previous formula can be written in general form as:

$$t_{int}^{/} = \frac{Q_{build} \cdot (t_{int} - t_{ext})}{Q_{build}^{dam}} + t_{ext}, \ ^oC \quad (16)$$

Q_{build} - summarized heat losses of the room enclosed by building envelopes without any damaged zone, W;

Q_{build}^{dam} - summarized heat losses of the room enclosed by building envelopes with some damaged zone, W.

3 Analysis

The widespead in Dnipropetrovsk for the eighties of XXth century standard composition of multi-layer wall was chosen as an exemple for calculation. Figure 2 shows the cross-section of the wall.

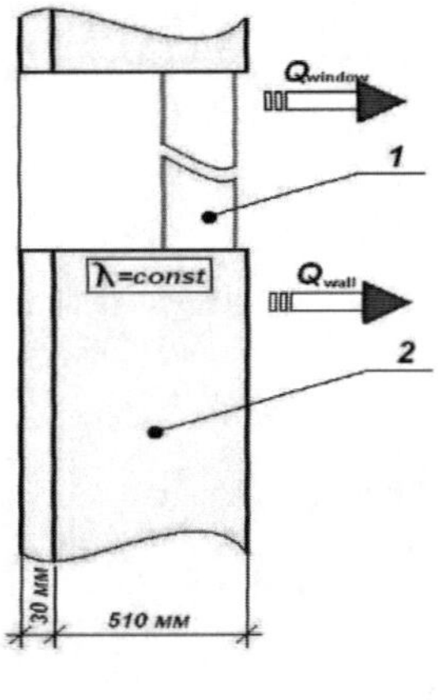

a)

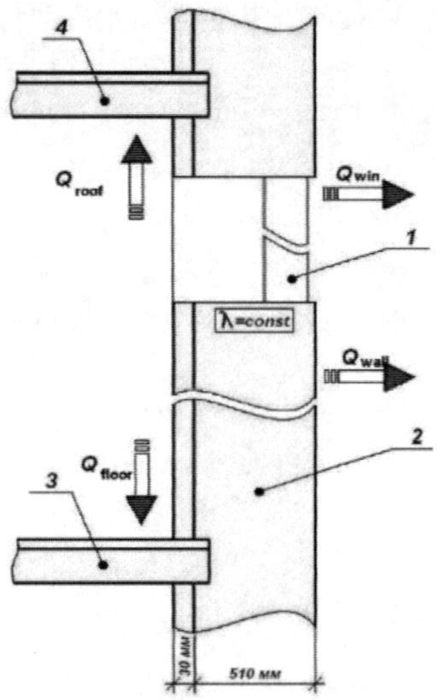

b)

Fig. 2. Variants of rooms enclosed by several different building envelopes:
a) external wall with an installed window; b) floor slab, roof slab, an external wall with an installed window.

Using (2) and (3) we can calculate the coefficient of heat transmission for the wall and thermal resistance on condition that the thickness of damaged area can take the next values $\delta^1_{dam} = 0\ mm$; $\delta^2_{dam} = 10\ mm$; $\delta^3_{dam} = 20\ mm$; $\delta^4_{dam} = 40\ mm$. The results of the calculation are shown in Table 1.

Table 1. Caracteristics of materials.

Thermal properties of the external wall	Thickness of the wall containing damaged area, mm			
	540 mm	530 mm	520 mm	510 mm
Thermal resistance of the wall, $\frac{m^2 \cdot K}{W}$	0,78	0,77	0,76	0,75
Coefficient of heat transmission, $\frac{W}{m^2 \cdot K}$	1,28	1,30	1,32	1,34

According to the obtained data, the coefficient of heat transfer and thermal resistance of the wall varies linearly [13] for $1,3\ \%$ per each $10\ mm$ of damaged area thickness.

Estimated inner air temperature in the room enclosed by an external wall with damaged zone and installed window was calculated using the expression (10). The results are given in Table 2 and shown on Fig. 3a.

Table 2. Estimation of interior airspace temperature on condition that external wall with the window encloses damaged area

$x_{dam} = \frac{F_{dam}}{F}$	Thickness of the damaged wall, mm			
	540 mm	530 mm	520 mm	510 mm
0	20,00	20,00	20,00	20,00
0,1	20,00	19,98	19,97	19,95
0,2	20,00	19,97	19,93	19,90
0,3	20,00	19,95	19,90	19,85
0,4	20,00	19,94	19,87	19,80
0,5	20,00	19,92	19,84	19,75
0,6	20,00	19,90	19,81	19,70
0,7	20,00	19,89	19,77	19,66
0,8	20,00	19,87	19,74	19,61
0,9	20,00	19,86	19,71	19,56
1	20,00	19,84	19,68	19,51

Estimated inner air temperature in the room enclosed by floor slab, roof slab, an external wall with damaged zone and installed window was calculated using the expression (15). The results are given in Table 3 and shown on Fig. 3b.

Table 3. Estimation of interior airspace temperature on condition that investigated room is enclosed by floor slab, roof slab, external wall with damaged zone and installed window.

$x_{dam} = \frac{F_{dam}}{F}$	Thickness of the damaged wall, mm			
	540 mm	530 mm	520 mm	510 mm
0	20,00	20,00	20,00	20,00
0,1	20,00	20,00	20,00	19,99
0,2	20,00	20,00	19,99	19,99
0,3	20,00	19,99	19,99	19,98
0,4	20,00	19,99	19,98	19,97
0,5	20,00	19,99	19,98	19,96
0,6	20,00	19,99	19,97	19,96
0,7	20,00	19,98	19,97	19,95
0,8	20,00	19,98	19,96	19,94
0,9	20,00	19,98	19,96	19,93
1	20,00	19,98	19,95	19,93

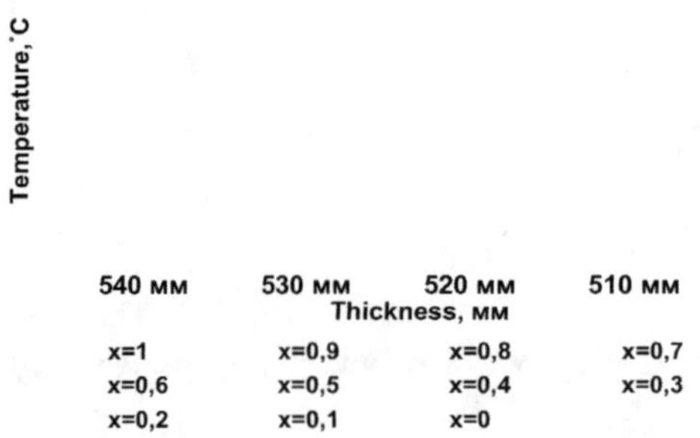

a)

https://doi.org/10.1051/e3sconf/20183201019

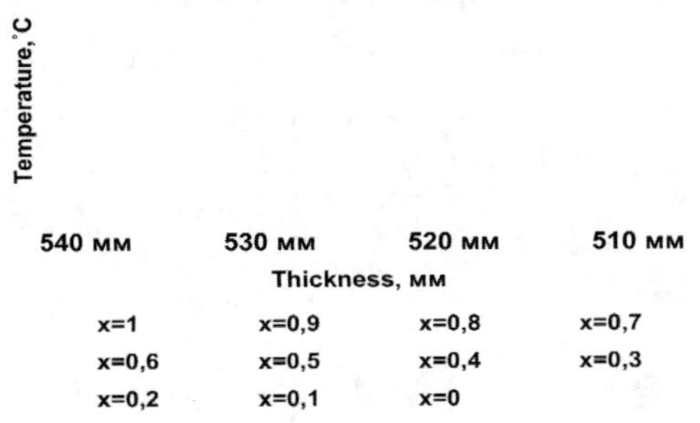

b)

Fig. 3. Evaluation of indoor airspace temperature depending on the damaged area change on condition that the room is enclosed by the following building envelopes: a) external wall with damaged zone and installed window; b) floor slab, roof slab, external wall with damaged zone and installed window

Analysis of the data given in Table 2, 3 and previous results presented in [4] leads to the statement that in the case of increasing of building envelopes number (taking into account only damages of the external wall) estimated parameters are decreasing, depending on calculated (basic) temperature, by:

- 2,45% provided that the room is enclosed by external wall with an installed window;
- 0,35% on condition that the room is enclosed by floor slab, roof slab, an external wall with an installed window.

The numerical analysis was carried out by ELCUT 6.3 workbench, a program which allows calculating a three-dimensional or a two-dimensional steady-state temperature distribution and heat transfer [15, 16]. The software is based on the finite element analysis using an appropriate grid. Properties of used materials are taken into consideration. Two basic models of rooms with various building envelopes (including damaged walls) were considered for simulations. Mentioned software allows evaluating the inner air temperature in the case of the basic variant without the damaged zone. Results are represented on the Fig. 3 a, c. Also, the variant which contains the damaged zone was investigated and the results are shown on Fig. 3 b, d.

According to the obtained results of the simulation for the both basic variants (with and without the damaged zone), we can state that there is a decreasing of the airspace temperature and the same effect was detected between the layers of building envelope. Table 4 represents temperatures on the external and inner surfaces of the building envelope and also it shows the temperature between the structural layers of the building envelope under the computer simulation. Since the data selection for the modelling was carried out in the place of the thickest damage, so we compare these data with the results of the calculations for the analogical thickness of the damaged zone from the Table 5.

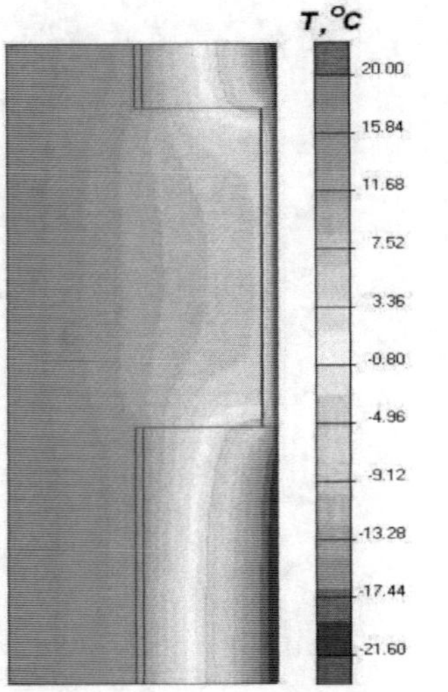

a)

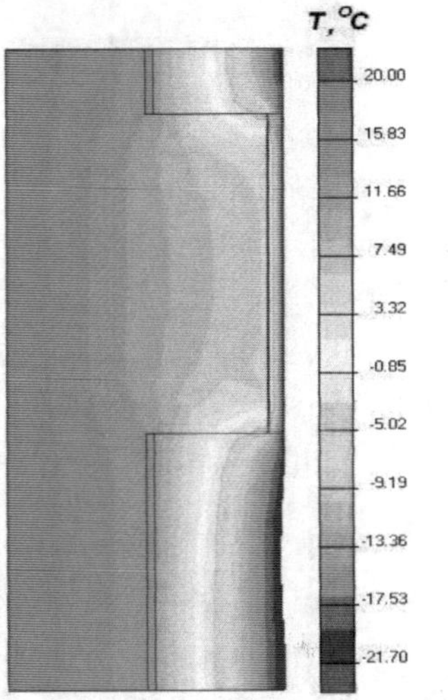

b)

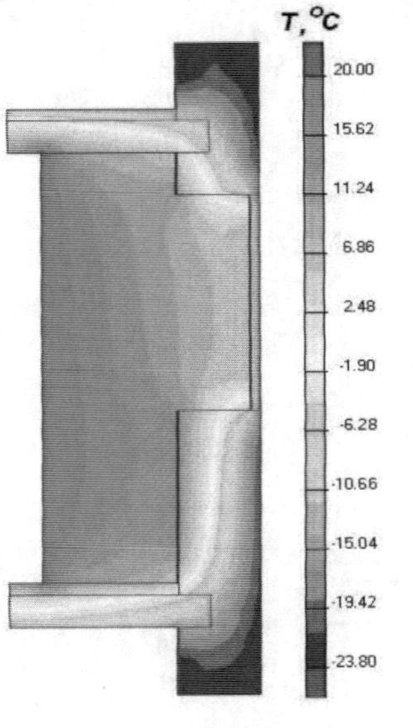

c)

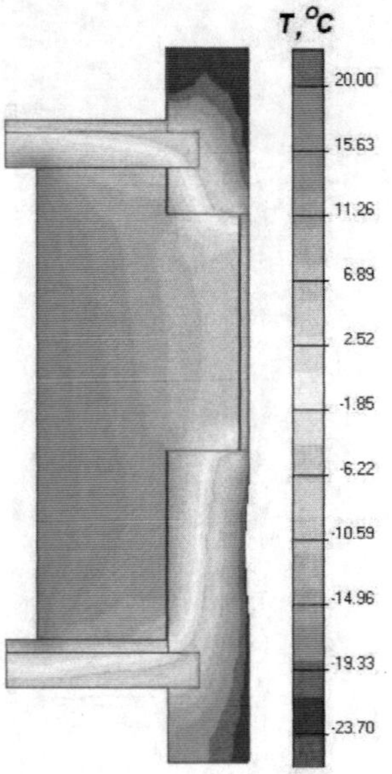

d)

Fig. 3. Modelling of the room which is enclosed by several multi-type building envelopes: a) external wall with an installed window; b) external wall which contains the damaged zone and the window; c) floor slab, roof slab, an external wall with window; d) floor slab, roof slab, an external wall with damaged zone and installed window.

Table 4. Estimated temperatures (using ELCUT 6.3) on the surfaces of the external wall layers taking into account the variation of the structural layers thickness caused by damages

Temperature values on the surfaces of the external wall layers	Thickness of building envelope with damaged zone, mm			
	Variant of Fig. 3a (540 mm)	Variant of Fig. 3b (510 mm)	Variant of Fig. 3c (540 mm)	Variant of Fig. 3d (510 mm)
Temperature on the inner surface of the wall, $\tau^{//}_{\text{int}}$, ^oC	13,95	13,72	13,49	13,26
Temperature between structural layers of the wall, $\tau^{//}_1$, ^oC	11,81	11,68	12,42	12,23
Temperature on the external surface of the wall, $\tau^{//}_{ext}$, ^oC	-21,5	-21,33	-21,35	-21,06

Table 5. Estimated temperatures on the surfaces of the external wall layers taking into account the variation of the structural layers thickness caused by damages.

Temperature values on the surfaces of the external wall layers	Thickness of the building envelope with damaged zone, mm	
	540 mm	510 mm
Temperature on the inner surface of the wall, $\tau^{/}_{\text{int}}$, ^oC	13,68	13,39
Temperature between structural layers of the wall, $\tau^{/}_1$, ^oC	11,64	11,25
Temperature on the external surface of the wall, $\tau^{/}_{зов}$, ^oC	-20,61	-22,48

The results represented in Table 4 and 5 are compared in Table 6.

Table 6. Part of divergence between the calculated results and the temperature values obtained using ELCUT 6.3

Temperature values on the surfaces of the external wall layers	% of divergence			
	Variant of Fig. 3a (540 mm)	Variant of Fig. 3b (510 mm)	Variant of Fig. 3c (540 mm)	Variant of Fig. 3d (510 mm)
Temperature on the inner surface of the wall, $\tau^{/}_{\text{int}}$, ^oC	1,9	2,4	1,4	0,9
Temperature between structural layers of the wall, $\tau^{/}_1$, ^oC	1,4	3,7	6,3	8,0
Temperature on the external surface of the wall, $\tau^{/}_{зов}$, ^oC	4,1	5,4	3,5	6,7

The calculated temperature values on the inner surface of the wall differ from the results of modelling by 2,4 %. This result doesn't exceed the tolerance.

The divergence of calculated results and computer modelling data (ELCUT 6.3) can be explained by the following:

- in case of two-dimensional simulation we don't take into account the part of damaged zone on the surface of plane wall x_{dam};
- in the case of mathematical modelling we take into account equal thickness for the building envelope in

damaged zone, whereas in the case of computer modelling mentioned parameter is varying.

Conclusions

This paper is devoted to the analytical investigation (for climite conditions of Dnipropetrovsk) of indoor airspace temperature variation provided that the room is enclosed by several multi-type building envelopes and one of them contains damaged zones. The obtained data reveals that the presence of damaged zones leads to decrease of temperature values at the border of each wall layer, therefore we observe the drop of indoor airspace temperature.

Taking into consideration the results presented in [4] we can state that in the case of increasing of building envelopes number (taking into account only damages of the external wall) estimated parameters are decreasing, depending on calculated (basic) temperature, by:

- 2,45% provided that the room is enclosed by external wall with an installed window;
- 0,35% on condition that the room is enclosed by floor slab, roof slab, an external wall with an installed window.

Results of mathematical and computer modelling (ELCUT 6.3) let us assume that its possible to evaluate the temperature change on external and inner surfaces of the building envelope and the temperature between layers on condition that the part of damaged zone on the surface of the plane wall x_{dam} is taken into account.

References

1. K.F. Fokin, *Heat transfer through the building envelope*. – 5-th ed– 256, (Moscow, 2006)
2. Building standard of Ukraine. ДСТУ Б В.2.6-101:2010 Methodes of thermal resistance determination for building envelopes. 84 p., (Kiev: Ministry of Con-s and Archit-re of Ukraine, 2010)
3. I. A. Kolesnik, V. O. Petrenko, I. L. Vetvitskiy, D. A. Vetvitskaya, *Analysis of the influence of the thermal characteristics of window units on the microclimate conditions in the heating season*. Const. Mat. Sc. Mec. Eng. **92.** – P. 67-73, (2016)
4. I. V. Golyakova, V. O. Petrenko, S. Z. Polischuk, A. S. Belikov, A. O. Petrenko, E. G. Kushnir, *Microclimate in rooms with local heat dissipation*. 120 p. (2016)
5. D. O. Lapshina, *Elaboration of energy saving and ecological means of microclimate normalization in chambers of mines with heat sources* E. E. M. of Ad. T. **2 (10).** - P. 32-39. (2016)
6. V. A. Pashynsky, O. A. Plotnikov, *The methodology of the probable estimating of temperature regime of the walls* Collection of sc. works of the Uk. State Ac. of Rail. Trans. **148 (2).** - P. 138-143. (2014)
7. V. V. Cherniavsky, A. M. Pashchenko, O. B. Borisenko, R. V. Lopatkov, *Analysis of the cracking of the decorative-plaster layer of the facade system depending on the temperature and humidity of the external air,* Coll. of Sc. W. of PNTU Y. Kondratyuk]: Sectoral engineering, construction. – 2010 **2.** - P. 147-151.
8. V. I. Deshko, N. A. Buyk *Influence of thermal protection of the building on parameters of thermal comfort* Collection of sc. works of the Uk. State Ac. of Rail. Trans. **153.** - P. 121-128. (2015)
9. K. Dikarev, A. Berezyuk, O. Kuzmenko, and A. Skokova, *Experimental and Numerical Thermal Analysis of Joint Connection «Floor Slab – Balcony Slabe» with Integrated Thermal Break*, En. Pr., **85**, pp. 184–192, (2016)
10. E. Zavadskas, S. Raslanas, A. Kaklauskas, *The selection of effective retrofit scenarios for panel houses in urban neighborhoods based on expected energy savings and increase in market value: The Vilnius case*, En. and Build. **4.** – P. 573-587. (2008)
11. M. Bojic, A. P. Parvedy, H. Boyer, *Optimization of thermal comfort in building through envelope design.* Int. Conf. ECOS 2012, pp.1-10, (2012)
12. V. O. Petrenko, A. O. Petrenko, I. V. Golyakova, *Factors that affect the microclimate in the room, which has defects of the enclosing structures and OVEC systems* Const. Mat. Sc. Mec. Eng. **93.** – P. 286-291, (2016)
13. V. Petrenko, K. Dikarev, A. Skokova, O. Kuzmenko, *Estimation of indoor temperatures on condition that building envelope is damaged* Rev. Rom. De ing. Civ. **1.** – P. 36-44, (2017)
14. SNiP 2.04.05-91 *Heating, ventilation and air conditioning.* Moscow: Stroiizdat, - 342 p., (1991)
15. L.V. Zinevich. *Solution of construction tasks using ELCUT software package* Coll. of Sc. W. of the XIV intern. Interuniv. Sc.-pr. Conf. of Y. Sc., D. St. and Post-grad. St. - Moscow: MSSU. 864p. (2011)
16. A.I. Mishichev, A.E. Martyanova. *Solution of heat conduction problems by the finite element method in the CAE-system ELCUT: Methodical instructions for studying CAD courses.* - Astrakhan: ASTU, - 39 p., (2001)
17. O. T. Voznyak, *Influence of indoor climate parameters on human heat exchange* Bull. of the Nat. Univ. "Lv. Polytech." **662**: Th. and Pract. of Const. - P. 84-88, (2010)
18. Estimated parameters of the microclimate of premises for designing and estimating of energy characteristics of buildings in relation to air quality, heat comfort, lighting and acoustics of buildings (EN 15251: 2007, IDT): DSTU B EN 15251: 2011. K .: - Minregion of Ukraine. - 33 p., (2012).
19. EN ISO 7730 European Standart. Moderate thermal environments- Determination of the PMV and PPD indices and specification of the conditions for the thermal comfort. CEN, (1995)
20. Ergonomics of the thermal environment. Analytical definition and interpretation of thermal comfort on the basis of calculation of PMV and PPD indices and criteria of local thermal comfort (EN ISO 7730: 2011, IDT): DSTU B EN ISO 7730: 2011- K .: - Minregion of Ukraine, - 64 p., (2012).

Wind tunnel study of natural ventilation of building integrated photovoltaics double skin façade

Sebastian Valeriu Hudişteanu[1], *Cătălin George* Popovici[1], and *Nelu-Cristian* Chereches[1,*]

[1]"Gheorghe Asachi" Technical University of Iaşi, Faculty of Civil Engineering and Building Services, Romania

Abstract. The paper presents a wind tunnel experimental analysis of a small-scale building model (1:30). The objective of the study is to determine the wind influence on the ventilation of a double skin façade channel (DSF) and the cooling effect over integrated photovoltaic panels. The tests were achieved by conceiving and implementation of an experimental program using a wind tunnel with atmospheric boundary layer. The effect of the wind over the ventilation of the horizontal channels of double skin façades is evaluated for different incident velocities. The results are generalized for the average steady state values of the velocities analysed. The experimental results put in evidence the correlation between the reference wind velocity and the dynamics of the air movement inside the double skin façade. These values are used to determine the convective heat transfer and the cooling effect of the air streams inside the channel upon the integrated photovoltaic panels. The decrease of the photovoltaic panels temperature determines a raise of 11% in efficiency and power generated.

1 Introduction

The complex effect of the wind over buildings is most often studied using wind tunnel simulations. The analysis is focused on the dynamic loads on the buildings facades [1, 2], snow accumulation on roofs [3] or natural ventilation of the indoor environment [4]. When buildings are exposes to the wind, some aerodynamic effects such as the bar, Venturi, stack, corner, slipstream or pyramid effect are generated [5].
The air velocity and direction has a random character, which represents the most important problem for studying its action on the buildings. Typically, for designing the wind effect, the mean values for velocity and direction are used. The probabilistic theories are used to determine the effects of the fluctuations, by knowing the statistical distributions specific of the wind velocities and pressures. The standard meteorological measurements are made using anemometers. The continuous and discontinuous recorded values are used for determining the 10 minutes average values, that is used to obtain the characteristic basic wind velocity [5].
Usually, the basic wind velocity, v_b, includes both the direction and season effects and it is measured at 10 m height, with 2% probability of annual exceeding the mean [5].
The mean wind velocity, $v_m(z)$, is also based on the v_b value, but varies with the height from the ground, z, with the roughness of the terrain and orographic particularities of the site [16].
The basic wind pressure, p_b, can be determined using the basic wind velocity, v_b:

$$p_b = \frac{\rho \cdot v_b^2}{2} \quad (1)$$

where, ρ – air density [kg/m^3].
Usually, the mean wind speed maintains its dynamic and random character, while the spatial fluctuations of the instantaneous velocity have also an important role. The instant velocity at the height z above the ground has two components:

$$v(z) = v_{med}(z) + v'(z) \quad (2)$$

where, $v_{med}(z)$ – mean velocity on time interval τ [m/s];
$v'(z)$ – fluctuating velocity [m/s].
The mean value is a characteristic of the wind profile, while the fluctuating component is used for measuring the turbulence intensity and quantifies the maximum and minimum velocities.
The peak velocity pressure, $q_p(z)$, becomes relevant for designing the wind action, because it includes the effects of the random wind speed and pressure fluctuations. The peak pressure is obtained by using an exposure factor, which increases the basic velocity pressure, p_b, about 3,4...4 times, proportional to the height above the ground and the turbulence intensity.
The wind velocity and direction have a spatial variation - in horizontal and vertical planes. The urban areas are characterized by lower wind velocities, but increased turbulence intensities, while the open fields are characterized by higher wind velocities and lower turbulence intensity – Fig. 1 [5].

* Corresponding author: chereches@tuiasi.ro

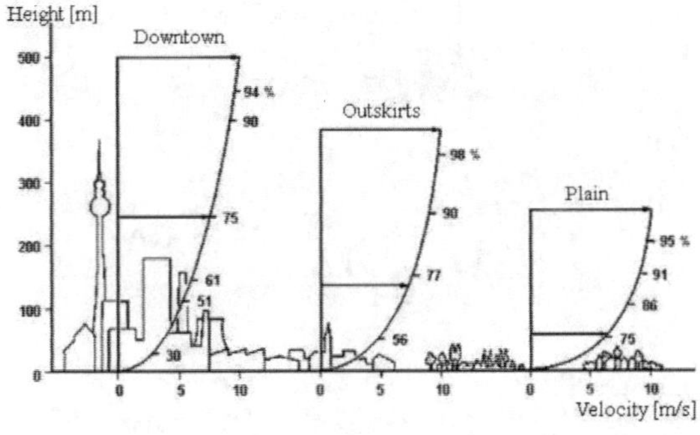

Fig. 1. Wind velocity in urban areas.

The wind effect on the ventilation of double skin façades is studied in literature [6-10, 18]. Recent studies are also focused on the building integrated photovoltaics system [1, 11-14]. These studies provide both qualitative and quantitative data concerning the influence of the wind on the natural circulation of the air inside double skin facades and cooling effect of airstreams [14].
The action of the wind causes a dynamic pressure, which is increasing with the height [5], and has a favourable effect for the ventilation of double skin facades. When the wind velocity is 10 m/s or higher, the flow becomes neutrally stable. Most often, the wind action upon the façades of the building determines a series of local pressures and suctions.
It is recommended to use simulations in wind tunnels with turbulent boundary layer in order to determine the wind action on the buildings [16]. These tests are distinguished by the similarity requirement for the model and prototype [16]. The main target of the studies is to identify the similarity model and the criteria necessary for simulation and also in detecting the similarity criteria that may be passed by, due to their irrelevance [5].
Usually, the analysis made in wind engineering, showed that the inertial forces prevail, and the flow is a turbulent one, at very high Reynolds numbers (*Re*). Also, for objects with sharp edges, the similarity is ensured by self-modelling, being insensitive to the *Re* numbers [5]. Some studies regarding the simulations in wind tunnel showed that a self-modelling condition is accomplished when *Re* numbers reach values of $(0.5...1.2)\cdot 10^5$ [5]. The self-modelling of the airflow over the bluff bodies simplifies physical models and avoid the inconveniences in satisfying the similarity requirements [15].
The current analysis is focussed in determining the wind effect over the ventilation of the vertical channels of double skin façades. The results for velocities are processed as average steady state values of 4 series of measurement of 30 seconds, resulting a total time of 2 minutes for each case. The frequency of measurement was set to 1000 values/second. The study was achieved by conceiving and implementing an experimental program in the atmospheric boundary layer wind tunnel.

2 Experimental setup

A reduced scale model (1:30) of a ventilated double skin façade is experimentally tested in a wind tunnel. A real building is a 3 stores building, 13.5 m height and 12 m width is modelled. The outer layer of the double skin façade system is realized from photovoltaic (PV) panels. The purpose of the study consists in determining the cooling effect of the air circulation when wind is incident on the façade of the building [5].
During the study, it is obtained with acceptable probability the air speed inside double skin façade, near the PV panels, during hot season. Also, a series of correlations between reference velocity and the velocities inside the channel are established.
The experimental measurements consist in obtaining the reference wind velocity, dynamic, static and total pressures, inside the channel of the double skin façade and in the external environment. The scale of the model, 1:30, was restricted by the dimensions of the transversal section of the wind tunnel, with 1.4 m height and 1.4 m width. Therefore, from the necessity of avoiding blockage of the airflow in the cross section, the transversal projection of the building was reduced to a maximum of 10% of the wind tunnel's one, resulting an area of 0.196 m^2. In these conditions, the building model has been constrained to the transversal dimensions of 0.40 m x 0.475 m, with an effective area of 0.19 m^2. The width of the channel of the ventilated façade is 0.01 m for the model, equivalent to a 0.3 m on the real building.
The building model is made of transparent plastic material (*Plexiglass*), and it has the components and dimensions presented in Fig. 2.

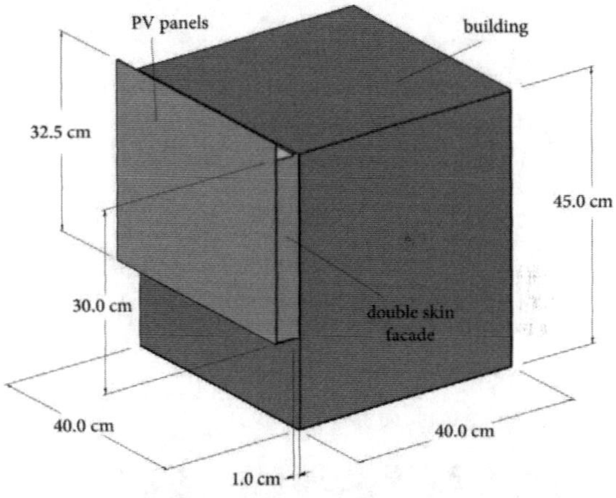

Fig. 2. Dimensions and components of the model for the building with double skin façade.

The model represents a building equipped with ventilated double skin façade system. The external glazing is considered to be replaced with photovoltaic panels, in order to achieve a building integrated photovoltaic system (BIPV). The present configuration of the double skin façade consists in three horizontal compartments on the full height and width of one façade and the air circulation due to wind action is supposed to cool the PV panels. The wind tunnel testing program is based on the assumption that for the full-scale buildings the air circulates horizontally, determined by the temperature gradient and wind action, avoiding the vertical overheating inside the channel. The main components and geometrical details of this configuration are displayed in Table 1 and Fig. 3 - Fig. 4.

Table 1. Double skin façade channel.

- 3 inlet sections (S_{in})
- 3 outlet sections (S_{ev})
- total area of inlet section: $S_{in} = 3 \text{ x } (10 \text{ x } 1) = 30 \text{ cm}^2$
- $S_{ev} = S_{in} = 30 \text{ cm}^2$

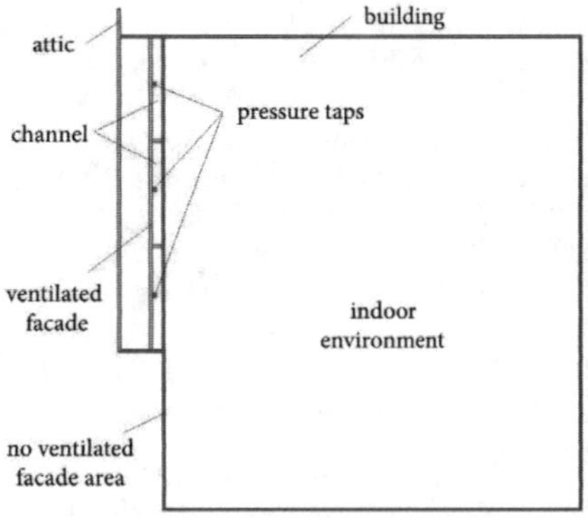

Fig. 3. Cross section through building model.

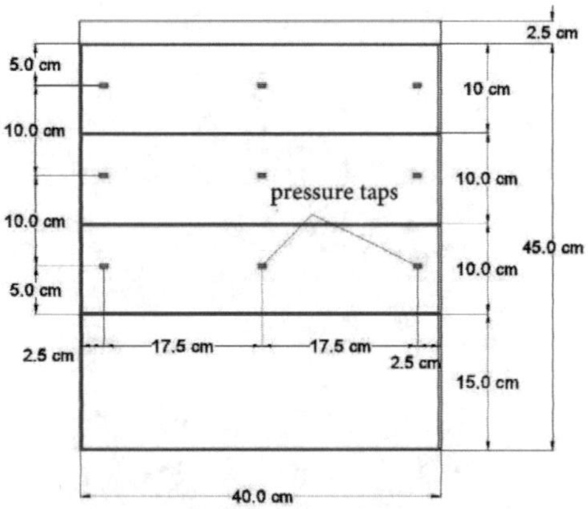

Fig. 3. Front view of the building model.

Fig. 4. Front view of the building model inside the wind tunnel – position of the pressure taps.

Fig. 5. Overview of the positioning of the model inside the wind tunnel and the rough elements for creating turbulence.

The wind tunnel is 10 m length and has rough elements that are creating the turbulent boundary layer, Fig. 5. The building model is placed at 6 m towards the turbulence grid of the tunnel.

The measurements of dynamic pressures and air speed are realized by a Pitot tube and a hot wire anemometer. Both measuring probes are placed at a 0.33 m height in the front of the building, equivalent to 10 m on the real scale. This point is named during the study, the reference velocity point, v_{ref}.

The direction of the wind in the reference point is normal to the double skin façade of the building model, as shown in Fig. 6. During the experimental study, the wind effect was tested for different reference velocities of 1, 2, 3, 4 and 5 m/s.

 https://doi.org/10.1051/e3sconf/20183201020

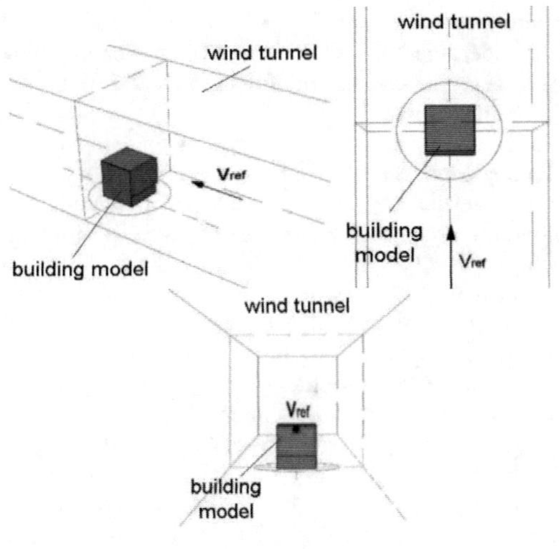

Fig. 6. Direction of the reference wind.

3 Results

The following results are based on the assumption that in case of this type of the building model, with sharp edges, it can be approximated that by assuring the geometric similarity, then the dynamic similarity is also performed. This hypothesis is available if the Reynolds numbers are superior to the turbulent flow limit.
In order to obtain a general view of the phenomenon, taking into account the self-modelling, the measured values are represented as dimensionless velocity coefficients:

$$c_v = \frac{v_m}{v_{ref}} \tag{3}$$

where, v_m – average velocity of the air inside the channel of the ventilated façade [m/s];

v_{ref} – average velocity of the air at the reference point [m/s].

These coefficients quantify the wind effect over the natural circulation of the air through the double skin façade channel. For reference velocities higher than 3 m/s, the self-modelling of the process is noticed and the coefficient determined using Eq. 3 tend to become constant, Fig 7. The self-modelling conditions were achieved and the results for the other velocities were neglected. Therefore, the c_v coefficients are the mean of these three velocities. Their values are very approximate, with differences compared to the mean of less than 30%.

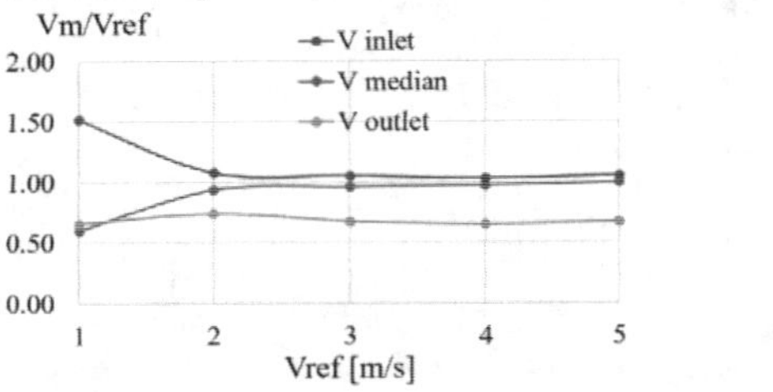

Fig. 7. Evolution of the c_v coefficient depending on the reference velocity v_{ref}.

The *Re* numbers to which the experiment was accomplished are synthesized in Table 2. It was noted that the values determined for reference velocities above 3 m/s confirm the previous assumption that the self-modelling is achieved for the studied model.

Table 2. Reynolds number depending on the reference velocity.

v_{ref} [m/s]	1	2	3	4	5
Re	26444	52889	79333	105778	132222

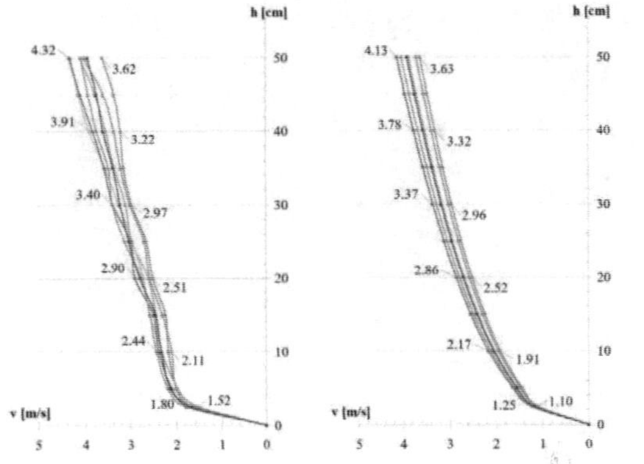

a) b)

Fig. 8. Comparison between vertical distribution of the reference velocity of experimental data (a) and exponential distribution law (b) at v_{ref}= 5 m/s.

The experimental results showed that the distribution of measured velocities is very close to the theoretical law, Fig. 8, with small differences for low level, at heights of maximum 5 cm from the horizontal plane. The match ratio of these distributions is presented in Table 3, for the reference velocity of 5 m/s. From 10 cm height above, the match is between 0.94 and 1.19 for the three reference velocities. This correspondence between experimental values and theoretical ones is very important and validates the results of the study.

Table 3. Match ration between the distribution of measured velocities and theoretical exponential law (v_{ref} = 5 m/s).

d [cm] / *h* [cm]	30	20	10	0	10	20	30
2,5	*1,39*	*1,45*	*1,50*	*1,37*	*1,40*	*1,48*	*1,41*
5	*1,38*	*1,37*	*1,29*	*1,32*	*1,28*	*1,37*	*1,37*
10	1,14	1,08	1,14	1,08	1,10	1,15	1,14
15	1,00	1,01	1,02	1,01	0,98	1,04	1,02
20	1,00	1,00	0,97	0,95	1,01	0,99	1,03
25	0,96	0,96	1,01	0,95	0,96	1,03	1,03
30	1,00	0,99	0,98	0,97	1,01	1,04	1,01
35	**1,00**	**1,00**	**1,00**	**1,00**	**1,00**	**1,00**	**1,00**
40	0,97	0,99	1,00	1,04	1,01	1,02	1,01
45	0,96	1,01	1,02	1,05	1,04	1,02	1,00
50	1,00	1,09	1,06	1,04	1,05	1,02	1,01

where, d – horizontal measurement (distance from the axis of the wind tunnel) [cm];

h – vertical measurement points (height above the floor of the wind tunnel) [cm].

The theoretical exponential law of vertical distribution of the wind velocities is:

$$v_z = v_g \cdot (\frac{z}{z_g})^\alpha = v_{10} \cdot (\frac{z}{z_{10}})^\alpha \qquad (4)$$

where, v_z – wind velocity at height z [m/s];

v_g – gradient velocity [m/s] ($v_g = v_{10}$);

z – calculation height [m];

z_g – gradient height [m] ($z_g = z_{10} = 10$ m);

α – coefficient of terrain.

The averaged values for c_v coefficients are represented in Fig. 9, in different measuring points.

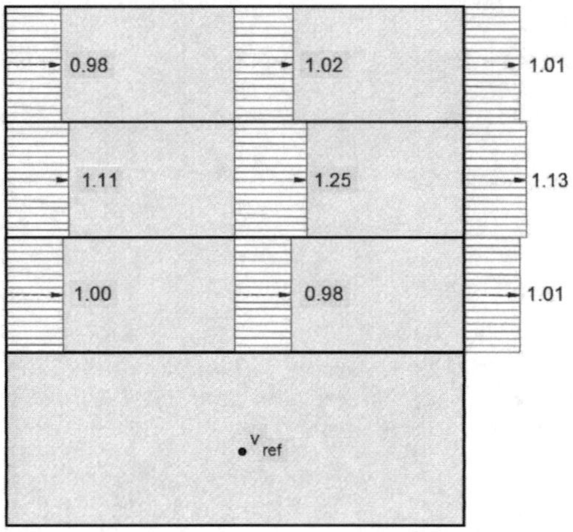

Fig. 9. c_v coefficients determined in measuring points.

The results regarding the c_v coefficients of the velocities inside the channels conducted to the analysis of the convective heat transfer coefficients obtained in the case of a real building, with the same geometrical characteristic as the modelled one.

Based on the c_v values, the convective heat transfer coefficients, h_c, inside double skin façade channel, near PV panels were calculated.

Taking into account the annual average velocities in real conditions, the convective heat transfer coefficients were calculated for reference velocities between 0,5 m/s and 3 m/s. Therefore, for this configuration of the double skin façade, the h_c coefficients reach the values according to Table 4.

Table. 4. Heat transfer coefficients, h_c, inside the double skin façade channel for c_v = 0,98...1,25.

v_{ref}	h_c [W/m^{2}K]		
[m/s]	Inferior	Median	Superior
0,50	2,30	**2,80**	2,38
1,00	4,01	**4,87**	4,14
1,50	5,54	**6,73**	5,72
2,00	6,98	**8,48**	7,20
2,50	8,34	**10,13**	8,61
3,00	9,65	**11,72**	9,96

For the usual air velocities of 0,5...3,0 m/s, the heat transfer coefficient inside the double skin façade channel achieve maximum values of about 2,8...11,72 W/m^{2}K in the median channel.

The analysis of the cooling impact over the integrated photovoltaic panels is based on the temperature effect over their efficiency [17]. A typical efficiency of the photovoltaic effect of 16 % is considered, for solar radiation of 1000 W/m^2. Given that in the present study the photovoltaics integration is realized on vertical surface (exterior layer of the double skin façade), a 500 W/m^2 nominal solar radiation can be considered [17]. In order to simplify the interpretation of the results, all data are related to a 1 m^2 monocristaline PV panel.

The power produced by the PV panel, at 500 W/m^2 and 25 °C is about 73.4 W/m^2. The influence of cooling is calculated by taking into consideration an overall effect of the temperature over the efficiency of -0.45%/C [17]. The solar radiation effect used in calculation has a value of 1,66%/100 W/m^2 at 500 W/m^2.

Therefore, for the base case, when the PV panel is integrated in the façade without ventilation, an operating temperature of 75 °C can be achieved. This temperature reduces the maximum power generated by PV panel to 77.5 % of the nominal one. In Table 5 is presented the effect of the cooling the PV panel for different reference velocities.

Table 5. Effect of cooling over the PV panel's efficiency

Case	t [°C]	[%] of P_N	η [%]	P_{gen} [W]	Increase over base case [%]
base case	75	77.5	11.37	14.22	-
v = 0.5 m/s	70.4	79.7	11.67	14.62	2.81
v = 1,0 m/s	62.6	83.4	12.19	15.30	7.59
v = 1.5 m/s	55.5	86.3	12.66	15.83	11.32

4 Conclusions

The experimental results have revealed the effect of the wind over the air circulation and velocities inside double skin façade channel in the near vicinity of the photovoltaic panels. The air flow inside the double skin façade is predominantly horizontal, which represents an important aspect in order to avoid the vertical overheating and extreme thermal draft effect during summer.

The results, quantified as c_v coefficients, showed that the average velocity of the air near the exterior glazing of the double skin façade is dependent on the reference velocity, v_{ref}, and has the same order of magnitude. The averaged values of the coefficients, c_v, in the median area of the façade, reach values between 1.11...1.25. For the inlet and outlet sections, these coefficients are laying between 0.59...0.91.

The double skin façade system is very suitable for building integrated photovoltaic solutions. Therefore, a combination of aesthetics and efficiency is achieved. The cooling effect of natural ventilation of channel by wind action revealed a possible increase of the generated electric power between 2.81…11.32 % compared to the base case.

References

1. Vasan N., Stathopoulos T., Wind Tunnel Assessment of the Wind Velocity Distribution on Vertical Façades, Proceedings of eSim, 2012, pp. 61-74.
2. Quirouette R., Arch B., La pression d'air et l'envelope du batiment, Ontario Association of Architects, 2004.
3. Axinte E., Teleman E. C., Băetu G., Pescaru R. A., Silion R., Studies for preventing the impact of climate changes upon built environment, The Bulletin of the Polytechnic Institute of Jassy, C-A Section, Tome: LXI (LXV), 2015, pp. 61-74.
4. Flamand O., Action du vent sur les ouvrages – Effets dynamiques, Universite de Nantes.
5. Axinte E., Pescaru R. A., Studiul acțiunii vântului în tunel aerodinamic, Editura Cermi, Iași, Romania, 2000.
6. Loua W., Huang M., Zhang M., Lin N., Experimental and zonal modeling for wind pressures on double-skin facades of a tall building, Energy and Buildings, Vol. 54, 2012, pp.179-191.
7. Kawai H., Nishimura H., Suzuki M., Oura Y., Field Measurement of Wind Pressure on a Double Skin with a Ventilator, EACWE 5 Florence, 12 pages, 2009.
8. Silva F.M., Gomes M.G., Rodrigues A.M., Measuring and estimating airflow in naturally ventilated double skin facades, Building and Environment, Vol. 87, 2015, pp. 292-301.
9. Tominaga Y., Blocken B., Wind tunnel experiments on cross-ventilation flow of a generic building with contaminant dispersion in unsheltered and sheltered conditions, Building and Environment, Vol. 92, 2015, pp. 452-461.
10. Barbosa S., Ip K., Perspectives of double skin façades for naturally ventilated buildings: A review, Renewable and Sustainable Energy Reviews, Vol. 40, 2014, pp. 1019-1029.
11. Agathokleous R.A., Kalogirou S.A., Double skin facades (DSF) and building integrated photovoltaics (BIPV): A review of configurations and heat transfer characteristics, Renewable Energy Vol. 89, 2016, pp. 743-756.
12. Sacht H., Bragança L., Almeida M., Caram R., Different module placements in a modular façade system for natural ventilation, Economics and Finance, Vol. 21, 2015, pp. 366-373.
13. Gaillard L., Ruedin, G. Giroux, Plantevit M., Kaytoue M., Saadon S., Ménézo C., Boulicaut J.-F., Data-driven performance evaluation of ventilated photovoltaic double-skin facades in the built environment, Energy Procedia, Vol. 78, 2015, pp 447-452.
14. Hudişteanu S.V., Popovici C.G., Mateescu T.D., Cherecheş N.-C., Efficiency analysis of BIPV systems for different locations in Romania, Energy Procedia, Vol. 112, 2017, pp. 404 – 411.
15. Neff D.E., Meroney N.R., Reynolds number independence of the wind tunnel simulation of transport and dispersion about buildings, Colorado State University, 20 pages, 1996.
16. Sacht H., Bragança L., Almeida M., Caram R., Study of natural ventilation in wind tunnels and influence of the position of ventilation modules and types of grids on a modular façade system, Energy Procedia 96 (2016) 953 – 964.
17. Popovici C.-G, Hudişteanu, S. V., Mateescu T. D., Chereches N.-C., (2016), Efficiency improvement of PV panels by using air cooled heat sinks, Energy Procedia, 85, 425-432.
18. Popovici C.-G., Cirlan V.V., Mateescu T.D., Chereches N.-C., Hudisteanu S.V., Influence of various angles of the venetian blind on the efficiency of a double skin façade, Energy Procedia 85 (2016) 416 – 424.

Study of an array of two circular jets impinging on a flat surface

Ștefan-Mugur Simionescu[1,*], *Nilesh* Dhondoo[1] and *Corneliu* Bălan[1]

[1]REOROMGroup, Department of Hydraulics, Hydraulic Machinery and Environmental Engineering, Faculty of Power Engineering, University Politehnica of Bucharest, Romania

Abstract. In this study, the flow characteristics of an array of two circular, laminar air jets impinging on a smooth solid wall are experimentally and numerically investigated. Direct visualizations using high speed/resolution camera are performed. The evolution of the vortical structures in the area where the jet is deflected from axial to radial direction is emphasized, as well as the interaction between the two jets. A set of CFD numerical simulations in 2D flow domains are performed by using the commercial software Fluent, in the context of Reynolds-averaged Navier-Stokes (RANS) modeling. The numerical resultsare compared and validated with the experiments. The vorticity number is computed and plotted at two different positions from the jet nozzle, and a study of its distribution gives a clue on how the jets are interacting with each other in the proximity of the solid wall.

1Introduction

Due to their practical importance, the study of impinging jets has received considerable attention in the last years. Of particular interest are the changes brought to the jet dynamics when modifying some important flow characteristics, as the position of the jet nozzle with respect to the impingement wall, shape of the nozzle, number of jet nozzles, wall roughness and structure.

The behaviour of impinging jets in the vicinity of solid walls has been extensively investigated experimentally. A survey of literature on flow characteristics of a single air jet impinging on a flat plate was made in 1970 by Gauntner, Livingood&Hrycak[1], focused mainly on basic flow statistics. Tahsini and Mousavi[2] provide a classification of different parameters that characterize impinging slot jets. The effects of the jet Reynolds number, jet confinement, nozzle-to-plate distance and turbulent intensity are discussed.

Later experimental studies brought up the phenomenon of heat transfer between a solid surface and a fluid jet, as the work of Jambunathan et al. [3] on single circular jet impingement.Hadziabdic and Hanjalic[4]showed results on turbulent heat transfer from an impinging slot/round jet. Geers et al. [5], [6], studied turbulent structures and heat transfer from arrays of impinging jets.

Despite significant progress in understanding macro- and microscopic phenomena occurring in impinging jets, many problems remain unsolved because of limitations in experimental measuring techniques. A solution to this problem appeared more than three decades ago, when numerical simulations of fluid flows became possible along with the advance on the market of powerful computing systems. Computational Fluid Dynamics (CFD), with its potential to solve the dynamics of a flow in both space and time, is nowadays used as a valuable instrument that provides information and even complements the experimental results.

The authors of the present study conducted a set of CFD simulations on the heat transfer between laminar impinging air jets and solid surfaces [7]. For turbulent flows, many numerical results can be found in the literature for this field of study, involving Reynolds-averaged Navier-Stokes (RANS) or Large Eddy Simulations (LES). Some significant examples are the studies by Jaramillo et al. [8] comparing the results given by different turbulence models, or Kalitzin et al. [9] studying the near-wall behaviour of RANS turbulence models.Other authors have developed new computational turbulence models, as in the case of Kubacki and Dick [10], [11] who used a k- ω based hybrid RANS/LES model for simulating plane impinging jets.

More recently Direct Numerical Simulations (DNS) of complex flows have become possible, solving the whole range of spatial and temporal scales of turbulence. Friedrich et al. [12] studied incompressible turbulent flows by using the same technique.

The aim of the present work is to study the time evolution of an array of two jets formed by circular orifices, impinging on a solid smooth wall, from both experimental and numerical point of view. The experiments provide a validation mean for the CFD results, which are then extended to extract new data. Detailed distributions of significant flow parameters in the jet impact region are inferred from the numerical tests.

* Corresponding author: stefan_simionescu@yahoo.com

 https://doi.org/10.1051/e3sconf/20183201021

2Theoretical aspects

For the case of a single unconfined jet impinging on a flat surface, most authors mention three main regions: free jet region, stagnation region and wall jet region [1], [3], [13]. For the array of two circular jets, each jet first develops as a free jet after emerging, then enters its stagnation region, close to the solid wall. After hitting the wall, each jet developsas a wall jet in all the radial directions where it is not influenced by the neighboring jet. Between the two jets, a constant mixing zone can be found,in form of two counter-rotating vortices.A view of two circular jets impinging on a solid wall is presented in Fig. 1, taken in the moments after the circular jets hit the wall.

Fig. 1. Visualization of an array of two circular jets impinging on a flat surface.

The flow characteristics of an array of two impinging jets depend upon a number of parameters: diameter of the two jets $d_1$and d_2, Reynolds number of each jet Re, nozzle-to-wall distance z, distance between the jet nozzles, structure and roughness of the impingement surface, jet configuration and orifice shape and profile. The Reynolds number Re of a circular jet is defined as $Re = \rho v d/\eta$, with ρ the fluid density, v mean velocity of the jet and η the fluid viscosity.
The flow regime presents a fundamental importance for the study of impinging jets. Comprehensive studies were written on the subject, but some still show contradictory results. A classification of circular gaseous jets, based on the Reynolds number Re is stated by Gardon and Akfirat[14], and separates laminar from turbulent jets at $Re \cong 1000$. This study distinguishes laminar jets for $Re < 1000$, transition jets or semi-turbulent jets for $1000 < Re < 3000$ and fully turbulent jets for $Re > 3000$.
The concept of vorticity is related to vortices and vortical flows[15], [16]. The nondimensionalvorticity numberWo is computed as the ratio between vorticity magnitude $\mathbf{\Omega}$and strain rate$\mathbf{D}$[16]. Whithin the present study it will be usedto quantify the existence vortices and vortical structures within the fluid domain.

3 Experimental set-up

As first step of the study, a set of direct visualizations was performed. The experimental facility used for the present study is sketched in Fig. 2. The two circular jet nozzles have a fixed 4.8 mm diameter, and are set at a distance equal to 4 times the jet diameter from each other. The impingement surface used for the tests is a

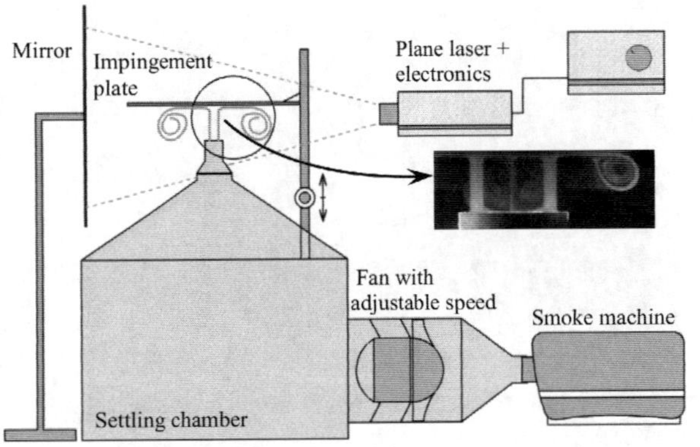

Fig. 2. Schematic view of the experimental set-up, with detail of the impingement zone.

smooth plexiglasplate. The nozzle-to-plate distance h/d is fixed for all the experiments, with a value of 4. The Reynolds number used for all the experimental and numerical tests is $Re = 200$. The settling chamber has a volume of$170\ dm^3$. The air is seeded with particles by use of a smoke machine Showtec Atmos 1000. The seeded air is sent inside the chamber and then through the jet nozzle by a fan with adjustable speed. A continuous laser with $532\ nm$ wavelength, equipped with a cylindrical lens to convert the beam in a planar sheet, is then lighting the impinging jet through its axis in order to make it visible. A mirror reflects back the laser beam to increase the lighting power in the area where the jets hit the wall. The dynamics and interaction of the impinging jets is recorded with a Nikon 1 J5 camera at 400 fps and a resolution of 800 x 296 pixels. The camera is placed perpendicular to the plane lighted by the laser sheet, so that it focuses the plane that passes through the center of the two nozzles.
The mean jet velocity is measured experimentally as the distance travelled by the fluid in a certain time period. The distance between the jet nozzle and the solid wall isrigorously set and is taken as reference, as for the time, it is obtained from the camera framerate.

4 Numerical study

In parallel with the experimental tests, a set of numerical simulations was performed, in order to validate and further analyze a case similar to the one experimentally tested. The numerical simulations were performed by using the commercial software ANSYS Fluent, in the context of Reynolds-averaged Navier-Stokes (RANS) modeling. As the Reynolds number used for all the simulations is low ($Re = 200$) the laminar model was used. The Fluent solver is based on a finite-volume solution of the mean momentum, energy and turbulent transport equations. The scheme selected to perform the computations was second order upwind.
The numerical cases were modeled in a two-dimensional flow domain, built and discretized by using the software program Gambit.

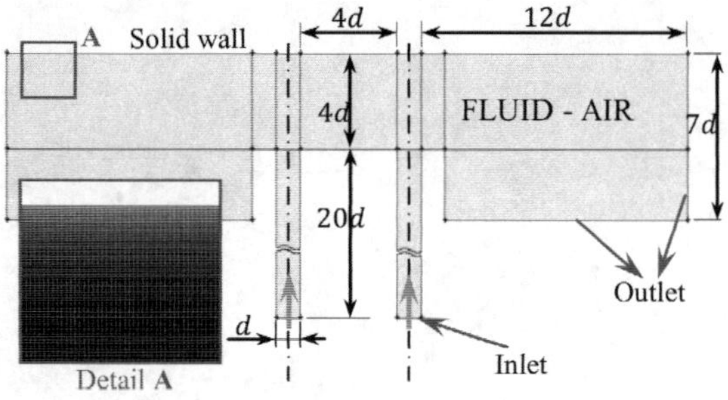

Fig. 3. Numerical flow domain and boundary conditions, with detail of the mesh.

A schematic view of the computational domain with boundary conditions is presented in Fig. 3. The outlet boundary is placed at a sufficiently large radial distance, so that the error from applying the outlet-pressure condition will not affect the region of interest. According to Dairay et al. [17], the radial expansion of the flow domain was chosen at $12d$. The mesh has 198000 cells, and was built with special attention in the region close to the impingement plate, where a further refinement was done.

5 Results and discussion

As the two jets emerge and approach the solid wall, its presence begins to affect the velocity field. A stagnation region appears close to the impingement point of each jet. Due to the presence of the impermeable wall, the axial velocity drops fast, at the same time with an increase of the static pressure [1]. After hitting the wall, the jets take radial direction and a wall-jet is formed downstream.

A time evolution of the two circular jets impinging on the flat surface is presented in Fig. 4. The turbulence induced between the two jets is highlighted by the two counter-rotating vortices found in this space. Perturbations present in the flow, as wall roughness, pressure pulsations etc., make these vortices change shape and position in time and dissipate a part of the energy from the jets.

The two jets start to influence each other from the first moments after emerging from the nozzles.So, they tend to move away from each other.This aspect is more visible in the numerical simulations, due to the limitations brought by the use of a two-dimensional flow domain instead of a physical three-dimensional one.

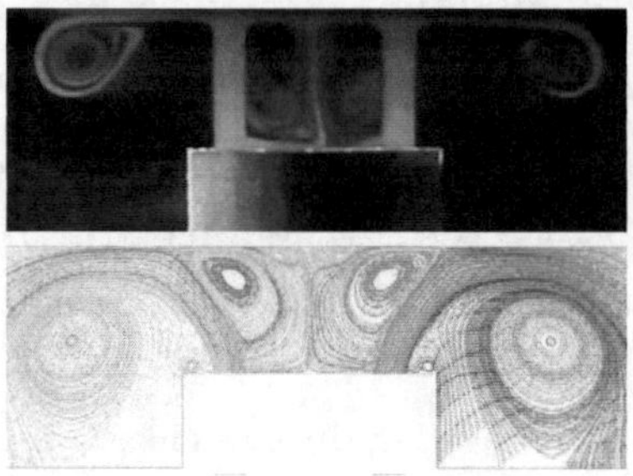

Fig. 5. Comparison between experimental and numerical results, at 400 ms after the jets emerge.

A direct flow visualization taken at 400 ms after the jets emerge from the nozzlesis compared with the corresponding numerical result in term of velocity pathlines – Fig. 5. A good agreement is found in terms of vortex core radial development, although for the numerical simulations the general shape of the jets is altered by the influence exerted by the two jets to each other. As the flow Reynolds number is low, the direct experimental visualizations show the two jets hitting the flat surface almost perpendicular. On the contrary, in the numerical resultthe jets tend to tilt and diverge as they approach the solid wall.

In Fig. 6a, the contours of vorticity number are shown at the time moment of 400 ms after jet emerge. In Fig. 6b, a vorticity number graph is plotted on the two lines shown: first continuous red line at a distance of $2d$ from the jet nozzle, and second dark blue line at a distance of $3d$ from the nozzle. This correlation between vorticity number contours in the domain and value on the two lines gives a clue on how the jets are interacting with each other,in the presence of the solid wall.

5 Conclusion

A set of flow visualizations on the behavior of an array of two circular, laminar air jets impinging on a smooth flat surface were conducted to show the non-stationary jet dynamics of the flow. The analysis of the three main zones crossed by the fluid (free jet, stagnation region and radial wall jet) show a specific dynamics for the case of flow emerging from two jet nozzles.

The stagnation region is the one with the highest velocity gradients, as the jet is constrained by the wall to take radial direction. Downstream of this region, the flow takes two different behaviors: forms a vortex ring that

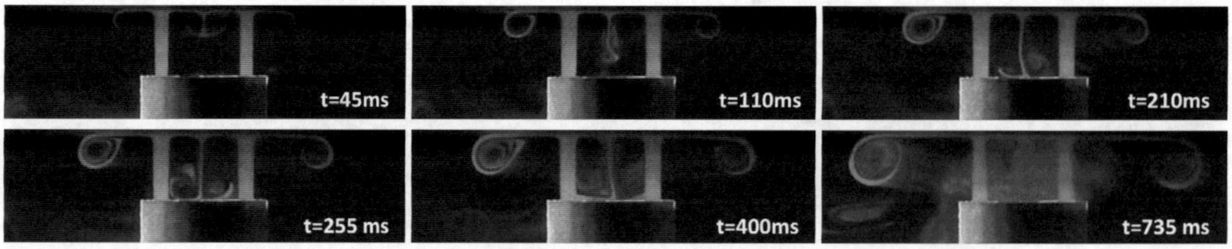

Fig. 4. Time evolution ofthe two circular jets impinging on flat surface:time measured from emergence of jets.

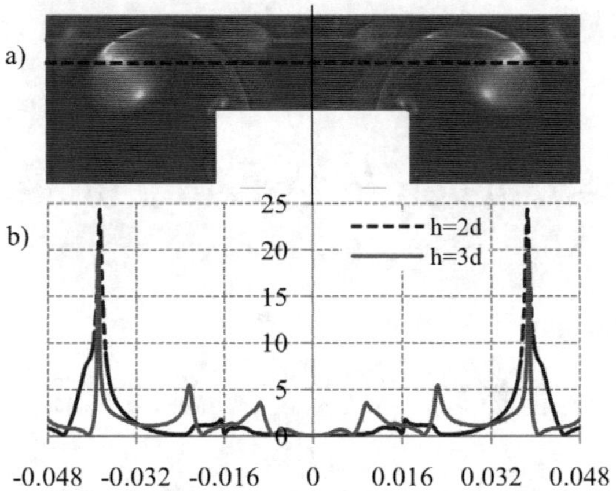

Fig. 6. Contours of vorticity number at 400 ms after jets emerge (a) and vorticity number on two lines at $h = 2d$ and $h = 3d$ from nozzle exit (b).

evolvesradially on the outer side of the two nozzles, as between the nozzles two counter-rotating vortices rollconstantly and dissipate a part of the jet energy.

In parallel with the experiments, a CFD study in the same flow conditions was performed in ANSYS Fluent. The evolution of the vortex rings observed in the numerical simulations, emerging after the jets hit the impermeable wall, agrees in terms of vortex radial expansion with the one found experimentally. After the validation is performed, a distribution of the vorticity number is plotted on two different lines.

After performing this work, the opportunity arises for new numerical simulations on the topic, using three-dimensional flow domains for a better reproduction of the physical processes. Also, a set of tests involving turbulent impinging jets can bring the opportunity for a further observation of the turbulence phenomena.

This work was partially supported by a grant of the Romanian National Authority for Scientific Research, CNCS, UEFISCDI, PHANTOM - Passive flow control for heat and mass transfer enhancement of impinging jets PN-III-P4-ID-PCE-2016-0758, and by a grant of the Romanian space agency ROSA, QUEST - Advanced air diffusion system of the crew quarters for the ISS and deep space habitation systems, STAR-CDI-C3-2016-577.

References

1. J. W. Gauntner, J. N. B. Livingood, and P. Hrycak, NASA Technical Note, Washington, DC (1970).
2. A. M. Tahsini and S. T. Mousavi, *World Acad. Sci. Eng. Technol.***6**, 1201–1205 (2012).
3. K. Jambunathan, E. Lai, M. A. Moss, and B. L. Button, *Int. J. Heat Fluid Flow***13**, 106–115 (1992).
4. M. Hadžiabdić and K. Hanjalić, *J. Fluid Mech.***596**, 221–260 (2008).
5. L. F. G. Geers, M. J. Tummers, and K. Hanjalić, *Exp. Fluids***36**, 946–958 (2004).
6. L. F. G. Geers, K. Hanjalic, and M. J. Tummers, *J. Fluid Mech.***546**, 255–284 (2006).
7. Ș.-M. Simionescu and C. Bălan, *Energy Procedia***85**, 481–488 (2016).
8. J. E. Jaramillo, K. Claramunt, J. Cadafalch, and C. D. . Pérez-Segarra, *Eur. Congr. Comput. Methods Appl. Sci. Eng. ECCOMAS*, 1–20 (2004).
9. G. Kalitzin, G. Medic, G. Iaccarino, and P. Durbin, *J. Comput. Phys.***204**, 265–291 (2005).
10. S. Kubacki and E. Dick, *Int. J. Heat Fluid Flow***31**, 862–878 (2010).
11. S. Kubacki, J. Rokicki, and E. Dick, *Int. J. Heat Fluid Flow***44**, 596–609 (2013).
12. R. Friedrich, T. J. Hüttl, M. Manhart, and C. Wagner, *Comput. Fluids***30**, 555–579 (2001).
13. N. Rajaratnam and K. A. Mazurek, *J. Hydraul. Res.***43**, 689–695 (2005).
14. R. Gardon and J. C. Akfirat, *Int. J. Heat Mass Transf.***8**, 1261–1272 (1965).
15. P. Brancher, J. M. Chomaz, and P. Huerre, *Phys. Fluids***6**, 1768-1774 (1994).
16. H. J. Lugt, *Introduction to vortex theory*. Vortex Flow Press (1996).
17. T. Dairay, V. Fortuné, E. Lamballais, and L. E. Brizzi, *Int. J. Heat Fluid Flow***764**, 362–394 (2015).

Preliminary research on virtual thermal comfort of automobile occupants

Tiberiu Horobet[1], *Paul Danca*[2], *Ilinca Nastase*[2], and *Florin Bode*[1,2*]

[1]Technical University of Cluj-Napoca, Dept. of Mechanical Engineering, Bd. Muncii 103-105, 400641, Cluj-Napoca, Romania
[2]Technical University of Civil Engineering Bucharest, Cambi Research Centre, Bd. Pache Protopopescu 66, Bucharest, Romania

Abstract. Numerical simulation of climate conditions in automotive industry for the study of thermal comfort had become more and more prominent in the last years compared with the classical approach which consists in wind tunnel measurements and field testing, the main advantages being the reduction of vehicle development time and costs. The study presented in this paper is a part of a project intended to evaluate different strategies of cabin ventilation for improving the thermal comfort inside vehicles. A virtual thermal manikin consisting of 24 parts was introduced on the driver seat in a vehicle. A heat load calculated for summer condition in the city of Cluj-Napoca, Romania was imposed as boundary condition. The purpose of this study was to elaborate a virtual thermal manikin suitable for our research, introduction of the manikin inside the vehicle and to examine his influence inside the automobile. The thermal comfort of the virtual manikin was evaluated in terms of temperature and air velocity.

1 Introduction

Computational Fluid Dynamics (CFD) methods are applied nowadays more and more in all stages related with conception of a motor vehicle. Among them, a particular interest for us represents the use of numerical simulations in predicting the thermal comfort of automobile occupants. In the last years regardless of the vehicle brand, an increased attention is oriented to the thermal comfort of the automobile occupants. The number of people spending a significant portion of their time in transportations is rising, and hence, there is an increasing demand for thermal comfort while traveling.

A more comfortable climate in automobile in almost all cases will work towards reducing driver's stress and thus contributes to a safer driving. In addition, today's demands for energy efficiency and performance, have led to an increased interest in investigating and analyzing the system and design requirements for good quality of the vehicular environment.

However, there are a few constraints which are making difficult this task: demands in increasing the fuel economy and tendency to use more glass in automobile design being among them. In order to achieve the thermal comfort inside the automobile a first step will be to evaluate the heat load that enter in the occupant's area. Is important that this evaluation to be obtained in the early stage of the vehicle design and this can be accomplished using CFD method, studies in the literature successfully applied this method [1-7].

The common interpretation of "thermal comfort" regards the state of a person that would express a feeling of wellbeing regarding the thermal conditions in an occupied space. However, thermal comfort is a term difficult to define and a universal definition of its meaning is almost impossible to obtain [8]. Nevertheless, it is the tangible interest of engineering applications that motivated a never-ending quest of quantitative models for estimating, predicting or classifying the state of thermal comfort in occupied spaces. Almost all these attempts are focusing on the physiological component. Most of them are relying on the assumption that the state of thermal comfort might be ensured when the heat produced in excess by the human body through its metabolic sources is dissipated in the environment and the thermoregulatory system intervention is infinitesimal. This way it is universally accepted nowadays that an environment is considered comfortable from the thermal point of view, when 80% to 90% of people from that environment do not express thermally dissatisfaction [9-12].

Achieving thermal comfort assume that both psychological and physiological factors are fulfilled with some degree. The air velocity, thermal radiation and temperature fields are among the most important factors that influence thermal comfort, thus in order to improve the design of a vehicle in terms of thermal comfort it is necessary to investigate the air-flow field and the temperature distributions inside the automobile.

* Corresponding author: florin.bode@termo.utcluj.ro

https://doi.org/10.1051/e3sconf/20183201022

2 Material and method

2.1 Geometry

In a previous study [4], we evaluated a simple approach of calibrating and validating a CFD model that reproduces the thermal environment and the flow dynamics inside a simple vehicular cabin. In this new study, a virtual manikin was installed in the automobile on the driver's seat. For this purpose, we studied tens of virtual manikins available on the web, we tried to pose them in driving position. We failed with almost all the geometrical models for the virtual manikins from different reasons. We succeed with one virtual manikin by reconstructing parts from his body, an image with the reconstructed virtual manikin in driving position is presented in Fig. 1a. The construction of the virtual manikin was made taking into account other real thermal manikins developed in our laboratory [13]. The virtual manikin was split in 24 parts (Fig. 1b) related with the anatomic parts of a real human.

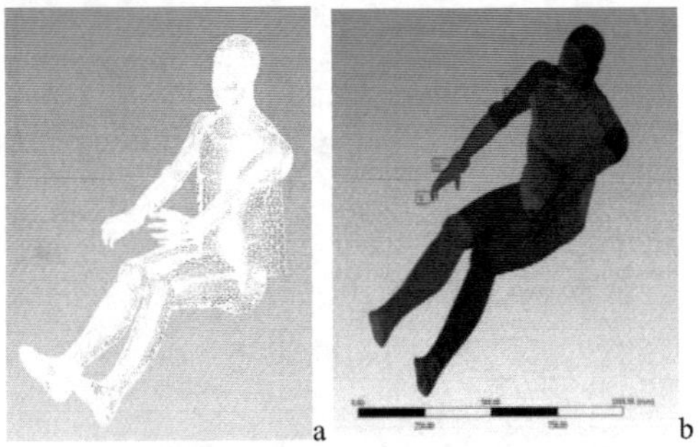

Fig. 1 Virtual thermal manikin used in this study

The studied vehicle geometry was built in SolidWorks. The virtual manikin was introduced into the vehicle studied in [4] with the purpose to simulate the presence of the driver (Fig. 2 and Fig. 3).

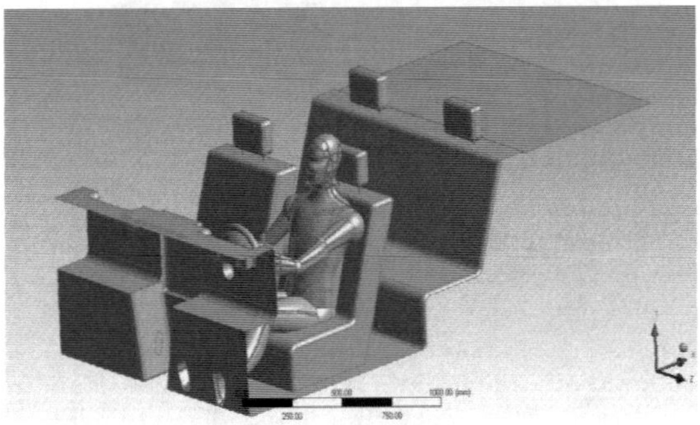

Fig. 2 Isometric view of the studied geometrical model

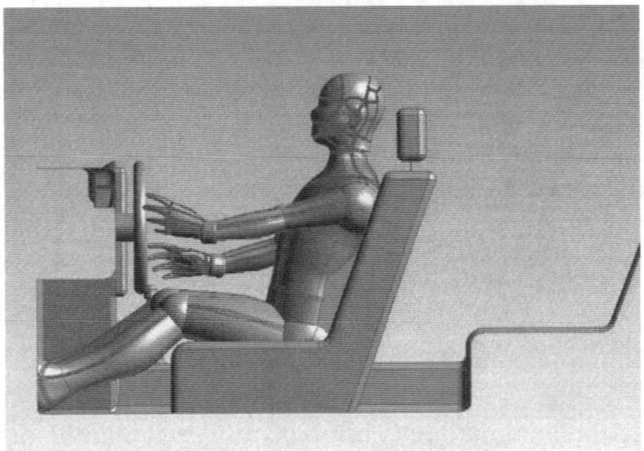

Fig. 3 Side view of the studied geometrical model

2.2 Mesh

The numerical grid was realized in Ansys Workbench. A fine mesh consisting of 6.5 million elements was created based on the experience from the previous studies where we performed the grid dependence test [4]. The boundary layer consists in five layers, with the first cell height of 0.75mm and a growth factor of 1.2.

Fig. 4 Tetrahedral mesh used for numerical simulation (different closeups)

2.3 Boundary conditions and setup

The introduction of air in the interior of the automobile was considered through the four air diffusers positioned on the dashboard of the vehicle using the appropriate airflow for the middle level of the air cooling system

 https://doi.org/10.1051/e3sconf/20183201022

($35m^3/h$ on each diffuser, air temperature: 18°C). The air jet is directed to the sides of the thermal manikin. Findings from previous studies considering the jet impacting into the surface have been taken into consideration in this study [14, 15].

Temperatures on the surface of the manikin body represent the skin temperature on different areas of the human body (Table 1, Fig. 5).

Table 1 Temperatures of the manikin parts

Body part	Foot	Calf	Thigh	Abdomen	Back and chest	Head	Arm	Hand
Temperature [°C]	28	31	32	33	35	37	33	35

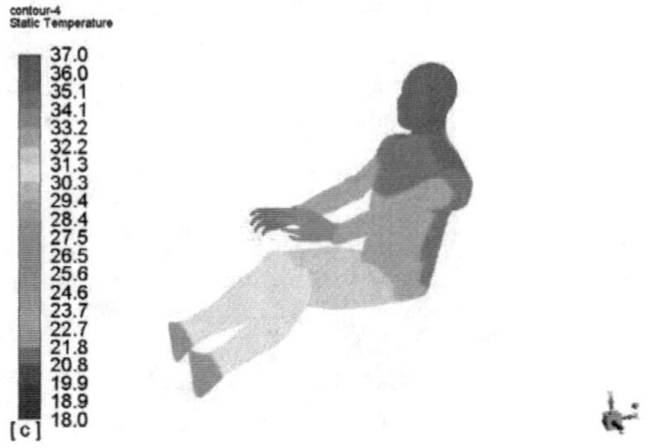

Fig. 5 Thermal manikin temperature distribution

A calculated radiative heat flux for the condition in the summer in the city of Cluj-Napoca, Romania was imposed on the windshield (951 W/m^2), left window (966 W/m^2), right window (792 W/m^2), lest doors (389 W/m^2), right doors (52 W/m^2), ceiling (908.6 W/m^2). A mention is that the automobile is heading west in this numerical simulation. The radiative heat flux for each orientation and part of the automobile was calculated using real climatic data for the city of Cluj-Napoca, Romania. An example of result for the solar radiative heat flux on the windshield of the studied automobile can be seen in Fig. 6.

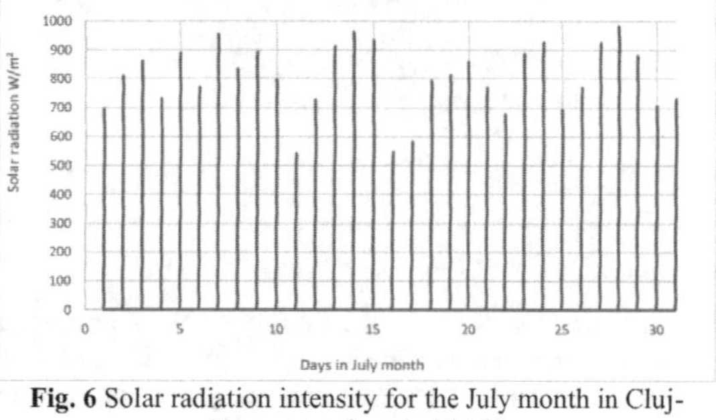

Fig. 6 Solar radiation intensity for the July month in Cluj-Napoca on the windshield

The numerical simulation was carried out in Fluent software [16]. The coupled algorithm was used for the pressure-velocity coupling. Also, a second order upwind scheme was considered for the calculation of the convective terms.

The turbulent model used was SST k-ω, a model which in the field of incompressible flow provides satisfactory results for both free flow and flow near the walls.

2 Results and discussions

As we already stated, the purpose of this study was to elaborate a virtual thermal manikin suitable for our research, introduction of the manikin inside the vehicle and to evaluate his influence inside the automobile.

The numerical validation of the vehicle geometry was studied in the mentioned article [4], a good agreement between the CFD model data and the measured data inside a real vehicle being obtained.

First stage of the numerical simulation was represented by the numerical simulation of the thermal plume of the virtual manikin. For this there was no flow rate from the air distribution system. The only source of heat was the virtual thermal manikin inside the automobile. Temperature field can be seen in Fig. 7a and velocity field in Fig. 7b.

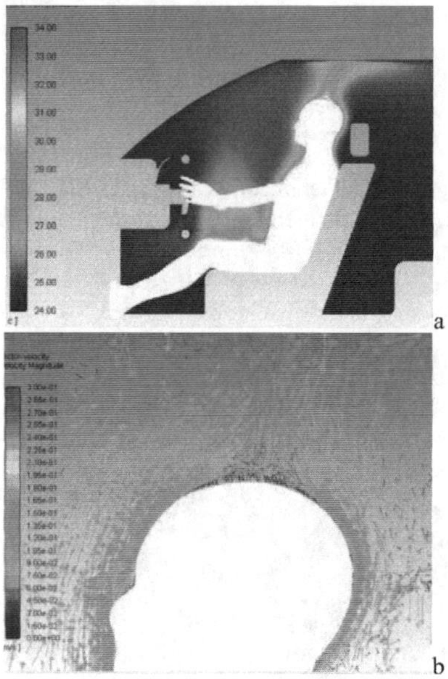

a

b

Fig. 7 Thermal plume for the virtual thermal manikin a. temperature field b. velocity vector field

In the second stage, the numerical simulation continued with the boundary conditions presented in the previous chapter. In Fig. 8 we have represented the temperature distribution in the sagittal plane of the virtual thermal manikin along with temperature distribution on the manikin body.

https://doi.org/10.1051/e3sconf/20183201022

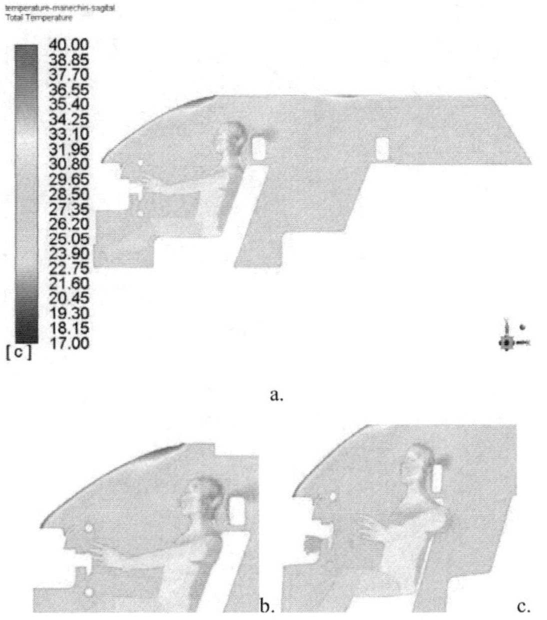

Fig. 8 Temperature distribution in the interior of the car in the sagittal plane of the virtual manikin

In Fig. 9 we can see the velocity field in the sagittal plane for the virtual thermal manikin.

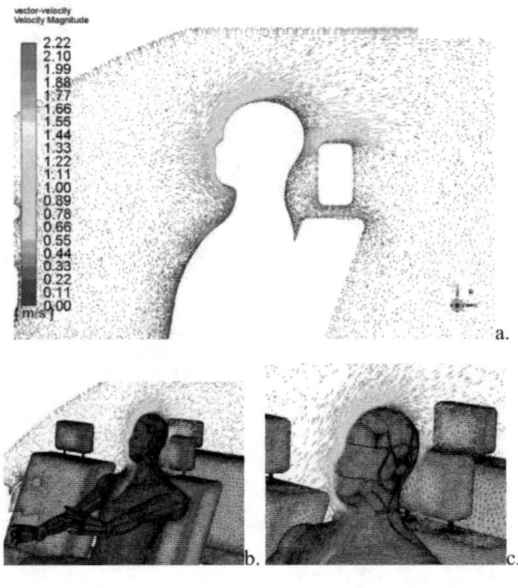

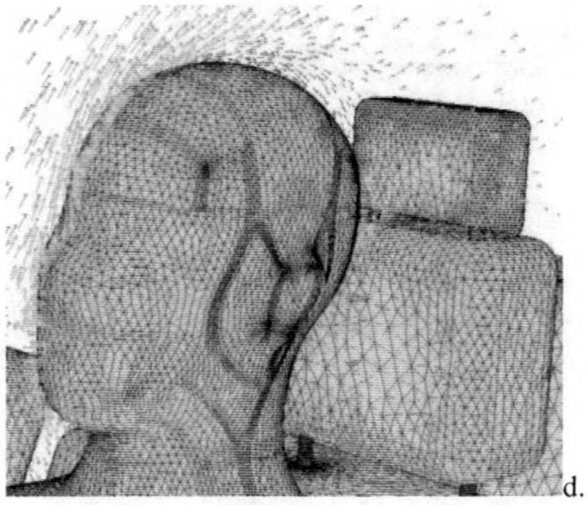

Fig. 9 Velocity field in the sagittal plane of the virtual manikin

3 Conclusions

This study continues the work started in [4] by introducing a virtual thermal manikin in the previously tested automobile interior. The study tries to evaluate a simple approach of a CFD model that models the thermal environment and the flow dynamics inside the interior of an automobile with a virtual thermal manikin present inside the vehicle. The main conclusion is that the virtual model is valid and is ready to be used in the next researches regarding thermal comfort in vehicular spaces.

This work was supported by Grant of the Romanian National Authority for Scientific Research, CNCS, UEFISCDI, Projects number: PN-II-PT-PCCA-2013-4-0569 Innovative strategies of HVAC systems for high indoor environmental quality in vehicle and a grant of the Romanian space agency ROSA, QUEST - Advanced air diffusion system of the crew quarters for the ISS and deep space habitation systems, STAR-CDI-C3-2016-577.

References

1. Moon, J.H., et al., *Thermal comfort analysis in a passenger compartment considering the solar radiation effect.* International Journal of Thermal Sciences, 2016. **107**: p. 77-88.
2. Kaynakli, O. and M. Kilic, *An investigation of thermal comfort inside an automobile during the heating period.* Applied Ergonomics, 2005. **36**(3): p. 301-312.
3. Li, W. and J. Sun, *Numerical simulation and analysis of transport air conditioning system integrated with passenger compartment.* Applied Thermal Engineering, 2013. **50**(1): p. 37-45.
4. Danca, P., et al., *On the Possibility of CFD Modeling of the Indoor Environment in a Vehicle.* Energy Procedia, 2017. **112**: p. 656-663.
5. Zhang, H., et al., *Studies of air-flow and temperature fields inside a passenger compartment for improving thermal comfort*

and saving energy. Part I: Test/numerical model and validation. Applied Thermal Engineering, 2009. **29**(10): p. 2022-2027.
6. Cheng, Y., J. Niu, and N. Gao, *Thermal comfort models: A review and numerical investigation.* Building and Environment, 2012. **47**(Supplement C): p. 13-22.
7. Bode, F., et al., *The influence of the Inlet angle of vehicle air diffuser on the thermal comfort of passengers.* ENERGY and ENVIRONMENT (CIEM), 2017 International Conference on, 2017: p. 442-446.
8. Slater, K., *Human Comfort.* Vol. 1. 1985.
9. ISO, E., *ISO 7730 - Ergonomics of the thermal environment —Analytical determination and interpretation of thermal comfort using calculation of the PMV and PPD indices and local thermal comfort criteria.* 2005.
10. ISO, *Ergonomics of the thermal environment - Evaluation of thermal environments in vehicles Part 3: Evaluation of thermal comfort using human subjects*, in *ISO 14505-3:2006.* 2006.
11. ISO, *Ergonomics of the thermal environment - Evaluation of thermal environments in vehicles - Part 2: Determination of equivalent temperature* in *ISO 14505-2:2006.* 2006.
12. ISO, *Ergonomics of the thermal environment - Evaluation of thermal environments in vehicles Part 1: Principles and methods for assessment of thermal stress*, in *ISO 14505-1:2007.* 2007.
13. Angel Dogeanu, et al., *Conception of a simplified seated thermal manikin for CFD validation purposes.* Romanian Journal of Civil Engineering, 2014. **5**(1).
14. Meslem, A., et al., *Comparison of turbulence models in simulating jet flow from a cross-shaped orifice.* European Journal of Mechanics - B/Fluids, 2014. **44**: p. 100-120.
15. Florin Bode, et al., *Numerical prediction of wall shear rate in impinging cross-shaped jet at moderate Reynolds number.* Scientific Bulletin - University Politehnica of Bucharest, Series D: Mechanical Engineering, 2014. **76**(12): p. 8.
16. FluentInc, *Fluent 18.2 Help.*

Comparison of indoor air pollutants concentration in two Romanian classrooms

Vasilica Vasile[1], Alina Dima[1], Elena Zorila[2,3], Andrei Istrate[2,3], Tiberiu Catalina[2,3]*

[1]INCD URBAN-INCERC Bucharest, Laboratory PFCH Pantelimon, 266, Bucharest, Romania
[2]Technical University of Civil Engineering Bucharest Blvd. Pache Protopopescu 66, Bucharest, Romania
[3]Babes-Bolyai University, Faculty of Environmental Science and Engineering 30, Fântânele Street, Cluj Napoca

Abstract. This paper investigates the air pollutions in space ventilated in two High School classrooms. The analysis consists of comparison of one classroom with hybrid ventilation system and another one stander-by classroom with natural ventilation. Several studies regarding indoor air quality during the experimental campaign have been done for VOC, CO_2, CO, other pollutants, keeping monitored for humidity and temperature. The experimental demonstrated that the highest value for CO_2 in stander-by classroom is 2691 ppm and in classroom with hybrid ventilation is 1897 ppm, while values for CO are 1.1 / 1.1 ppm and VOC 0.14 / 0.06 ppm, better use hybrid ventilation.

1 Introduction

Air quality of indoor environments draw considerable attention from the public as well as from the researchers [1] because relatively higher pollutant concentrations in combination with longer time spent indoors can result in higher exposures and can produce adverse health effects (Ugranli et al., 2015, Zhong et al., 2017) [2, 3]. All over the world [3], indoor monitoring of air compounds is an ongoing challenge (Ho et al., 2016) [4], numerous studies being in progress, whether it is office or dwelling buildings or schools [3, 5-9, 10].

Educational institutions are one of the most studied indoor environments with a focus on kindergartens and primary or high schools, because of the high density population potentially vulnerable who are still growing, making them especially susceptible to the effects of pollution (Ugranli et al., 2015, Zhong et al., 2017) [2, 3].

Our previous studies relating the monitoring of indoor air from different type of spaces showed high values of the carbon dioxide concentrations in the most of cases. For example, the average values of carbon dioxide concentration monitored during three experimental campaigns in residential spaces, first in March-May 2014, ranged between 898 ppm and 1,306 ppm, the second in September-October 2014, with average values between 811 ppm and 3,111 ppm and the third in March-April 2015, with values between 1,128 ppm and 2,490 ppm (Vasile et al., 2016) [5-9]. It appears that in all three monitored residential spaces there are high concentrations of carbon dioxide, which exceeded up to three times the limit allowed in indoor air. These high values of the concentrations of carbon dioxide in the space of a kitchen are justified since the cooking activities results in generation of carbon dioxide by combustion. It is interesting to note that, the same high concentrations are found in the bedrooms leading to the conclusion that these spaces are deficient in terms of ventilation and of the assurance of necessary fresh air. Also, the average concentration values of carbon dioxide, monitored in office spaces during March-April 2012 (National Project PN 09 14 04 02, 2012) [11] were between 661 ppm and 1,018 ppm, those recorded in the period of February-March 2015, ranged between 888 and 1,668 ppm (Vasile et al., 2017) [12], and those monitored in November 2015, inside the spaces for the educational process in higher education (chemistry lab), were between 912 ppm and 990 ppm (Vasile et al., 2016) [5-9], over the permissible limit of 800 ppm [10].

Average concentrations of carbon dioxide between 1,720 ppm and 4,302 ppm (National Project PN 09 14 04 02, 2015) [13, 14], two to five times the tolerable limit of 800 ppm, as determined by OSHA [10], were recorded in the investigated educational sites (kindergartens, secondary and high schools) during September-October 2015, and when talking about spaces where children are working, ages ranging from 3 to 17 years, have a particularly worrying situation with a strong negative impact on the educational process, which must be an alarm signal. Carbon dioxide present in indoor air can have negative effects on the concentration and attention of children, thus affecting the learning process. It can also cause a drowsiness or headache.

Due to their abundant nature, Volatile Organic Compounds (VOCs) are crucial parameters for assessing air quality both internally and externally environment. This class of compounds includes a wide range of air pollutants with a significant impact on atmospheric air and human health (Kumar et al., 2014, Lerner et al., 2012, Matsumoto et al., 2010) [15, 16, 17]. The chemical

* Corresponding author: tiberiu.catalina@gmail.com

variety of a VOC group constitutes the cause of negative impact factor on human health, ranging from noncancerous to carcinogenic effects (Kumar et al., 2014; Lerner et al., 2012; Srivastava et al., 2005; Ramirez et al., 2012) [15, 16, 18, 19].

2 Study case

The study presents the results obtained from the monitoring campaign conducted at "Mihai Viteazul" High School from Bucharest, Romania and is based on the comparative analysis of two different classrooms, one equipped with a mixed hybrid air ventilation system (I-17) and the other one does not have an air ventilation system (I-16). These two classrooms are located on the first floor of the building, consisting of ground floor and two levels, put into use in 1926.

Fig. 1. Classroom with air ventilation system and detail of the equipment's location.

The two classrooms are oriented to the south and the volume of this was 299.88 m^3. The interior finishes were consisted of: layered laminate parquet for flooring, water dispersion paint for wall and vinyl polychloride joinery with thermo-insulating glass for windows. The heating of spaces is normally carried out by the centralized heat supply system using steel radiators with a length of 2,000 mm. During of entire measuring period, the heating system was not active. The furniture elements are specific to classrooms and are made of MDF (medium density fibreboard) panels. Moreover, in the space equipped with the air ventilation system (I-17), are present the electronic equipment's (computer, projector) and decorative plants. The number of occupants varied in these two classrooms, reaching a maximum value of 30 persons.

For the first classroom (I-17) with ventilation systems, the main component of the ventilation system was the Xpelair GX9 fan with application for schools. The diameter of the ventilator fan is of 266 mm in reversible range and with enclosed external rotor motors. The system is controlled by an automatic module that allow the modification of the fresh air flow from 150 to 600 m^3/h. During the experimental measurements the fresh air flow was set at 450 m^3/h or approx. 20 m^3/h/student. The extraction of the air is realized using an extraction outlet mounted in the classroom door.

Fig. 2. Photos with the window fan + hygro-adjustable grids and the extraction outlets in the door

The proposed system is functioning only if the hygroscopic air vents mounted in the windows frame don't allow sufficient fresh flow. These air vents are Eha2 Aereco hygro-adjustable grids and six of them were installed allowing a maximum fresh air flow of 180-200 m^3/h depending on the pressure difference indoor/outdoor and indoor humidity.

2.1. Monitoring method

The volatile organic compounds, benzene, toluene and formaldehyde have been selected based on their effects on human health. Previous studies mention that the breathing benzene can cause the onset of illness, accompanied by slight irritation of the eyes and mucous membranes. If benzene is aspirated directly into the lungs it can lead to acute haemorrhagic pneumonitis. Major concerns benzene toxicity includes anaemia and acute leukaemia. Also, benzene is a known carcinogen while toluene is toxic at high concentrations (Khoder, 2006) [20]. Sources of benzene concentrations may be paints, furniture wax, lubricants and glues (Kumar et al., 2014) [15]. Formaldehyde is the simplest and most common aldehyde found in the environment and is perhaps the most important pollutant of the indoor environment due to its frequency of occurrence and the irritating potentially it has over the occupants (Occupational Health Guideline for Formaldehyde) [21], being toxic and irritating to the respiratory tract, eyes and skin and carcinogenic to humans at high concentrations (Khoder, 2006) [20]. Indoor concentrations of formaldehyde vary and are influenced by temperature, humidity, ventilation rate, building age, presence of combustion sources, type of indoor activity, etc.

All the measurements were performed in accordance with a standardized method based on the standards EN1525 and World Health Organisation -Guidelines for Indoor Air Quality: Selected Pollutants), using a monitoring protocol, which includes the specifications regarding the location area of equipment's installation (adjacent from the ventilation system) and the setting of the chemical compounds and the monitoring interval.

The monitoring process of benzene and toluene was conducted using a real time detection method with GrayWolf DirectSense IQ-610 portable data-logging detector. In this case, the device is based on a photo-ionization detector (PID) for VOCs (Vasile et al., 2011) [22].

For detection of formaldehyde present in the indoor air, we used a gas detector whose principle of operation is based on photoelectric photometry that involves the existence of a standard pill containing the test paper treated with special chemicals, an illuminating agent and a light beam. When gas is blown onto the standard pill, the paper emits illumination by chemical reaction and this causes the paper to change colour. The amount of colour change is determined by the level of formaldehyde exposure and the time of exposure. In this campaign, the sampling time used for formaldehyde monitoring was 15 minutes. In every classroom, the level of formaldehyde concentration was equal (<0.01 ppm) and below the permissible limit.

The volatile organic compounds have been measured continuously throughout the course hours starting with 7:41 AM, and recording the concentrations every minute.

Along with the measurements of the volatile organic compounds, have also been monitored indoor environmental parameters by measuring carbon monoxide and carbon dioxide concentrations, temperature and humidity. We chose to monitor the concentration of carbon monoxide next to carbon dioxide because an accumulation of this odourless, colourless gas in indoor air can be toxic for human body due to the affinity for combination with hemoglobin and forming carboxyhemoglobin (COHb) which leads to disrupting oxygen transport. This parameter was considered an indicator of indoor air quality. In this case the source of carbon monoxide is only from outdoor air into the indoor environment. This parameter was considered an indicator of indoor air quality. In this case the source of carbon monoxide is only from outdoor environment.

Indoor air parameters monitoring was carry out by using several measuring principles, presented in table 1. Before the starting of the measurements, it was performed the calibration of the equipment.

Table 1. Measuring principles for detection indoor air parameters

Parameter	Measuring principle	Range	Accuracy
Carbon monoxide (CO_2)	Electrochemical	0÷500 ppm	±2 ppm
Carbon dioxide (CO)	NDIR	0÷10,000 ppm	±50 ppm
Relative humidity	Capacitive probe	0÷100 %	±2% ÷ ±3%
Temperature	Thermal resistance Pt100	-25° ÷ +70°C	±0.3°C

3 Results and interpretation

The contribution of this solution for air ventilation is also observed in the recorded values for the concentrations of the total volatile organic compounds, the mean values being lower compared to the non-ventilated space. Also, in our previous experimental campaigns in kindergartens, TCOV concentration had the highest average value of 0.52 ppm (minimum 0.23 ppm – maximum 0.85 ppm), in isobutylene units 1,193.349 μg/m^3 (527.82-1,950.659 μg/m^3), while in secondary or high schools, it recorded average values of 1.11 ppm (2,547.329 μg/m^3), with a minimum of 0.18 ppm (413.08 μg/m^3) and a maximum of 1.85 ppm (4,199.649 μg/m^3) (Vasile et al., 2016c) [9].

The mean concentration of TVOC recorded in our previous campaign in office spaces was between 0.03 ppm and 0.3 ppm (69.1-690.8 μg/m^3 isobutylene units, at 24 °C and 1 atm (Vasile et al., 2015) [23, 24], in residential spaces, the values ranged between 0.16 ppm and 0.36 ppm (Vasile and Dima, 2015) [23, 24], and in a chemistry lab was recorded an average of 0.25 ppm (0.20 ppm minimum - maximum 0.35 ppm), isobutylene units 573.72 μg/m^3 (minimum 458.97 μg/m^3 - maximum 803.21 μg/m^3) (Vasile et al., 2016a) [7]. Our results are comparable with the similar studies conducted in earlier years. Recently, Kumar et al., 2014 [15] reported the mean concentration of TVOC as 465.8 μg/m^3 (145.3–1,503.2 μg/m^3) in winter and 321.8 μg/m^3 (90.7–1,100.9 μg/m^3) in summer for indoor air of the library of Jawaharlal Nehru University, New Delhi, a range of 100 and 538 μg/m^3 for TVOC concentrations was reported in indoor air of Japanese University (Hori et al., 2012) [25]. Sarkhosh et al., 2012 [26] measured the concentrations in range from 113.4 to 486.3 ppm (227–973 μg/m^3 isobutylene units) in photocopy centres and Chan et al., (2009) [27] recorded TVOC concentrations in new hotels guest rooms of the factory region which varied between 416 and 2,900 μg/m^3.

Benzene concentration of 0.1 ppm recorded in this campaign is at the minimum level registered in kindergartens (between 0.1 ppm and 0.5 ppm) (National Project PN 09 14 04 02, phase no. 22, 2015) [13] and another secondary high schools from Bucharest (between 0.1 ppm and 0.9 ppm).

Our previous studies showed levels of benzene mean concentration of 0.3 ppm (958.4 μg/m^3) in residential spaces and 0.2 ppm (638.9 μg/m^3) in office spaces (Vasile et al., 2016b) [8], within PELs, and 0.13 ppm (415.30 μg/m^3), with a minimum of 0.10 ppm (319.59 μg/m^3) and a maximum of 0.20 ppm (638.94 μg/m^3) in a chemistry lab (Vasile et al., 2016a) [7], but were higher than those reported in indoor air of Egyptian offices (4.32 ppb) (Khoder, 2006) [20], or in the library of Jawaharlal Nehru University, New Delhi (7.2-12.2 μg/m^3) (Kumar et al., 2014) [15], or in indoor air of offices from NCSR "Demokritos", Aghia Paraskevi, Athens (15.3-17.6 μg/m^3) (Saraga et al., 2011) [28].

In the classroom without air ventilation system has been recorded the presence of toluene with an average value of 0.1 ppm and a maximum concentration of 0.2 ppm compared to classroom with ventilation solution, where this compound was not present in indoor air.

Since the classrooms are similarly from the dimensional point of view and from the finishing products, we can consider a constant contribution of these parameters, the comparison of the recorded values based on the influence of the space ventilation solution.

At the end of the monitoring program it was observed that all the specific volatile organic compounds are below of the permissible exposure limits. However, from a comparative point of view, it can be notice the presence of benzene and toluene in classroom without air ventilation system versus the classroom with ventilation solution. This concentration of the specific organic compound is also reflected in the concentration values of total volatile organic compounds.

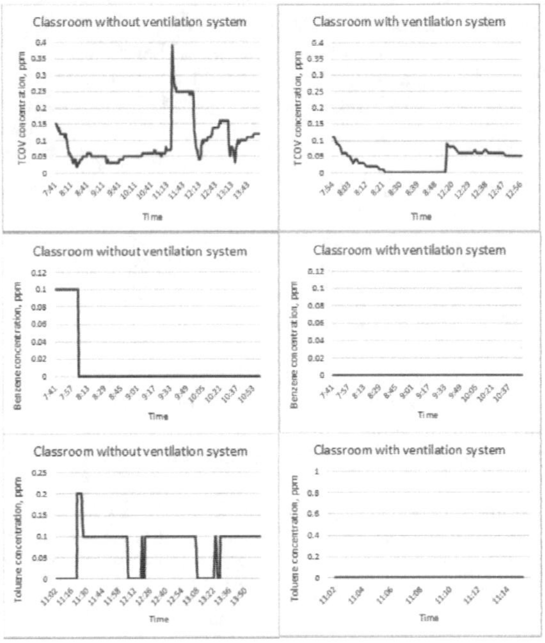

Fig. 3. Comparison between non-ventilated and ventilated classroom for TCOV, Benzene and Toluene.

Table 2. Monitored compound and environmental conditions.

The monitoring period: April 4th- April 5th 2017	Monitored compound and environmental conditions Average value (min/max)							
	TCOV, ppm	Benzene, ppm	Toluena, ppm	Formaldehyde, ppm (15 min)	CO_2, ppm	CO, ppm	T, °C	RH, %
Classroom with ventilation system (I-17)								
07:54–08:53 AM	0.02 (0/0.11)	0	unmeasured	< 0.01	1,521 (880/2,379)	1.1 (0.9/1.4)	24.6 (22.8/25.8)	35.2 (33.6/37.0)
12:17–12:56 AM	0.06 (0.05/0.09)	unmeasured	0		1,897 (1,710/2,026)	0.7 (0.5/0.8)	25.1 (23.9/25.8)	35.1 (33.3/38.9)
Classroom without ventilation system (I-16)								
07:41 AM–08:03 AM	0.13 (0.10/0.15)	0.1*	unmeasured		1,317 (1,139/1,564)	1.1 (1.0/1.2)	25.3 (24.3/26.0)	35.7 (34.9/37.0)
08:04 AM–10:59 AM	0.05 (0.02/0.08)	0	unmeasured	< 0.01	1,872 (1,000/2,567)	0.6 (0.4/1.1)	27.0 (24.5/28.0)	34.0 (30.8/35.9)
11:02 AM–14:03 AM	0.14 (0.03/0.39)	unmeasured	0.1 (0/0.2)		2,691 (1,530/3,492)	0.5 (0.1/0.8)	27.2 (25.5/28.2)	37.9 (30.7/42.4)
Permissible exposure limits (PELs), ppm	-	1	200	Class E_1 (≤ 0.041)	800	9	-	-

4 Conclusions

Within this article we have realized experimental campaign with the purpose to compare indoor air pollutants in case of two ventilation strategies. Two identical classrooms as shape, volume, about the same number of students, orientation and furniture was compared. The difference in comparing indoor air pollutants was the using of hybrid ventilation (natural ventilation with humidity sensitive air vents added with window fan) for one classroom and only natural ventilation for other classroom. In this way, it has been demonstrated that can be obtained good air quality using hybrid ventilation. After the analysis of the measured data using professional equipment with GrayWolf Direct Sense IQ-610 portable data-logging detector we can conclude that it is necessary the introduction even more of fresh air to reduce CO_2 limit at 800 ppm in both classroom, more acceptable values obtaining using hybrid ventilation. The same for VOC, CO and for other pollutants have been obtained good permissible exposure values using hybrid ventilation. In these conditions, using hybrid ventilation to achieve more suitable limits for indoor air pollutants concentration can allow more fresh airflow rates.

Acknowledgements

The research article is supported by the project ID P_37_229, SMIS 103427, Contract No. 22/01.09.2016, with the title „Smart Systems for Public Safety through Control and Mitigation of Residential Radon linked with Energy Efficiency Optimization of Buildings in Romanian Major Urban Agglomerations SMART-RAD-EN" supported by the Competitiveness Operational Program 2014-2020, POC-A.1- A.1.1.4 -E- 2015 competition.

References

[1]. C. Soviany, Embedding Data and Task Parallelism in Image Processing Applications, PhD Thesis, Technische Universiteit Delft, 2003

[2]. T. Ugranli, M. Toprak, G. Gursoy, A.H. Cimrin, S.C. Sofuoglu, Indoor environmental quality in chemistry and chemical engineering laboratories at Izmir Institute of Technology, Atmospheric Pollution Research 6, 147-153, 2015

[3]. L. Zhong, F.C. Su, S. Batterman, Volatile organic compounds (VOCs) in conventional and high performance school buildings in the U.S., Int.J.Environ.Res.Public Health 14, 100-118, 2017

[4]. S.S.H. Ho, Y. Huang, S.C. Lee, J. Cao, Challenges on field monitoring of indoor air quality in China, Indoor and Built Environment, available on https://www.researchgate.net/publication, 2016

[5]. V. Vasile, H. Petran, A. Dima, C. Petcu, Indoor air quality – a key element of the energy performance of the buildings, Jarek Kurnitski, Energy Procedia - Sustainable Built Environment Tallinn and Helsinki Conference SBE16 Build Green and Renovate Deep, 972 pag., ISSN 1876-6102, pag. 277-284, available at www.sciencedirect.com, 2016

[6]. V. Vasile, A. Dima, M. Ion, Indoor air quality in built environment subject to vibrations generated by non-seismic sources, Conferinţa de cercetare în construcţii, economia construcţiilor, urbanism şi amenajarea teritoriului. Rezumate ale lucrărilor, ISSN 2343-7537, indexată în Ulrich's, ProQuest şi Europa World of Learning/Routledge, disponibil la http://pub.incd.ro/, 2016
[7]. V. Vasile, A. Dima, M. Ion, Indoor air quality in built environment subject to vibrations generated by non-seismic sources, Conferinţa de cercetare în construcţii, economia construcţiilor, urbanism şi amenajarea teritoriului. Rezumate ale lucrărilor, ISSN 2343-7537, indexată în Ulrich's, ProQuest şi Europa World of Learning/Routledge, disponibil la http://pub.incd.ro/, 2016a
[8]. V. Vasile, A. Dima, M. Ion, Monitoring of indoor air pollution in residential and office spaces, 16th International Multidisciplinary Scientific GeoConference SGEM 2016,www.sgem.org, SGEM2016 Conference Proceedings, ISBN 978-619-7105-64-3 / ISSN 1314-2704, June 28 - July 6, 2016, Book4 Vol. 2, 467-474 pp, 2016b
[9]. V. Vasile, A. Dima, M. Ion, Children exposure to volatile organic compounds in indoor air of Bucharest educational spaces, Proceedings of The 8th Edition of European Exhibition of Creativity and Innovation – EUROINVENT 2016, Romanian Inventors Forum, Iaşi, 492 pag., pag. 429-430, disponibil la http://www.euroinvent.org/cat/e2016.pdf, 2016c
[10]. Ocuppational Safety & Health Administration (OSHA) Standards – Permissible Exposure Limits (PELs)
[11]. The project PN 09 14 04 02, Concepts of making green buildings, in embodiments and of environmentally friendly materials, with low energy, phase no.7, May 2012
[12]. V. Vasile, A. Dima, M. Ion, Monitoring of the inorganic pollutants in built indoor environment, Rev.Chem.(Bucharest), 68 (1):85-89, available at http://www.revistadechimie.ro/, 2017
[13]. The project PN 09 14 04 02, Concepts of making green buildings, in embodiments and of environmentally friendly materials, with low energy, phase no.22, November 2015
[14]. The project PN 09 14 04 02, Concepts of making green buildings, in embodiments and of environmentally friendly materials, with low energy, phase no.23, December 2015
[15]. A. Kumar, B.P. Singh, M. Punia, D. Singh, K. Kumar, V.K. Jain, Assessment of indoor air concentrations of VOCs and their associated health risks in the library of Jawaharlal Nehru University, New Dehli, Environ Sci Pollut Res, 21:2240-2248, 2014
[16]. J.E. Lerner, E.Y. Sanchez, J.E. Sambeth, A.A. Porta, Characterization and health risk assessment of VOCs in occupational environments in Buenos Aires, Argentina, Atmos Environ, 55:440-447, 2012
[17]. K. Matsumoto, R. Mizuno, M. Igawa, Volatile organic compounds in ambient aerosols, Atmos Res, 97:124-128, 2010
[18]. A. Srivastava, B. Sengupta, S.A. Dutta, Source Apportionment of ambient VOCs in Delhi City, Sci Total Environ 343(1-3):207-220, 2005
[19]. N. Ramirez, A. Cuadras, E. Rovira, F. Borrull, R.M. Marce, Chronic risk assessment of exposure to volatile organic compounds in the atmosphere near the largest Mediterranean industrial site, Environ Int, 39:200-209, 2012
[20]. M.I. Khoder, Formaldehyde and Aromatic Volatile Hydrocarbons in the Indoor Air of Egyptian Office Buildings, Indoor and Built Environment, 15, pp 379-387, 2006
[21]. Ocupational Health Guideline for Formaldehyde – National Institute for Occupational Safty and Health, US Department of Health and Human Services
[22]. V. Vasile, A. Cioacă (Dima), Metode şi tehnici experimentale pentru determinarea emisiilor provenite din materialele de construcţii, Urbanism, Arhitectură, Construcţii, 2(2):63-66, 2011
[23]. V. Vasile, A. Dima, C. Stamate, Indoor air pollution and their associated health risks, Conference Importance of Place – Conference Proceedings, Sarajevo, 21-23 oct.2015, CICOPBH, Sarajevo, vol.3, no.1, pp.277-291, 2015
[24]. V. Vasile, A. Dima, Indoor environment – air quality and the risks on human health, Studia UBB Ambientum, LX, 1-2, pp.137-145, 2015
[25]. H. Hori, S. Ishimatsu, Y. Fueta, T. Ishidao, Evaluation of a real-time method for monitoring volatile organic compounds in indoor air in a Japanese university, Environ Health Prev Med. doi:10.1007/s12199-012-0319-1, 2012
[26]. M. Sarkhosh, A.H. Mahvi, M.R. Zare, Y. Fakhri, H.R. Shamsolahi, Indoor contaminants from hardcopy devices: characteristics of VOCs in photocopy centers, Atmos Environ 63:307–312, 2012
[27]. W. Chan, S. Lee, Y. Chen, B. Mak, K. Wong, C. Chan, Indoor air quality in new hotels' guest rooms of themajor world factory region, Int J Hosp Manag 28:26–32, 2009
[28]. D. Saraga, S. Pateraki, A. Papadopoulos, C. Vasilakos, T. Maggos, Studying the indoor air quality in three non-residential environments of different use: a museum, a printery industry and an office, Building and Environment 46, pp 2333-2341, 2011

E3S Web of Conferences **32**, 01024 (2018) https://doi.org/10.1051/e3sconf/20183201024
EENVIRO 2017

Urban sound energy reduction by means of sound barriers

Vlad Iordache**and Mihai Vlad* Ionita

Research Center CAMBI, Technical University of Civil Engineering of Bucharest, Romania

Abstract. *In urban environment, various heating ventilation and air conditioning appliances designed to maintain indoor comfort become urban acoustic pollution vectors due to the sound energy produced by these equipment. The acoustic barriers are the recommended method for the sound energy reduction in urban environment. The current sizing method of these acoustic barriers is too difficult and it is not practical for any 3D location of the noisy equipment and reception point. In this study we will develop based on the same method a new simplified tool for acoustic barriers sizing, maintaining the same precision characteristic to the classical method. Abacuses for acoustic barriers sizing are built that can be used for different 3D locations of the source and the reception points, for several frequencies and several acoustic barrier heights. The study case presented in the article represents a confirmation for the rapidity and ease of use of these abacuses in the design of the acoustic barriers.*

1 Introduction

The noise protection represents today one of the main design requirements in order to meet international LEED and BREAM standards as well as the Romanian Ministry of Development design option F requirement. Today, several types of building services equipment (chillers, heat pumps, compressor, VRVs, Rooftops, and others) alongside their role in preserving the indoor comfort of the serviced building, they become sound pollution vectors and sometimes leading to the violation of the noise protection requirements in urban environment [1]

The sound energy emitted by the equipment in the environment and leads to an increase of the acoustic pressure above the background value. The higher the acoustic pressure in a specific reception point the higher the sound energy arrived in that reception point. The amount of energy arrived at the reception point depends on the length of the sound wave propagation path: the further the reception point, the smaller the energy arrived, the smaller the acoustic pressure and finally the smaller the acoustic pressure level.

Therefore, engineers and architects strive to find solutions to reduce the sound energy arrived at specific locations in the urban environment and consequently reduce the noise level at that location. The installation acoustic barriers in-between the noise generation equipment and the reception point represents a noise protection solution. The design of these acoustic barriers consists in determining the position, length and height of this acoustic barrier. The height of the acoustic barrier is a design element that is most often done using the Maekawa diagram [2]. Although, other theories have been developed depending on the type of relief, the shape of the acoustic barrier [3], its thickness or the presence of an inclined cover [4] [5] as well as the type of noise source (point or line) [6], however Makawa Diagram remains today the most widely used method.

This acoustic barrier sizing procedure based on Maekawa's theory represents a method hardly exploitable by engineers and architects due to the complexity of calculations required to be performed for the entire spectrum according to national noise protection requirements [7]. In this study we propose a new and simplified method for the sizing of an acoustic barrier, method based on the same theory. This new design tool will lead, in a faster and easier way, to the same results. Today there are no simplified calculation abacuses for this sizing procedure neither in international norms nor in research literature. The aim of this study is to produce several abacuses in order to rapidly determine the correct height for these acoustic barriers. These abacuses should be relevant to different 3D geometries and different locations of the reception point (that is intended to be protected).

The paper presents the method for the construction of the abacus for the rapid calculation of acoustic barrier for all frequencies and a study case that aims to understand the applicability and the ease of the new design method.

2. Method

In this paragraph we will present the analyzed geometry and the method used for the construction of the calculation abacus for fast noise level decrease.

The analyzed geometry at urban level (**Fig. 1**) consists of three different elements:

- a noise source (chiller, pump, fan or other noisy equipment) that will be considered in our analysis as a stationary noise source, placed at 1 m above the ground;
- an acoustic barrier located at 1m away from the noise source. Its height is from the ground level up to the design height. Such an acoustic barrier should exceed the altitude of the noise source. In this study we aim to examine five different heights for the acoustic barrier:

* Corresponding author: viordach@yahoo.com

a) 1m representing the lower limit, b) 2m, c) 3m, d) 4m and e) 5m. From a structural viewpoint, the barrier has a specific mass greater than 10 kg/m^2, so that the main sound wave is the diffracted wave (the wave passing around the acoustic barrier) and not the wave perpendicular on the acoustic barrier crossing the barrier.

- the area protected by the acoustic barrier is the area behind this acoustic barrier that is not directly visible from the noise source (the shaded area in Figure 1). For any point in this area, the sound wave generated by the noise source suffers is diffracted at the top of the acoustic barrier. In this study, we analyze five different heights of the acoustic barrier (five different geometries) and for each height the protected area will be different.

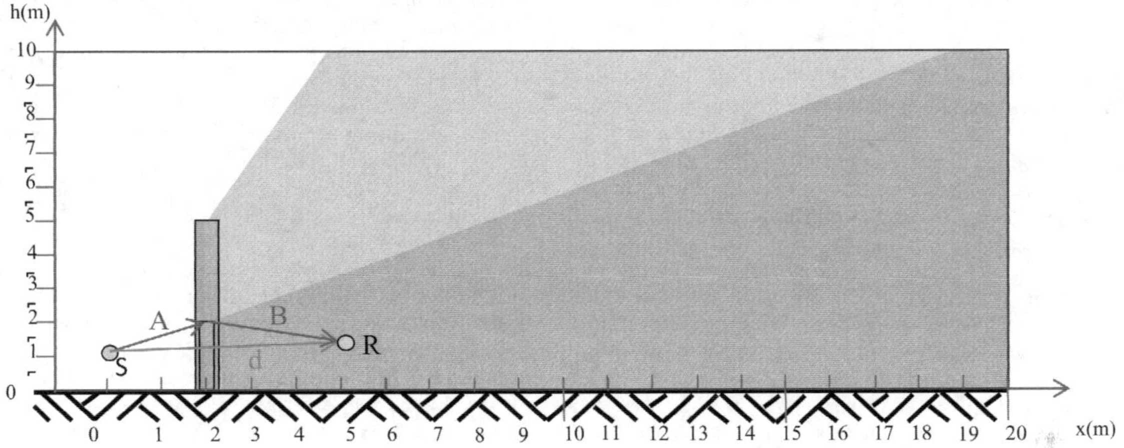

Fig.1. 2D geometry for two acoustic barrier heights (2m and 5m) and the corresponding protected area

When no acoustic barrier is mounted the noise level in the protected area is due to the sound wave propagated directly from the noise source (S) to the reception point (R), covering a distance d (m) (**Fig. 1**). After the installation of the acoustic barrier, the noise level is due to the diffracted wave, (the wave generated by the noise source (S), and passing around the acoustic barrier at its highest point to reach the receiver point (R). Thus, the new distance covered by the diffracted sound wave, A+B (m), where A (m) is the distance from the source to the highest point of barrier and B (m) is the distance from the highest point of the barrier to the receiver, is longer than if there were no acoustic barrier.

To determine the sound energy decrease due to a certain acoustic barrier, firstly we determine the Fresnel's number N (-) (Equation 1) [8].

$$N = 2 \cdot \frac{A + B - d}{\lambda}, \qquad (1)$$

where λ *(m)* is the sound wave length, (variable with the frequency). Further, the Maekawa theory will lead to assessment of the noise level attenuation in the reception point ΔLp *(dB),* compared to the case when there was no acoustic barrier, as a function of the Fresnel's number, N.

In urban environment the noise source is an HVAC equipment that generates sound energy on all frequencies of the audible spectrum. Thus the Fresnel's number and further the noise level attenuation are different from one frequency to another.

The urban noise protection norms in Romania [7] set sound pressure level maximum limit values for each frequency at both the street level [1] and the building facade level [9]. Thus the sizing of this type of urban acoustic barriers should simultaneously fulfill all the requirements for all the frequencies.

Therefore, the sizing of an acoustic barrier during the design phase becomes a highly complex procedure given the 3D geometry of the environment where sound wave propagates and the analysis that should be carried out for all frequencies.

This higher complexity of the calculation procedure has negative consequences upon the design solution. We wish to generate a new tool for acoustic barrier sizing for architects and engineers. The purpose of this study is to generate abacuses for rapid calculation of noise attenuation at the receiving point and this new tool will represent an essential advancement in this field.

3. Results

The calculation method presented in the previous paragraph was applied for each geometric configuration (each panel height). The receiver point was placed between 1m to 20m behind the acoustic barrier (1m horizontal grid spacing) and between the ground level to 10m height (0.5m vertical grid spacing). For all these reception points, the calculation method was applied, calculating the parameters: distance *A*, distance *B*, distance *d*, distance difference δ, Fresnel's number *N*, and noise attenuation ΔLp, for each frequency. For these receiver positions inside this grid where the receiver is

* Corresponding author: viordach@yahoo.com

not protected by the acoustic barrier (outside the shaded area), a zero noise attenuation value was imposed (negative Fresnel's number is irrelevant for our application).

In order to obtain the abacuses for rapid calculation of the noise level attenuation, these attenuations were calculated for each receptor position inside the matrix, each height of the acoustic barrier and each frequency: 125Hz and 250Hz (low frequencies), 500Hz and 1000Hz (median frequencies) and 2000Hz and 4000Hz (high frequencies).

In **Fig. 2**, corresponding to the 125 Hz frequency, we can compare the effect of the five heights of acoustic barriers. We considered the HVAC equipment of a house representing the noise source while the facade of the closest building (block of flats) is the receptor. In-between the two buildings, at 2m from the noise source the acoustic barriers are placed. In the case of the 1m height barrier (**Fig. 2**a), the protected area is the narrowest and therefore such an acoustic barrier is appropriate for receiver points placed on the ground or below ground level (depending on the relief). The protection provided by this 1m height barrier is 5-6 dB at ground level, representing a low noise protection.

For the 2m height barrier (**Fig. 2**b), the protected area is larger and the attenuation of ground level noise is approximately 8-10 dB. For the 3m height barrier (**Fig. 2**c), the protected area is wider, and the attenuation of ground level noise is 12-15 dB. For the 4m high barrier, the attenuation of the ground level noise increases to 14-17 dB, and for the 5m barrier height it reaches 16-19 dB.

As a general trend it is noted that the higher the height of the barrier the larger the protected area and the higher the noise level attenuation.

The minimum values of the sound level attenuation (about 5dB) are found on the separation line between the area protected by the acoustic barrier (shaded area in Figure 1) and the unprotected area. Above this separation line, the attenuation rapidly descends to 0dB (zone corresponding to negative Fresnel's number). In our study we considered the effect of the acoustic barrier is considered to be null.

The location where the sound level attenuation is highest is in the immediate vicinity of the barrier behind it. At this location, for the 1 m high barrier the attenuation is 7dB, while for the 5 m barrier the attenuation reaches 20dB.

If we consider a reception point at a horizontal distance of 20m from the noise source and at a height of 5m (located on the first floor of the block of flats) it is observed that the 1m height barrier does not provide any protection whatever the frequency. But the 2m barrier provides a 6.3dB sound level attenuation at 125Hz (**Fig. 2**b), a 3m barrier assures a 10dB attenuation, a 4m barrier assures an attenuation of 13dB, and the 5m barrier provides a 15dB attenuation.

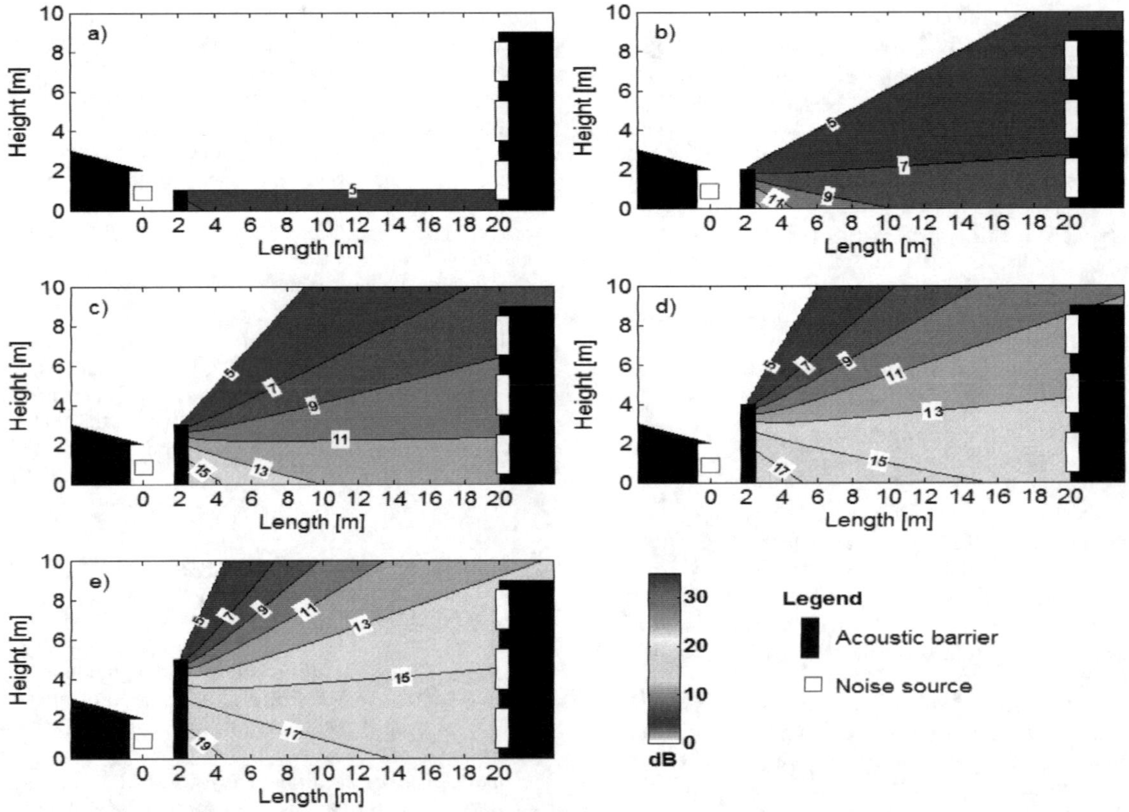

Fig. 2. Noise level attenuation at 125 Hz for an acoustic barrier height of: a) 1m; b) 2m; c) 3m; d) 4m; e) 5m

* Corresponding author: viordach@yahoo.com

 https://doi.org/10.1051/e3sconf/20183201024

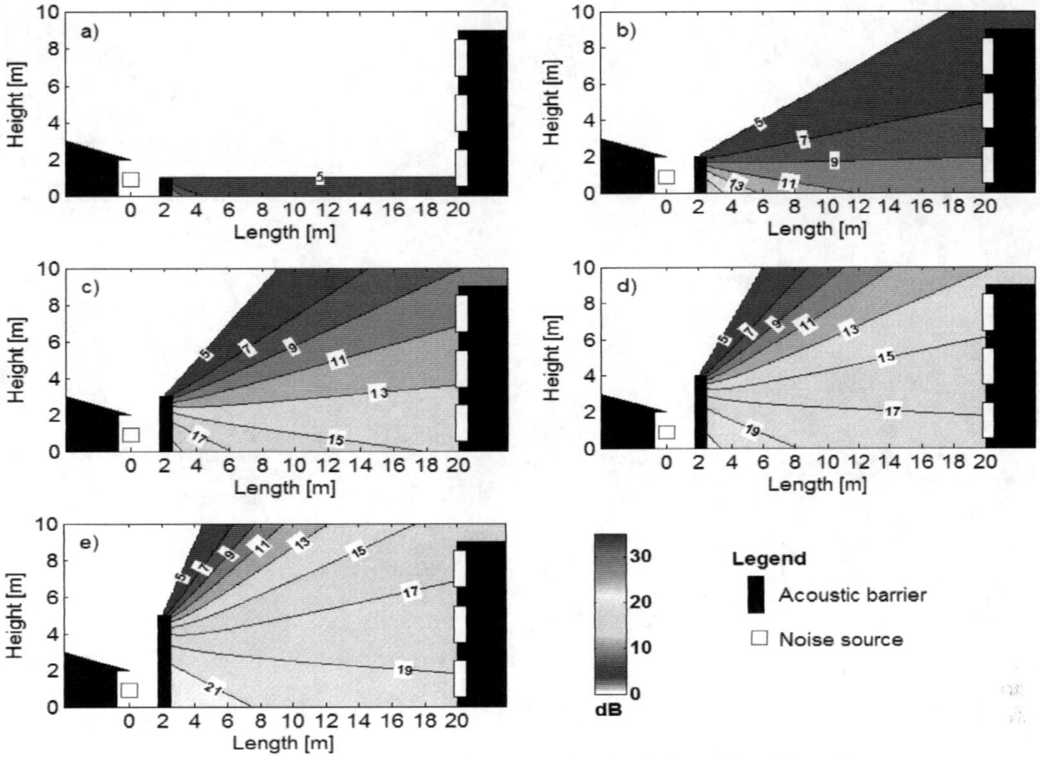

Fig. 3. Noise level attenuation at 250 Hz for an acoustic barrier height of: a) 1m; b) 2m; c) 3m; d) 4m; e) 5m

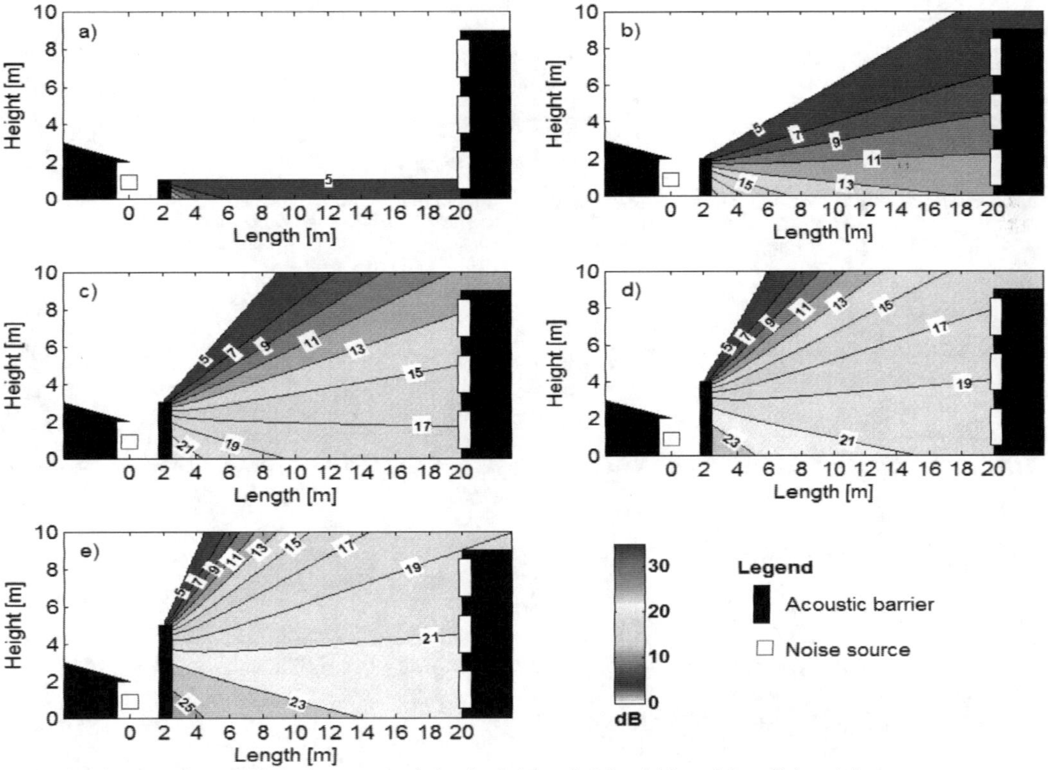

Fig. 4. Noise level attenuation at 500 Hz for an acoustic barrier height of: a) 1m; b) 2m; c) 3m; d) 4m; e) 5m

* Corresponding author: viordach@yahoo.com

https://doi.org/10.1051/e3sconf/20183201024

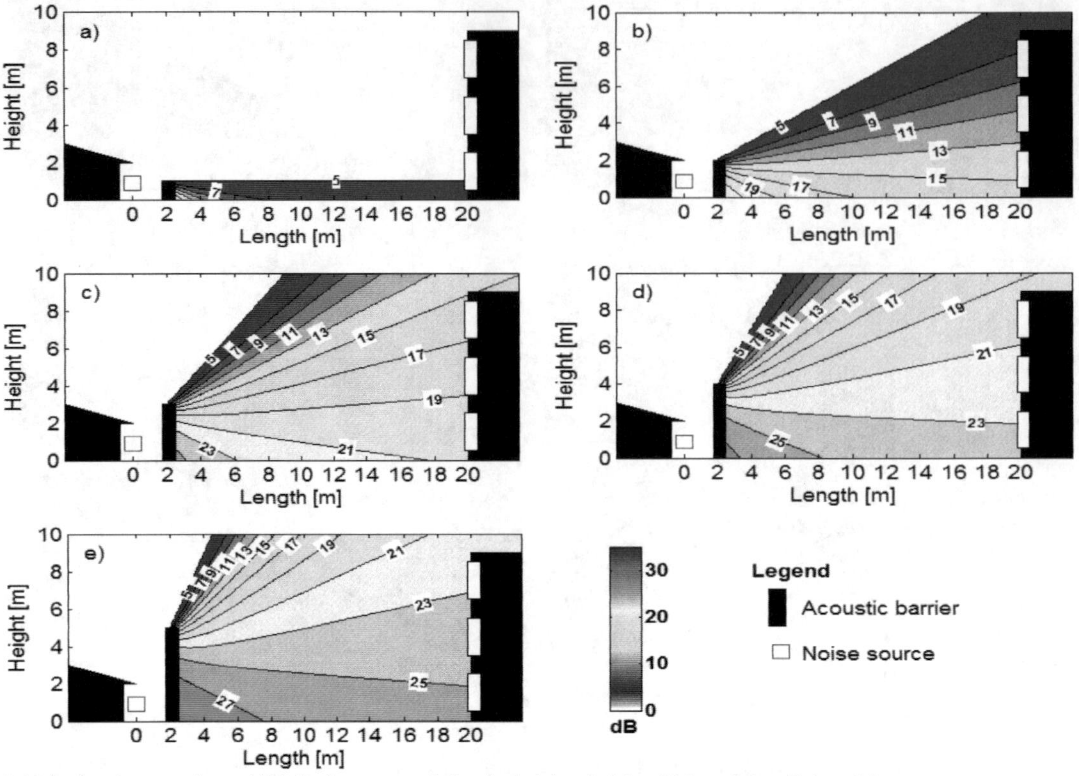

Fig. 5. Noise level attenuation at 1000 Hz for an acoustic barrier height of: a) 1m; b) 2m; c) 3m; d) 4m; e) 5m

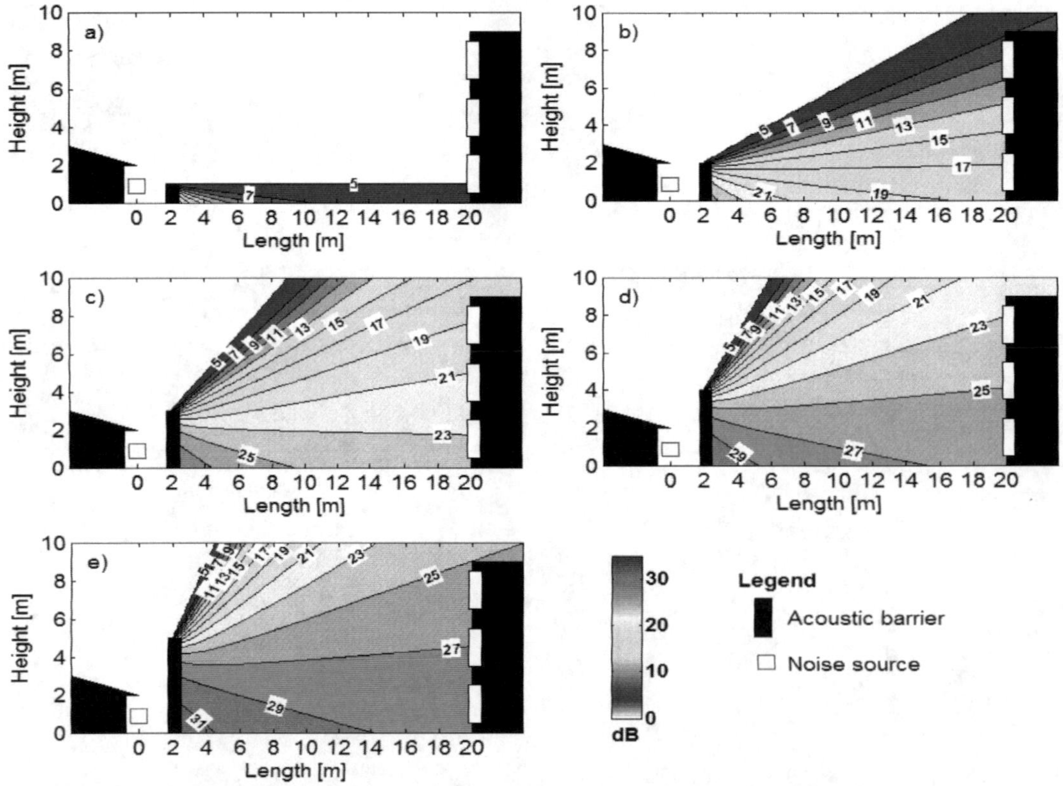

Fig. 6. Noise level attenuation at 2000 Hz for an acoustic barrier height of: a) 1m; b) 2m; c) 3m; d) 4m; e) 5m

* Corresponding author: viordach@yahoo.com

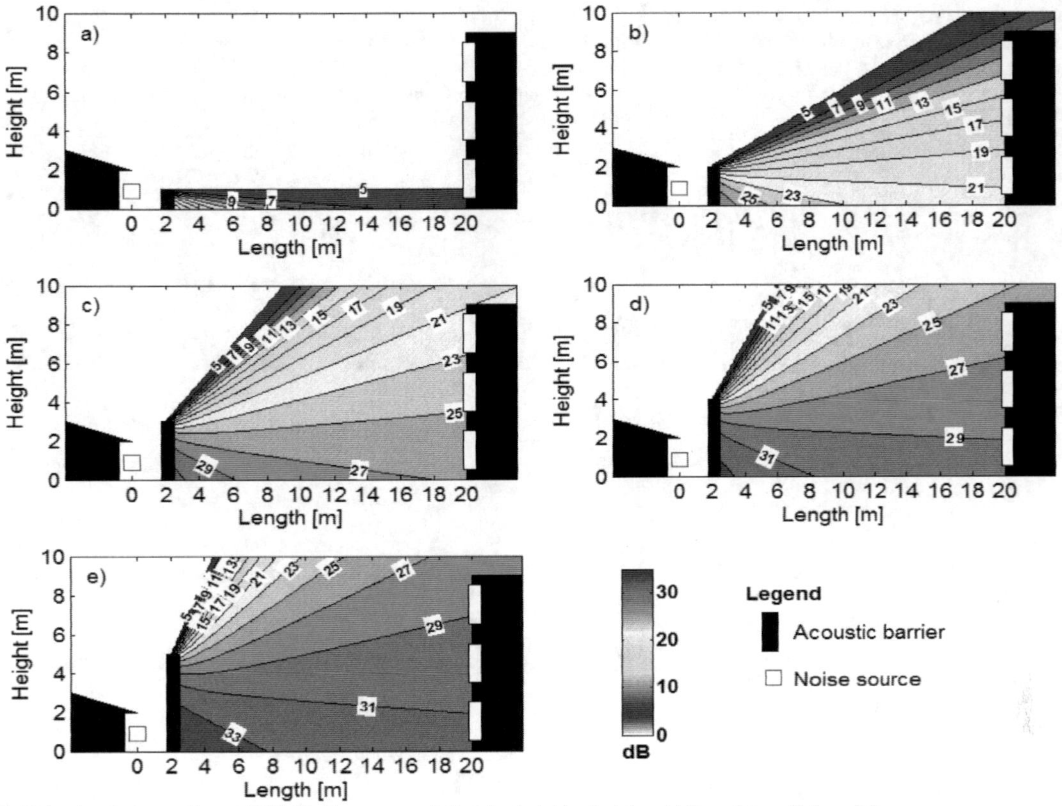

Fig. 7. Noise level attenuation at 4000 Hz for an acoustic barrier height of: a) 1m; b) 2m; c) 3m; d) 4m; e) 5m

In **Fig. 3**, corresponding to the 250Hz frequency, the same general tendency is observed: the higher the barrier height, the higher the protected area and the higher the sound level attenuation. The maximum attenuation corresponding to the 5m height acoustic barrier is about 23dB, higher than the maximum attenuation corresponding to the 125 Hz frequency (20dB). For a receiver located 18 m behind the acoustic barrier and 2 m high, the 2m height barrier sound level attenuation at 250 Hz is about 7.5 dB.

In **Fig.4 - 7**, corresponding to frequencies 500Hz, 1000Hz, 2000Hz and 4000Hz respectively, the same general tendency is observed: the higher the barrier height, the higher the protected area and the higher the sound level attenuation. The maximum attenuation is met behind the acoustic barrier for all frequencies.

Moreover, the sound level attenuation is variable as a function of the frequency: the higher the frequency the higher the attenuation. Thus, for 4000Hz the highest attenuation is about 34dB compared to only 20dB corresponding to 125Hz.

4. Study case

This chapter exemplifies how these simplified abacuses can be used to correctly design the height of an acoustic barrier. We shall consider the case of a rooftop ventilation equipment placed of the flat roof of an office building (marked "OB" in **Fig. 8**), which represents the sound source (marked with "S" in **Fig. 8**) placed at 2m from the office building attic. The reception point (marked with R in **Fig. 8**) is represented by the window of the nearby apartment building (marked "AB" in **Fig. 8**) where the façade maximum allowed noise level is 45 dB at 1000Hz. The distance between the rooftop and the analyzed window is 10m and the acoustic pressure level on the façade of the apartment building is 59.5dB, thus it is 14.5 dB over the maximum allowable limit. We note the higher the distance between the receiver and the sound source, the smaller the noise level in the receiver point. This is due to the dispersion of the sound energy emitted by the rooftop all-around into a sphere shape and thus the further the reception point from the source the higher the surface area of the sphere where the energy is dispersed (**Fig. 8**a).

An acoustic barrier would be installed close to the attic of the office building (blue line in **Fig. 8**) at 2m away from the noise source and at 8 m from the reception point. This barrier would alter this spherical energy dispersion behind the acoustic barrier.

We will consider the initial height of the acoustic barrier to be 1m. Thus for a 1000Hz frequency we will use the abacus in **Fig. 5**a and the sound level attenuation is about 6dB. Consequently, the noise level after the installation of the acoustic barrier is about 54dB (**Fig. 8**b). The attenuation on the façade of the apartment building is variable (lower acoustic pressure levels for lower floors of the apartment building façade), but for our reception point the noise protection condition is not fulfilled. For the case of the 2m height acoustic barrier

* Corresponding author: viordach@yahoo.com

 https://doi.org/10.1051/e3sconf/20183201024

(attenuation abacus in **Fig. 5**b) the attenuation is 15 dB, and thus the acoustic prresure level is 44.5dB (**Fig. 8**c) and the maximum limit condition is fulfilled.

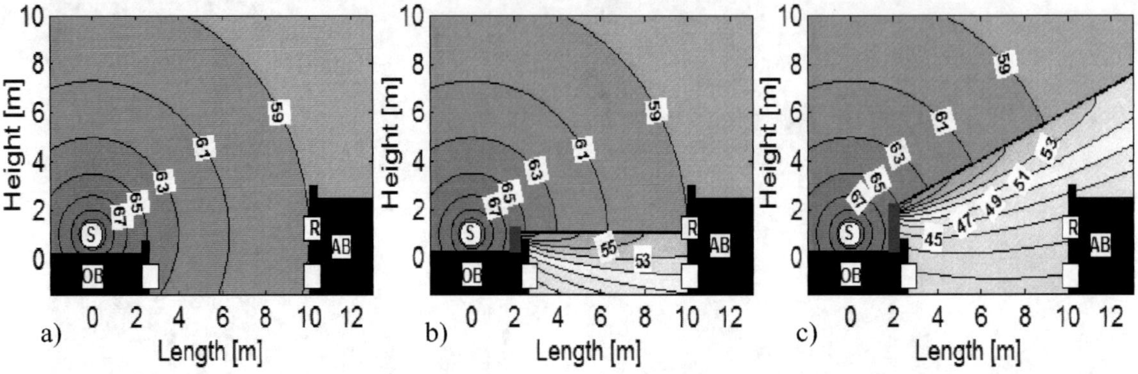

Fig. 8. Noise level comparison at 1000 Hz betwen three situations: a) no acoustic barrier, b) acoustic barrier height 1m, c) acoustic barrier height 2m

5. Conclusions

The Maekawa method of calculating the noise attenuation characteristic of an acoustic barrier has been applied to several receptor positions in the shaded area. The results were used to create simplified abacuses for the sizing of the acoustic barrier height. The abacuses were built for five types of barrier heights from 1m to 5 m and for different frequencies from 125 to 4000 Hz.

The study case proves that these abacuses essentially contribute to simplifying the height sizing calculation of an acoustic barrier while maintaining the same precision as the classical method.

Moreover, this study also represents a validation of this approach in order to achieve other simplifying methods for other geometries and special cases where the classical method can also become difficult to apply.

Acknowledgement

This study is supported by Romanian Authority for Scientific Research and Innovation, CNCS/CCCDI - UEFISCDI, through research project PN-III-P2-2.1-PED-2016-1951.

References

[1] SR 10009:2017 - Acustică. Limite admisibile ale nivelului de zgomot din mediul ambiant.

[2] Z. Maekawa, "A simple estimation method for noise reduction by various shaped barriers," *Revista de Acustica,* vol. XXI, pp. 32-37, 1990.

[3] K. W. W. H. Kasess CH, "Deriving correction functions to model the efficiency of noise barriers with complex shapes using boundary element simulations.," *Applied Acoustics,* vol. 102, pp. 88-99, 2016.

[4] U. J. Kurze and G. S. Anderson, "Sound attenuation by barriers," *Applied Acoustics,* vol. 4, no. 1, pp. 35-53, 1971.

[5] Y. M. Lam, "Using Maekawa's chart to calculate finite length barrier insertion loss," *Applied Acoustics,* vol. 42, no. 1, pp. 29-40, 1994.

[6] Z. Maekawa, "Noise reduction by screens," *Applied Acoustics,* vol. 1, no. 3, pp. 157-173, 1968.

[7] "Normativ C125:2012 - Normativ privind acustica în constructii şi zone urbane".

[8] R. Peng and L. Tang, "Narrowing effect of focused polychromatic laser with small Fresnel number," *Optics Communications,* pp. 2707-2710, 2011.

[9] "SR 6161-1:2008 Acustica în construcţii. Partea 1: Măsurarea nivelului de zgomot în construcţii civile. Metode de măsurare".

* Corresponding author: viordach@yahoo.com

AUTHOR INDEX